Lecture Notes in Control and Information Sciences

Edited by M. Thoma and A. Wyner

For information about Vols. 1–61 please contact your bookseller or Springer-Verlag.

Lecture Notes in Control and Information Sciences

Edited by M. Thoma and A. Wyner

120

L. Trave, A. Titli, A. Tarras

Large Scale Systems: Decentralization, Structure Constraints and Fixed Modes

Springer-Verlag
Berlin Heidelberg GmbH

Authors
Louise Trave
André Titli
Ahmed Maher Tarras

Laboratoire d'Automatique et d'Analyse
des Systèmes du Centre National
de la Recherche Scientifique
Toulouse
France

ISBN 978-3-540-50787-1

Library of Congress Cataloging in Publication Data

Trave, L. (Louise)
Large scale systems : decentralization structure constraints and fixed modes
L. Trave, A. Titli, A. Tarras.
(Lecture notes in control and information sciences ; 120)
Bibliography: p.
Includes indexes.
ISBN 978-3-540-50787-1   ISBN 978-3-540-46085-5 (eBook)
DOI 10.1007/978-3-540-46085-5
1. System theory. 2. Control theory.
I. Titli, André. II. Tarras, A. (Ahmed). III. Title. IV. Series.
Q295.T73   1989   88-35984

2161/3020-543210

PREFACE

The growing dimensions and complexity of the present day technological, environment and societal processes is one of the foremost challenges to system theory. Determining a solution for the problems arising in large scale systems may become either very uneconomical or even impossible if using the classical mathematical tools developed for system analysis and control. The main reason is that classical theories are not built for dealing with high dimensionality models. Now, the essential characteristics of large scale systems are a huge number of input and output variables on subsystems which are generally geographically distributed.

These new features involve large and complex models, though the modelling problem may not be solvable. Moreover, we must face economical and reliability problems related to the information transfer between control stations.

For these reasons, the decomposition, aggregation and model reduction technics have received considerable attention in the last ten years. A great deal of theoretical and practical results concerning their applications have been obtained in the area of stability and decentralized control. In particular, the problem of stabilization and pole placement with decentralized dynamic compensation is of great practical interest. Despite the numerous advances around this problem, which are materialized by a large number of papers, there is none synthetical survey work exclusively concerned with this problem and the various other ones which are relevant.

The main objective of this book is to provide such global survey by presenting the present day results which can be used for :

- the analysis of stabilizability and pole placement under decentralized constraints,
- the determination of a control policy solving the problem of stabilization or pole placement when decentralized dynamic compensation fails (preserving a decentralized scheme of control or minimizing the cost associated to the information transfer),
- the design of these prespecified control laws.

By this way, this book supplies the tools for building a methodology which brings a solution to the complete problem of control in the context of large scale systems. Moreover, the last part of the work takes into account parametric and structural robustness constraints.

Chapter I presents an overview of the well-known results around the problem of stabilization and pole assignment of linear time-invariant dynamic systems subjected to centralized control (no structural constraints). The fundamental concepts of controllability and observability are introduced and they are extended to the concepts of structural controllability and observability, which are of major practical interest in the study of large scale systems. Indeed, these properties are established from the system structure and they do not depend on the particular configuration of the parameters' values. In this framework, the problem reduces to one of binary nature that allows for the application of graph-theoretic concepts. This approach is thus specially adequate for large scale systems.

The necessary and sufficient conditions for the existence of a solution to the problem of stabilization and pole assignment are presented for the following two cases :

- centralized state feedback
- centralized output feedback

They are stated in terms of the controllability and observability properties of the system.

It is clear that a good understanding of the concepts of controllability and observability is indispensable before proceeding to the study of the above problem with structural constraints on the control. This problem is introduced in Chapter II.

In the control of large scale systems whose essential characteristic is their high dimensionality, conventional techniques fail to give reasonable solutions with reasonable computational efforts. The classical control theory generally stands on the assumption of a centralized information pattern ; i.e., all the information on the system is available at a given center, generally a geographical position, where all the calculations can be carried out.

For most large scale systems, this centralization assumption does not hold due to the geographical distribution of the information. This new constraint leads to economical and reliability problems related to the information transfer. This implies that the control system should be made of a number of local controllers that are only allowed to use part of the whole information in order to generate part of the whole control. In particular, the design of feedback controllers requires restrictions on the particular system output-input pairs that the controller can connect.

When no transfer of information between the different local stations is allowed, this yields to a decentralized scheme of control. When some but not all transfers (those of minimum cost for example) are allowed we obtain a nonstandard reduced information pattern.

It is clear that the decentralized control scheme is the most economically advantageous since no transfer of information for one geographical location to another is required : system inputs are assigned to a given set of local controllers (stations), which observe only local system outputs. This is the reason the first studies for the problem of stabilization and pole assignment were investigated within a decentralized control scheme. The results refering to this study are presented in the first part of the chapter. These results were then extended to the more general case of arbitrarily structurally constrained control which is considered in the second part of the chapter.

The main concept to deal with this kind of problem is the new notion of fixed modes which is fundamental in the study of stabilization and pole placement with structurally constrained dynamic compensation. Indeed, the existence of a solution depends critically on the properties of this finite set of numbers. The presence of unstable fixed modes indicates that stabilization is impossible while the presence of any sort of fixed modes rules out arbitrary pole placement.

Due to the theoretical and practical importance of the notion of fixed modes, the whole Chapter 3 is concerned with their characterization. The number of different characterizations which can be found in the scientific literature is impressive. Moreover, every one of them is expressed in terms of its own authors definitions and introduced in a different way. With the main objective of classification, they are presented in two groups, the time-domain and the frequency-domain characterizations. A particular attention is given to show the existing equivalences. The analysis of every one of them allows us to point out the conditions for the existence of fixed modes and to give a deap inside into their interpretation related to their origins. The different types of fixed modes are outlined : no structurally fixed modes which need a quantitative analysis of the system and structurally fixed modes for which a structural approach is more suitable (representation of the system by a graph, use of general concepts of graph theory).

From the above analysis, the different results concerning the problem of decentralized stabilization and pole placement are unified and expressed in terms of the different types of fixed modes.

Whereas Chapter 2 makes clear that the choice a priori of a feedback control structure (decentralized for example) can generate some problems if it gives rise to the existence of fixed modes, Chapter 3 provides all the necessary tools to analyse and explain the situation.

As a natural consequence, the following chapters are concerned with the different methods which are available to determine an acceptable control policy such that stabilization or pole placement is possible.

In the context of large scale systems, the objective is to preserve a decentralized control structure (Chapter 4) or to determine a new control structure such that the cost of the information transfer is minimum (Chapter 5).

Chapter 4 presents a class of decentralized time-varying controllers which allow to avoid fixed modes provided that they are not from a structural origin. Several kinds of time-varying laws are examined and compared.

An original approach is then developed : it consists in using vibrational control. Vibrational control can be useful in the cases where conventional control methods (based on feedback or feedforward principles) do not apply because of lack of measurements. It is shown that vibrational control as a stabilization method is compatible with the decentralization constraints and that it solves the problem in presence of unstable fixed modes. From another point of view, a particular application of vibrational control is presented that constitutes a method for the design of decentralized time-varying control laws.

When the implementation of time-varying or non-linear control laws appears to be too difficult or when we are in presence of structurally fixed modes, we must consider the problem of determining a new control structure which minimizes the cost of the information transfer. This problem, which is of immense practical interest, is the subject of Chapter 5.

In fact, relaxing the structural constraints on the control seems to be the most convenient way to solve the stabilization or pole placement problem in presence of fixed modes. The purpose of Chapter 5 is to present the different available methods for the design of an appropriate feedback control structure. These methods are based on the different characterizations of fixed modes which are presented in Chapter 3 : one can be more appropriate than another depending on the type of situation we are dealing with. Roughly speaking, two types of situations can occur :

- the system is physically partitioned in several stations, due for example to different geographical locations of the inputs and outputs. In this case, it is clear that the decentralized structure would be the most appropriate. Our goal is thus to determine the minimal information exchanges between stations which get rid of fixed modes. The optimality criterion can be chosen as the number of feedback links between two different stations or as the cost associated with the implementation of these feedback links,

- either the system does not reflect a prespecified partitioning or the stations present the particularity that the cost of local feedbacks is not neglectable with respect to the cost of feedbacks between two different stations. In these cases, we want to determine the minimal control structures (if several) for which the system has no fixed modes. They do not generally involve all the local feedbacks.

Note that the problem resulting from the second situation is more general. Indeed, we are brought back to the first problem by setting to zero the costs associated to the local feedbacks. Therefore, all the methods which are presented in this general framework can also be used for the particular case.

As a logical following to the determination of adequate feedback control structures, Chapter 6 considers the problem of the synthesis of feedback gains under structural constraints. It provides an overview of appropriate near-optimal design techniques. Considerations on the robustness of such controllers are also included, in the sense that uncertainties due to parameter variations or external disturbances are considered. The effects of structural perturbations (failure of sensors or actuators, line cuts...) are studied later.

Chapter 7, and the last one, approaches indeed the problem of structural robustness. The chapter extends the results concerning decentralized or structurally constrained control systems to systems subjected to structural perturbations, mamely failures of sensors or actuators or cuts of lines implementing feedback-loops. The notions of structurally robust control and structurally robust modes are introduced. Several ways to conclude on the robustness of a control, knowing its prespecified structure, are presented. They appear as an extension of well-known results derived from fixed modes characterizations. At last, a graph-theoretic algorithm is presented to determine the information pattern of a robust regulator with minimum cost.

Through all the book, every result is illustrated by small significative examples which make easier their understanding. Moreover, the appendices contain a collection of packages corresponding to some important algorithms presented in the book (evaluation of fixed modes, determination of their type, calculation of constrained structure feedback matrices...).

This book is the consequence of an intensive research activity of several years in the area of analysis and control of large scale systems and more particularly in decentralized control in the Laboratoire d'Automatique et d'Analyse des Systèmes (LAAS).

The authors would like to express their gratitude to all the colleagues of their research group and to the Director of the LAAS, Professor A. COSTES, for their scientific and financial support.

The authors sincerely thank Miss C. FABRE for typing all these pages and Mr. E. LAPEYRE-MESTRE for the drawings of this book.

Toulouse, January 1989

Louise TRAVE
André TITLI
Ahmed TARRAS

TABLE OF CONTENTS

CHAPTER 4. DECENTRALIZED STABILIZATION IN PRESENCE OF NON
STRUCTURALLY FIXED MODES

CHAPTER 5. CHOICE OF FEEDBACK CONTROL STRUCTURE TO AVOID FIXED
           MODES

CHAPTER 6. DESIGN TECHNIQUES - PARAMETRIC ROBUSTNESS

CHAPTER I

CENTRALIZED CONTROL :

STABILIZATION AND POLE ASSIGNMENT

1.1. - INTRODUCTION

The general introduction pointed out that, very often, large scale systems control problems are characterized by structurally constrained feedback patterns. Before taking into account these new requirements, this chapter presents an overview of the well-known results concerned by the problem of stabilization and pole assignment of a linear time-invariant dynamic system subjected to centralized control (no structural constraints). The fundamental concepts of controllability and observability are introduced and extended to the concepts of structural controllability and observability which are of a major practical interest in the study of large scale systems. Indeed, these properties are established from the system structure and do not depend on the particular configuration of the parameters' values. In this framework, the problem reduces to one of binary nature that allows for the application of graph-theoretic concepts. This approach is thus especially adequate for large scale systems.

The necessary and sufficient conditions for the existence of a solution to the problem of stabilization and pole assignment are presented for the following two cases :

- centralized state feedback
- centralized output feedback

They are stated in terms of the controllability and observability properties of the system.

It is clear that a good understanding of the concepts of controllability and observability is indispensable before proceeding to the study of the above problem with structural constraints on the control.

1.2. - CONTROLLABILITY AND OBSERVABILITY (FOS-77) (KAI-80)

Consider the linear time-invariant dynamic system described by the following state-space model :

$$\begin{cases} \dot{x}(t) = A\ x(t) + B\ u(t) \\ y(t) = C\ x(t) \end{cases} \tag{1.2.1}$$

where $x(t) \in R^n$, $u(t) \in R^m$ and $y(t) \in R^r$ are the state, input and output vectors respectively, and A, B and C are invariant matrices of appropriate dimensions.

1.2.1. - Stability (WIL-70)

<u>Definition 1.1.</u> The autonomous system (1.2.1) (i.e. with $u(t) = 0$) is <u>stable</u> if for any given value $\epsilon > 0$, there exists a number $\delta_1(\epsilon, t_0) > 0$ such that :

$$\| x(t_0) \| < \delta_1 \implies \| x(t) \| < \epsilon \qquad \text{for all } t > t_0$$

The autonomous system (1.2.1) is <u>aymptotically stable</u> if :

(i) - it is stable

(ii)- $\forall\ x(t_0)$, $\quad x(t) \xrightarrow[t \to \infty]{} 0$

It is well-known that the solution of the equations (1.2.1) with $u(t) = 0$ is given by :

$$x(t) = e^{A(t-t_0)}\ x_0$$

Given $\{\lambda_1, \ldots, \lambda_n\}$ the set of eigenvalues of A, system (1.2.1) with $u(t) = 0$ is asymptotically stable if and only if all the eigenvalues of A have a negative real part.

In the opposite case, the state space X can be split into the stable subspace X_S which is generated by the set of eigenvectors associated with the stable eigenvalues and the unstable subspace X_U which is generated by the set of eigenvectors associated with the unstable eigenvalues. For $x(t_0) \in X_S$, the system response converges toward zero. For $x(t_0) \in X_U$, the system response diverges.

1.2.2. - Controllability

<u>Definition 1.2.</u> A state x_1 is said to be <u>controllable</u> at time t_0 if for every initial state x_0 defined at time t_0, there exists a control $u(t)$ that transfers the system from the state x_0 to the state x_1 in a finite time $(t_1 - t_0)$.

If every state of the system is controllable whatever t_0 may be, the system is said to be "completely controllable" or just controllable. For system (1.2.1), this is equivalent to stating that the pair (A, B) is controllable.

The necessary and sufficient conditions for system (1.2.1) to be controllable are stated in the following theorem :

<u>Theorem 1.1.</u> The system (1.2.1) is controllable if and only if either of the two following conditions holds :

1. - <u>Kalman's criterion</u> (KAL-62). The columns of the controllability matrix :

$$\Phi_C = (B, AB, \ldots, A^{n-1}B)$$

generate a space of dimension n, i.e., rank $\Phi_C = n$.

2. - <u>Popov-Belevitch-Hantus criterion</u> (KAI-80). There exists $j \in \{1, \ldots, m\}$ such that all the scalar products $\langle w_i, b_j \rangle$ are non zero, where b_j is the jth column of B and w_i, (i=1, ..., n), are the left eigenvectors of A.

Note that Kalman's criterion is not "minimal". More often than not, it will turn out that the matrix :

$$\Phi_C = (B, AB, \ldots, A^{\nu-1}B)$$

is of rank n for some ν less than n. The smallest such ν, say ν_c, will then be called the <u>controllability index</u>.

Popov-Belevitch-Hantus criterion may be more convenient in some cases since it may be restated in the following form :

The system (1.2.1) is controllable if and only if :

$$\text{Rank } (\lambda I - A \ B) = n \qquad \forall \ \lambda \in \sigma (A) \qquad \qquad (1.2.2)$$

where $\sigma (.)$ denotes the set of eigenvalues of (.).

In this new form, this criterion introduces the definition of a controllable pole (eigenvalue of A) as a pole for which condition (1.2.2) holds.

The components of every state $x \in X$ of the system can be partitioned such that : $x = x_c \oplus x_{unc}$ where $x_c \in X_C$ and $x_{unc} \in X_{UNC}$. X_C (X_{UNC}) is the controllable (uncontrollable) subspace generated by the eigenvectors associated to the controllable (uncontrollable) eigenvalues of A. It can then be shown that the equations (1.2.1) can take the form :

$$\begin{bmatrix} \dot{x}_c \\ \dot{x}_{unc} \end{bmatrix} = \begin{bmatrix} A_{11} & A_{12} \\ 0 & A_{22} \end{bmatrix} \begin{bmatrix} x_c \\ x_{unc} \end{bmatrix} + \begin{bmatrix} B_1 \\ 0 \end{bmatrix} u$$

$$y = \begin{bmatrix} C_1 & C_2 \end{bmatrix} \begin{bmatrix} x_c \\ x_{unc} \end{bmatrix}$$

where it appears that the components of X_{UNC} are not connected to the input.

From this point of view, Popov-Belevitch-Hantus criterion gives a deep insight into the controllability properties of the system. Kalman's criterion is suitable for checking the global controllability only.

1.2.3. – Observability

Definition 1.3. A state $x(t_0)=x_0$ is said to be <u>observable</u> at time t_0 if it can be determined from the knowledge of the input u(t) and of the output y(t) over a finite interval of time (t_0,t_1).

If every state of the system is observable whatever t_0 may be, the system is said to be "completely observable" or just observable. For system (1.2.1), this is equivalent to stating that the pair (C,A) is observable.

The necessary and sufficient conditions for system (1.2.1) to be observable are stated in the following theorem :

Theorem 1.2. The system (1.2.1) is observable if and only if either of the two following conditions holds :

1. – <u>Kalman's criterion</u> (KAL-62). The rows of the observability matrix :

$$\Phi_O = \begin{bmatrix} C \\ CA \\ \vdots \\ CA^{n-1} \end{bmatrix}$$

generate a space of dimension n, i.e., rank Φ_O = n.

2. - <u>Popov-Belevitch-Hantus criterion</u> (KAI-80). There exists j $\in$ {1, ..., r} such that all the scalar products $\langle c_j, v_i \rangle$ are non zero, where c_j is the jth row of C and v_i, (i=1, ..., n), are the right eigenvectors of A.

As for controllability, it will generally turn out that the matrix :

$$\Phi_O = \begin{bmatrix} C \\ CA \\ \vdots \\ CA^{\nu-1} \end{bmatrix}$$

is of rank n for some ν less than n. The smallest such ν, say ν_o, will then be called the <u>observability index</u>.

In the observability case, Popov-Belevitch-Hantus criterion may be restated in the following form :

The system (1.2.1) is observable if and only if :

$$\text{Rank} \begin{bmatrix} \lambda I - A \\ C \end{bmatrix} = n \qquad \forall \lambda \in \sigma(A) \tag{1.2.3}$$

and an observable pole is defined as a pole for which condition (1.2.3) holds.

The components of every state of the system can be partitioned such that $x = x_o \oplus x_{uno}$ where $x_o \in X_O$ and $x_{uno} \in X_{UNO}$. X_O (X_{UNO}) is the observable (unobservable) subspace generated by the eigenvectors associated to the observable (unobservable) eigenvalues of A. The decomposition of the system with regard to observability puts equations (1.2.1) in the following form :

$$\begin{bmatrix} \dot{x}_o \\ \dot{x}_{uno} \end{bmatrix} = \begin{bmatrix} A_{11} & 0 \\ A_{21} & A_{22} \end{bmatrix} \begin{bmatrix} x_o \\ x_{uno} \end{bmatrix} + \begin{bmatrix} B_1 \\ B_2 \end{bmatrix} u$$

$$y = \begin{bmatrix} C_1 & 0 \end{bmatrix} \begin{bmatrix} x_o \\ x_{uno} \end{bmatrix}$$

where it is clear that the components of X_{UNO} are not connected to the output.

The obvious analogy between theorems 1.1 and 1.2 points out the duality between the concepts of controllability and observability. Two systems are called dual if they are defined respectively by the equations :

$$S : \begin{cases} \dot{x} = A\,x + B\,u \\ \\ y = C\,x \end{cases} \qquad S^* : \begin{cases} \dot{x}^* = A'\,x^* + C'\,u^* \\ \\ y^* = B'\,x^* \end{cases}$$

These systems are such that, if S is controllable, S* is observable and vice versa. It is thus possible to check the observability of a system by examining the controllability of the dual system.

1.2.4. – Kalman's canonical form (KAL-62)

In view of paragraphs 1.2.2 and 1.2.3 , it follows that the state-space X can be decomposed into four subspaces such that :

$$X = X_1 \oplus X_2 \oplus X_3 \oplus X_4$$

where :

$X_1 = X_C \cap X_{UNO}$ (controllable and unobservable subspace)
$X_2 = X_C \cap X_O$ (controllable and observable subspace)
$X_3 = X_{UNC} \cap X_{UNO}$ (uncontrollable and unobservable subspace)
$X_4 = X_{UNC} \cap X_O$ (uncontrollable and observable subspace)

Kalman (KAL-62) showed that there exists a real, regular transformation matrix such that the system (1.2.1) can be put in the following canonical form :

$$
\begin{bmatrix} \dot{x}_1 \\ \dot{x}_2 \\ \dot{x}_3 \\ \dot{x}_4 \end{bmatrix} = \begin{bmatrix} A_{11} & A_{12} & A_{13} & A_{14} \\ 0 & A_{22} & 0 & A_{24} \\ 0 & 0 & A_{33} & A_{34} \\ 0 & 0 & 0 & A_{44} \end{bmatrix} \begin{bmatrix} x_1 \\ x_2 \\ x_3 \\ x_4 \end{bmatrix} + \begin{bmatrix} B_1 \\ B_2 \\ 0 \\ 0 \end{bmatrix} u \qquad (1.2.4)
$$

$$
y = \begin{bmatrix} 0 & C_2 & 0 & C_4 \end{bmatrix} \begin{bmatrix} x_1 & x_2 & x_3 & x_4 \end{bmatrix}'
$$

illustrated by figure 1.1 :

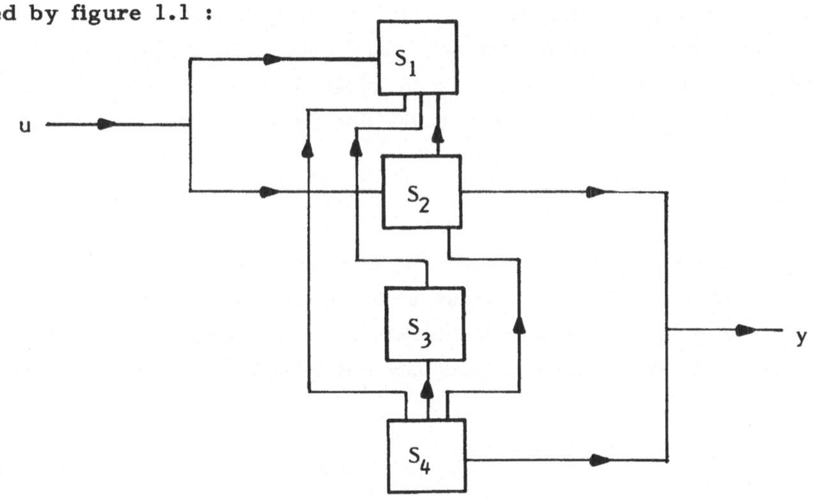

Fig. 1.1. : Canonical decomposition of a linear time-invariant system

Starting from the canonical form, the transfer function matrix of the system is :

$$
W(p) = \frac{Y(p)}{U(p)} = C_2 \, [pI - A_{22}]^{-1} \, B_2 \qquad \text{(p : Laplace variable)}
$$

in which only the simultaneously controllable and observable poles are present.

Note that the poles of the system corresponding to the eigenvalues of A_{11}, A_{22} and A_{44} (the non simultaneously uncontrollable and unobservable poles) verify the condition :

$$
\text{rank} \begin{bmatrix} \lambda I - A & B \\ C & 0 \end{bmatrix} = n
$$

easily derived from the Popov-Belevitch-Hantus criterion (1.2.2) and (1.2.3).

1.2.5. - Practical importance of the concepts of controllability and observability (FOS-77)

It is now interesting to examine the consequences of the existence of uncontrollable and unobservable modes on the behaviour of the system. These few following remarks consider several cases and point out the practical importance of the concepts of controllability and observability.

<u>Remark 1.1.</u>   As shown in paragraph 1.2.3, an uncontrollable mode is not connected to the input. The response associated with such mode will thus evolve in time independently of the control input, whether in open-loop or closed-loop configuration. Its evolution will depend only on the mode dynamics and the corresponding initial conditions.

<u>Remark 1.2.</u>   Consider that an uncontrollable mode is unstable. If it is observable, the unstability will appear at the output and will thus be detectable. Nevertheless, the fact that the mode is uncontrollable excludes all possibility of stabilizing the system. What is required is not a control law but a modification of the system structure.

<u>Remark 1.3.</u>   Consider now the case for which an unstable mode is not observable. The unstable dynamics of this mode will not appear on the output, since we have seen in paragraphe 1.2.4 that unobservable modes are not connected to the output. The system may thus be observed as stable. Nevertheless, the internal unstability of the system may come to either a break-up of the system or the appearence of a non linear function (saturation) so that the linear model is no longer valid.

These remarks strength the importance of having criteria which allow the detection of uncontrollable and unobservable modes when considering system control problems.

1.2.6. - Stabilization and pole assignment (BRA-70) (WON-67)

The problem of stabilization is formulated as follows : given the unstable system (1.2.1) (i.e., having some poles with positive real parts), find a controller of the form :

$$u = K \, y \hspace{4cm} (1.2.5)$$

such that the closed-loop system :

$$\dot{x}(t) = (A + BKC)\, x(t) \tag{1.2.6}$$

is stable ; i.e. every eigenvalue of the closed-loop dynamic matrix $(A + BKC)$ has a negative real part.

1.2.6.a. - State feedback control

First consider the case for which every state of the system (1.2.1) can be measured, what can be expressed by :

$$C = I_n \text{ (Identity matrix of order nxn)}$$
$$y = x$$

The feedback control then takes the form :

$$u = K\,x \tag{1.2.7}$$

System (1.2.1) is stabilizable using such a control law if and only if the unstable subspace X_U (see § 1.2.1) is included in the controllable subspace X_C (see § 1.2.2) and every pole of (1.2.1) can be arbitrarily assigned if and only if (1.2.1) is controllable. This is clearly understandable from the definition of controllability.

However, more often than not, the states are not directly available from the measurements and additional conditions are required.

1.2.6.b. - Output feedback control

With a control law of the form (1.2.5), using the Kalman's canonical form of (1.2.1) (see § 1.2.4), the closed-loop system is described by :

$$
\begin{bmatrix} \dot{x}_1 \\ \dot{x}_2 \\ \dot{x}_3 \\ \dot{x}_4 \end{bmatrix}
=
\begin{bmatrix}
A_{11} & A_{12}+B_1KC_2 & A_{13} & A_{14}+B_1KC_4 \\
0 & A_{22}+B_2KC_2 & 0 & A_{24}+B_2KC_4 \\
0 & 0 & A_{23} & A_{34} \\
0 & 0 & 0 & A_{44}
\end{bmatrix}
\begin{bmatrix} x_1 \\ x_2 \\ x_3 \\ x_4 \end{bmatrix}
\tag{1.2.8}
$$

It is thus apparent that the closed-loop system is stable if and only if :

(i) the unstable subspace X_U is included in the controllable subspace X_C and in the observable subspace X_O ; i.e., the eigenvalues of A_{11}, A_{33} and A_{44} are stable. Stated in a different way, this is equivalent to have all the unstable poles controllable and observable.

(ii) there exists a matrix K such that $(A_{22} + B_2 K C_2)$ is stable, what is expressed by :

$$F_2 = K \ C_2 \quad \text{and} \quad \text{rank} \ (C_2, \ F_2) = \text{rank} \ C_2$$

where F_2 is a stabilizing state feedback for the second subsystem that always exists since the second subsystem is controllable.

When considering arbitrary pole assignment, condition (i) is replaced by :
(i*) the system (1.2.1) is controllable and observable.

When condition (ii) cannot be verified, a dynamic output feedback control of the form :

$$\begin{cases} \dot{z}(t) = S \ z(t) + R \ y(t) \\ u(t) = Q \ z(t) + K \ y(t) + v(t) \end{cases} \tag{1.2.9}$$

is required. Then, condition (i) is sufficient (and necessary) to stabilize the system (1.2.1) and arbitrary pole assignment is possible if and only if (1.2.1) is controllable and observable (condition (i*)). In this latter case, Brasch and Pearson (BRA-70) showed that the minimal order of the required dynamic compensator to achieve pole assignment is :min (ν_c-1, ν_o-1), where ν_c and ν_o are the controllability and observability indices, respectively.

1.2.7. - Origins of uncontrollable and unobservable modes

In order to show the mechanism of uncontrollability and unobservability, consider the following simple example of a single-variable system taken from (FOS-77) :

$$\dot{x} = \begin{bmatrix} -1 & 0 & 0 & 0 \\ 0 & -2 & 0 & 0 \\ 1 & 1 & 1 & 0 \\ 1 & 1 & 0 & -3 \end{bmatrix} x + \begin{bmatrix} -2 \\ 3 \\ 0 \\ 0 \end{bmatrix} u$$

$$y = \begin{bmatrix} 0 & 0 & 0.5 & 0.5 \end{bmatrix} x \tag{1.2.10}$$

that can be represented by the block-diagram of figure 1.2 :

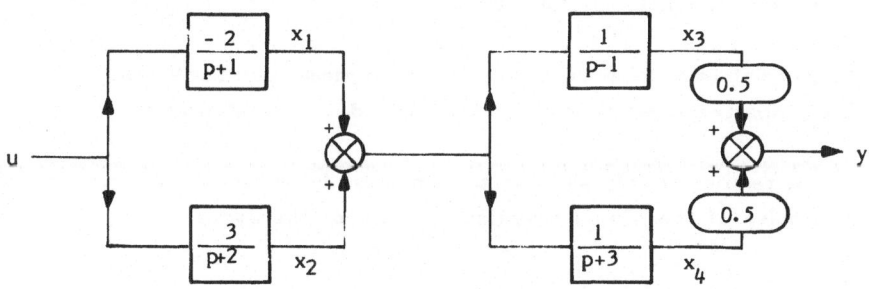

Figure 1.2.

Using conditions (1.2.2) and (1.2.3), it can be easily checked that the mode $\lambda_1=1$ is uncontrollable and $\lambda_2=-1$ is unobservable :

$$\text{rank } [\lambda I- A \ B]_{\lambda_1 =1} \quad = \text{ rank } \begin{bmatrix} 2 & 0 & 0 & 0 & -2 \\ 0 & 3 & 0 & 0 & 3 \\ -1 & -1 & 0 & 0 & 0 \\ -1 & -1 & 0 & 4 & 0 \end{bmatrix} = 3 < 4$$

$$\text{rank } \begin{bmatrix} \lambda I- A \\ \\ \\ C \end{bmatrix}_{\lambda_2 =-1} \quad = \text{ rank } \begin{bmatrix} 0 & 0 & 0 & 0 \\ 0 & 1 & 0 & 0 \\ -1 & -1 & -2 & 0 \\ -1 & -1 & 0 & 2 \\ 0 & 0 & .5 & 0.5 \end{bmatrix} = 3 < 4$$

This system is composed of two subsystems in cascade and can be rewritten in the equivalent form of figure 1.3. :

Figure 1.3.

where it appears that the uncontrollability and the unobservability result from the cancellation of a pole of one transfer function by a zero of the other and vice versa. In fact, if the cancellation is of the form :

1 - "zero upstream, pole downstream", it induces uncontrollability as for $\lambda_1=1$.
2 - "pole upstream, zero downstream", it induces unobservability as for $\lambda_2=-1$.

From this result, it follows that the uncontrollable and unobservable modes do not appear as poles of the global transfer function of the system :

$$y = \frac{1}{(p+2)(p+3)} \, u$$

Indeed, the uncontrollable and unobservable modes are "transparent" with respect to the input-output relation since the first ones are disconnected from the input and the second ones are disconnected from the output.

It is clear that this situation arises from the special configuration of the parameters' values in the matrices of the model of the system (1.2.10). Introducing a perturbation ε on the parameters a_{31} and a_{32}, the dynamic matrix becomes :

$$A \quad = \quad \begin{bmatrix} -1 & 0 & 0 & 0 \\ 0 & -2 & 0 & 0 \\ 1+\varepsilon & 1+\varepsilon & 1 & 0 \\ 1 & 1 & 0 & -3 \end{bmatrix}$$

The configuration of two subsystems in cascade is preserved as shown in figure 1.4 :

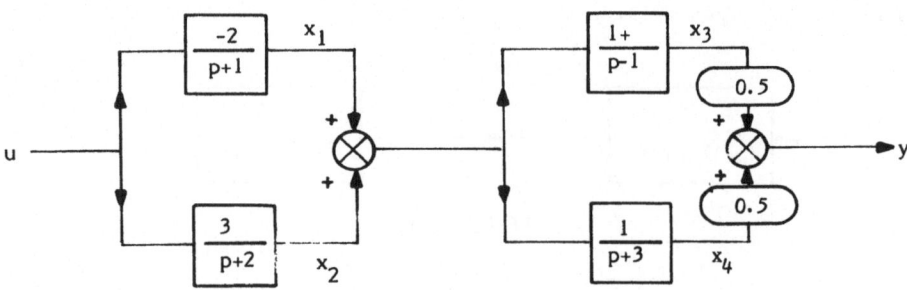

Figure 1.4

13

or equivalently in figure 1.5 :

Figure 1.5

In figure 1.5, it is clear that the pole-zero cancellation giving rise to the unobservability no longer occurs for $\epsilon \neq 0$. Therefore, this unobservability can be avoided by changing the values of some adequate components of the system.

Consider now the case for which a_{31} and a_{41} in (1.2.10) are fixed zeros, reflecting the internal interconnection structure of the system (i.e., these zero entries cannot be perturbed by any change of the components values). In this situation, the system is represented by figures 1.6 and 1.7 below :

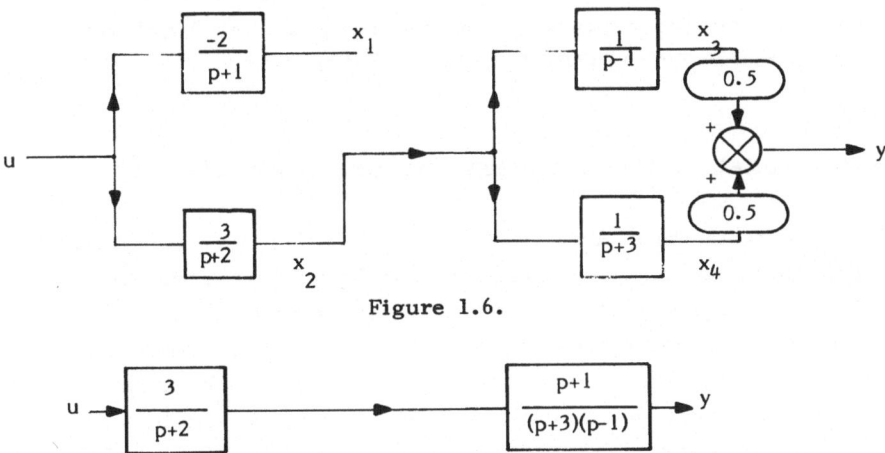

Figure 1.6.

Figure 1.7.

and its transfer function is :

$$y = \frac{3(p+1)}{(p+2)(p+3)(p-1)}$$

where the pole $\lambda_2 = -1$ is absent. This result is not surprising since Figure 1.6 shows that the block $-2/p+1$ is disconnected from the output. $\lambda_2 = -1$ is therefore an unobservable mode.

The interesting point of this discussion is that the pole $\lambda_2=-1$ will not appear in the transfer function and will remain unobservable however the parameters of the system are changed. This situation arises from the interconnection structure of the system. Such a pole is called <u>structurally unobservable</u>. A similar discussion could be done for uncontrollability coming to the concept of <u>structurally uncontrollable</u> poles.

These concepts present a great physical interest. Indeed, practically, most of the entries of A, B and C in (1.2.1) are known with the approximation of some errors of measurements (it is especially true for large scale systems). Only some of these entries are known with 100 percent precision : this happens for the entries that are equal to zero and reflect the internal interconnection structure of the system. The concepts of structural controllability and structural observability then makes the meaning of controllability and observability more complete from the phy-sical point of view.

The following paragraph presents the mathematical formulation of these pro-perties and some results expressing the necessary and sufficient conditions for a system to be structurally controllable and observable.

1.3. - STRUCTURAL CONTROLLABILITY AND OBSERVABILITY

These concepts were first introduced by Lin in 1974 (LIN-74). His results provide the conditions for structural controllability (and by duality for structural observability) of single-variable systems in a graph-theoretic approach. They were extended to multi-variable systems by Shield and Pearson in 1976 (SHI-76) but in a purely algebraic approach. In the same year, Glover and Silverman (GLO-76) pre-sented a simple algebraic solution and derived a recursive algorithm for determining structural controllability which utilizes Boolean operations only. This paragraph presents these results and establishes the equivalence between the graph-theoretic and the algebraic approaches.

Consider the linear time-invariant dynamic system in the form :

$$\begin{cases} \dot{x}(t) = A\ x(t) + B\ u(t) \\ y(t) = C\ x(t) \end{cases}$$

$$(1.3.1)$$

where $x \in R^n$, $u \in R^m$ and $y \in R^r$. A, B, and C are invariant matrices of appro-priate dimensions in which the entries are either fixed zeros of undeterminate (arbi-trary). Such a system is called a <u>structured system</u> and A, B and C are <u>structured matrices</u> (SHI-76). It is worth noting that the model (1.3.1) refers therefore to a class of systems.

<u>Definition 1.4.</u> The triple $(\overline{C}, \overline{A}, \overline{B})$ has the same structure as (is <u>structurally equivalent</u> to) the triple (C, A, B) of the same dimensions if for every fixed zero entry of the matrix $(\overline{C}\ \overline{A}\ \overline{B})$, the corresponding entry of the matrix $(C\ A\ B)$ is also a fixed zero, and vice versa.

Obviously, if the triple $(\overline{C}, \overline{A}, \overline{B})$ is structurally equivalent to the triple (C, A, B), then the pairs $(\overline{C}, \overline{A})$ and $(\overline{A}, \overline{B})$ are structurally equivalent to the pairs (C, A) and (A, B), and the inverse is also true.

<u>Definition 1.5.</u> A pair (A, B) $((C, A))$ is structurally controllable (observable) if there exists a pair equivalent to (A, B) $((C, A))$ which is controllable (observable) in the usual sense.

1.3.1. - Structural controllability

1.3.1.a. - Algebraic approach (SHI-76)

<u>Definition 1.6.</u> The matrix $(A\ B)$ has form I if there exists a permutation matrix P satisfying :

$$ P'\ [A\ B] \begin{bmatrix} P & 0 \\ 0 & I \end{bmatrix} = \begin{bmatrix} A_{11} & 0 & 0 \\ A_{21} & A_{22} & B_2 \end{bmatrix} $$

with A_{11} of order txt, $1 \leqslant t < n$. We assume that B_2 has at least one nonzero element.

The matrix $(A\ B)$ has form II if :

$$ gr(A\ B) < n $$

where gr (.) denotes the generic rank[1] of (.).

---

[1]Consider a structured matrix $\overline{M}$ with ν arbitrary entries. Then the parameter space R^{ν} is associated with $\overline{M}$ such that every data point $d \in R^{\nu}$ defines a matrix $M = \overline{M}(d)$. Conversely, a structured matrix $\overline{M}$ is associated with every matrix M such that $M = \overline{M}(d)$ for $d \in R^{\nu}$. The <u>generic rank</u> of M or of $\overline{M}$ is defined as follows :

$$ gr(M) = gr(\overline{M}) = \max_{d \in R^{\nu}} \{ rank\ \overline{M}\ (d) \} $$

<u>Theorem 1.3</u> (SHI-76). The pair (A, B) is structurally controllable if and only if the two following conditions both hold :

1 - the matrix (A B) is not of form I.
2 - the matrix (A B) is not of form II.

1.3.1.b. - Graph-theoretic approach (LIN-74)

The structured nature of the system (1.2.5) naturally comes to associate a digraph.

A <u>digraph</u> is a pair D=(V,E), where V is a set of vertices and E is a set of edges (v_j, v_i) directed from the vertex v_j to the vertex v_i. If $(v_j, v_i) \in E$ then v_j is said to be <u>adjacent</u> to v_i and v_i adjacent from v_j. The adjacency relation defines a square binary matrix $\overline{R}=(\overline{r}_{ij})$ called the <u>adjacency</u> (or interconnection) <u>matrix</u> $\overline{r}_{ij}=1$ if and only if $(v_1, v_i) \in E$. A sequence of edges $\{(v_1, v_2), (v_2, v_3),...,(v_{k-1}, v_k)\}$ where all the vertices are distinct, is called a <u>path</u> from v_1 to v_k. When v_k coincides with v_1, then the path is a <u>cycle</u>. A digraph D is said to be <u>acyclic</u> if it contains no cycles. If there is a path from v_j to v_i, then we say that v_i is <u>reachable</u> from v_j. Similarly, a subset $V_i \subset V$ is reachable from a subset $V_j \subset V$ if every vertex in V_i is reachable from some vertex in V_j. We say that $V_j \subset V$ reaches V_i if every vertex in V_j reaches at least one vertex in V_i. Like adjacency, the reachability relation on V can be represented by a binary matrix $R=(r_{ij})$ such that $r_{ij}=1$ if and only if v_i is reachable from v_j.

A pair of vertices of D are said to be <u>strongly connected</u> if they are reachable from each other. A maximal subgraph of D in which every pair of vertices are strongly connected is called a <u>strong component</u> of D.

The strong components of D are uniquely determined. If D has N strong components $D_1,...,D_N$, they can be ordered such that for some $1 \leqslant k \leqslant N$, none D_i, (i=1,...,k), is reachable from any other D_j, j≯i. If the corresponding permutation is performed on the adjacency matrix of D, it appears then in a block-triangular form where every block in the diagonal is the adjacency matrix of a strong component.

Let M be a rectangular pxq structured matrix and define the digraph D=(V,E) associated to M as the digraph whose adjacency matrix is M', obtained from M by adding (q-p) zero rows.

<u>Definition 1.7.</u> A digraph D=(V,E) contains a <u>dilation</u> if and only if there exists a set S ⊂V' (V'⊂ V is the set of vertices which are adjacent from at least one vertex in V) of K vertices such that the set T(S)⊂ V of vertices that are adjacent to a vertex of S contains no more than (K-1) elements. The dilation is denoted by {S,T(S)}.

This definition is a generalization of the concept of dilation introduced by (LIN-74) in the context of structural analysis for single-variable systems.

In practice, D contains a dilation if a set of K rows can be found in the adjacency matrix of D such that there are no more than (K-1) columns with nonzero entries in the submatrix formed by these K rows.

Given the triple (C,A,B), one defines a digraph Γ =(U ∪ X ∪ Y,E) where {U= $u_1,\ldots,u_m$} is the set of input vertices, X={$x_1,\ldots,x_n$} is the set of state vertices and Y={$y_1,\ldots y_r$} is the set of output vertices. E is the set of edges such that (u_i, x_j)∈ E if and only if b_{ji}≠0 in B, (x_i,x_j) ∈ E if and only if a_{ji}≠0 in A and (x_i,y_j)∈ E if and only if c_{ji}≠0 in C.

In this way, the digraph Γ completely describes the structure of the system.

<u>Definition 1.8</u> (SIL-78). The system (C,A,B) is said to be <u>input reachable</u> (or input connectable (DAV-77a)) if X is reachable from U and <u>output reachable</u> (or output connectable (DAV-77a)) if X reaches Y.

Although Lin's work (LIN-74) dealt only with single-input systems, it can be directly extended to multi-input systems.

Consider the pair (A,B) and its associated digraph Γ_1=(U ∪ X,E_1) obtained from Γ by deleting the set of output vertices and every edge from X to Y. Then, we have the following result :

<u>Theorem 1.4.</u> The pair (A,B) is structurally controllable if and only if the two following conditions both hold :

1 - (A,B) is input reachable.
2 - Γ_1 does not contain a dilation.

1.3.1.c. - **Equivalence of the two approaches**

It is now worth verifying that theorems 1.3 and 1.4 are equivalent. Consider the first parts of the theorems.

If matrix (A B) has form I, by definition 1.6, there exists a permutation matrix P satisfying :

$$P' [A\ B] \begin{bmatrix} P & 0 \\ 0 & I \end{bmatrix} = \begin{bmatrix} A_{11} & 0 & 0 \\ A_{21} & A_{22} & B_2 \end{bmatrix}$$

with A_{11} of order txt, $1 \leqslant t < n$.

The first equation in (1.3.1) is thus rewrited with a partitioned state vector :

$$\begin{bmatrix} \dot{X}_1 \\ \dot{X}_2 \end{bmatrix} = \begin{bmatrix} A_{11} & 0 \\ A_{21} & A_{22} \end{bmatrix} \begin{bmatrix} X_1 \\ X_2 \end{bmatrix} + \begin{bmatrix} 0 \\ B_2 \end{bmatrix} u$$

and the graph associated with the pair (A,B) in this form is given in Figure 1.8.

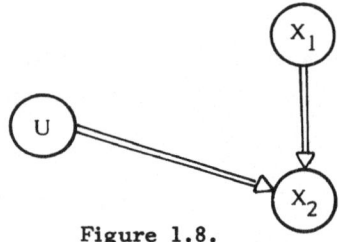

Figure 1.8.

It is clear that the state vertices in X_1 are not input reachable. The inverse also holds : if the graph of a pair (A,B) contains non-input-reachable state vertices, the matrix (A B) can be brought to form I. The first parts of the theorems 1.3 and 1.4 are therefore equivalent.

To point out the equivalence of the second parts of the theorems, we need the following result due to Shield and Pearson (SHI-76).

Theorem 1.5. (SHI-76). Assume a rectangular matrix M of order pxq with $p \leqslant q$. gr(M) < t for some t, $1 \leqslant t \leqslant p$, if for some k in the range $q-t < k \leqslant q$, M contains a zero submatrix of order (p+q-t-k+1)xk.

Theorem 1.5 applied to M=(A B) of order nx(n+m) obviously leads to the following corollary :

<u>Corollary 1.1.</u> : gr(A B) <n (i.e, (A B) has form II) if for some k in the range
m< k ≤ n+m, (A B) contains a zero submatrix of order (n+m-k+1)xk.

Making the change of variables : K=n+m-k+1, corollary 1.1 becomes :

gr(A B) <n if for some K in the range 1 ≤ K <n+1, (A B) contains a zero
submatrix of order Kx(n+m-K+1), which is equivalent, from definition 1.8, to have
the digraph associated with the pair (A,B) contains a dilation. This is illustrated by
Figure 1.9 with an example :

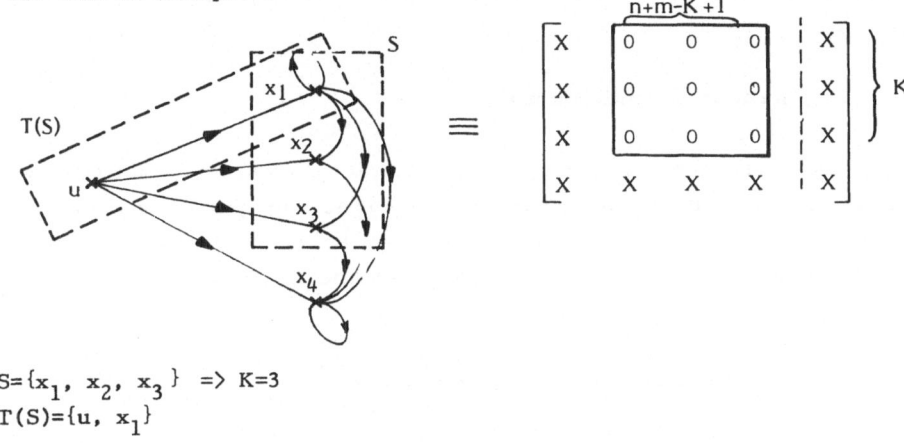

$S=\{x_1, x_2, x_3\}$ => K=3
$T(S)=\{u, x_1\}$

Figure 1.9

1.3.2. - General results on structural controllability and observability

As far as the dual concept of structural observability is concerned, a similar
study could be done and the general results are stated in the following theorems :

1.3.2.a. - Algebraic approach

<u>Theorem 1.6.</u> The system (1.2.5) represented by the triple (C,A,B) is structurally
controllable and observable if and only if the following conditions hold :

1 - (A,B) is not of form I.

2 - (A', C') is not of form I.

3 - gr(A B)=n (i.e, (A B) is not of form II).

4- $gr\begin{bmatrix} A \\ C \end{bmatrix}$=n (i.e, (A' C') is not of form II).

1.3.2.b. - Graph-theoretic approach

Theorem 1.7. The system (1.2.5) represented by the triple (C,A,B) is structurally controllable and observable if and only if the following conditions hold :

1 - (A,B) is input reachable.
2 - (A,C) is output reachable.
3 - The digraph associated with the pair (A,B) does not contain a dilation.
4 - The digraph associated with the pair (A', C') does not contain a dilation.

1.3.3. - Computational considerations

The most practical way to check structural controllability and observability of a given system (C,A,B) is to combine the algebraic and graph-theoretic approaches. Indeed, the input and output reachability can be easily tested and on the other hand, a lot of efficient algorithms exist to determine the generic rank of a matrix.

1.3.3.a. -Input and output reachability conditions

The adjacency matrix of the digraph associated with (C,A,B) has the following form :

$$
\bar{R} = \begin{array}{ccc} X & U & Y \\ \left[\begin{array}{ccc} \bar{A} & \bar{B} & 0 \\ 0 & 0 & 0 \\ \bar{C} & 0 & 0 \end{array}\right] & \begin{array}{c} X \\ U \\ Y \end{array} \end{array} \qquad (1.3.2)
$$

where $\bar{A},\bar{B}$ and $\bar{C}$ are obtained by replacing every undeterminate (arbitrary) entry in A,B and C by 1.

The reachability matrix, which has the same dimension as $\bar{R}$ above is given by :

$$
R = \begin{array}{ccc} X & U & Y \\ \left[\begin{array}{ccc} E & F & 0 \\ 0 & 0 & 0 \\ G & H & 0 \end{array}\right] & \begin{array}{c} X \\ U \\ Y \end{array} \end{array} \qquad (1.3.3)
$$

R can be computed directly from $\bar{R}$ as shown in (SIL-78).

By the sole observation of R, we can conclude on the reachability conditions (SIL-78) : (C,A,B) is input reachable if and only if F has no zero rows, it is output reachable if and only if G has no zero columns, and it is input-output reachable if and only if H has neither zero rows nor zero columns.

1.3.3.b. - Generic rank conditions

A number of authors have suggested alternative algorithms for finding the generic rank of a structured matrix. Several algorithms were shown to be wrong ((SHI-76), (BUR-81), (DAV-77a)), revealing a tricky problem.

Among the efficient algorithms, we mention the one by Prescott and Pearson (PRE-81) (JAM-83) based on combinatorial mathematics (LIU-68). The basic idea is that the evaluation of the generic rank of an nxm structured matrix is equivalent to the determination of the maximum number of "nontaking roots" that can be placed on an nxm chessboard, where the zeros of the matrix are interpreted as forbidden positions.

Another interesting algorithm was presented recently by Johnston et al. (JOH-84) that is based on the detection in a systematic manner of any dilation. The initial matrix is first brought back to a triangular block-structure form where different disjoint blocks appear in the diagonal. The blocks that have a greater number of rows than columns indicate a possible rank deficiency, which is determined by a simple procedure examining the position of the nonzero entries.

We mention also that an APL routine to compute the generic rank of a matrix is provided in (EVA-84). It is based on the fact that the generic rank of a matrix is equal to the sum of the dimensions of the nonzero diagonal blocks constituting the maximal permutation matrix.

The algorithm that we present here (see the next section) is the one in (TRA-86a) which is specially suitable for application to large scale systems. Using an appropriate reordering of the matrix, the global generic rank condition is expressed in terms of conditions refering to smaller dimensioned matrices. The checking is thus carried out in a sequential way by using, at each step, a slightly modified version of the algorithm in (JOH-84).

1.3.3.c. - A sequential algorithm to conclude on structural controllability and observability of large scale systems (TRA-86a)

When dealing with large scale systems, checking the conditions for structural controllability and observability may cause computational difficulties and require high calculation time. This section presents the work in (TRA-86a) which takes advantage of the inherent structure of a class of large scale systems which can be decomposed into hierarchically ordered interconnected irreducible subsystems. The underlying idea is to achieve computational and conceptual simplifications by subsequently solving each subsystem. This approach has been used to analyse systems stability in (SEZ-81c) (MIC-78) and for estimators and controllers design in (PIC-83b). Although (PIC-83b) uses the property that each subsystem is structurally controllable and observable, the conditions guaranteing these properties have not been yet derived within this perspective. This problem is considered in (TRA-86a) and sequential conditions for structural controllability (observability) of the overall system are provided. An iterative algorithm is proposed and its improvements with respect to an algorithm using a global approach are discussed. An illustrative example is presented.

1. Preliminaries

In this section, some new concepts of graph theory (TRA-86a) are introduced.

Definition 1.9. A G-dilation (Generalized dilation) is defined as any pair of sets $\{S \subset V', T(S) \subset V\}$ such that each vertex of $T(S)$ is adjacent to some vertex of S. The order of a G-dilation is $d_s = \text{card } S - \text{card } T(S)$ where card * denotes the number of elements in the set *.

Obviously, a dilation is a G-dilation whose order is positive. Then, it is clear that $gr(M) = p$ (full generic rank) if and only if the order of all the G-dilations contained in D is $\leqslant 0$.

We say that a G-dilation $\{S, T(S)\}$ of order d_s is maximal if, for all proper subsets $L \subset (V'-S)$, the G-dilation $\{S \cup L, T(S \cup L)\}$ is of order $\neq d_s$.

Theorem 1.8. (TRA-86a). If the digraph D associated with a full generic rank matrix M contains k G-dilations $\{S_i, T(S_i)\}$, (i=1,...,k), of order 0, then the maximal G-dilation of order 0 $\{\Omega, T(\Omega)\}$ is unique and "includes" all the G-dilations of order 0, i.e., $S_i \subset \Omega$, $T(S_i) \subset T(\Omega)$, (i=1,...,k).

2. Sequential conditions for structural controllability

Consider the system (1.2.1) represented by the triple (C,A,B) and its associated digraph $\Gamma = (U \cup X \cup Y, E)$ as defined in section 1.3.1.b. The following theorem summarizes the results provided in (LIN-74) and (SHI-76) :

Theorem 1.9. The system (1.2.1) is structurally controllable (observable) if and only if the two following conditions both hold :

    1- (A,B) ((C,A)) is input (output) reachable (SIL-77)
    2- gr(A B) = n (gr(A' C')= n)

From the computational point of view, checking the above conditions requires the evaluation of the reachability matrix of the digraph associated with (A,B) ((C, A)) and the determination of the generic rank of a matrix whose dimension is in the range of the order of the system. Although efficient algorithms are available (see (SIL-78) (BOW-76)) for the reachability matrix and (JOH-84) (MOA-80) for the generic rank), some computational difficulties may arise when dealing with large scale systems.

The system is decomposed into those subsystems for which the state variables correspond to the strong components of $\Gamma_X = (X, E_x)$, obtained from Γ by deleting U, Y, and the corresponding edges. If the strong components are adequately ordered and after the corresponding permutation of the state variables, system (1.2.1) takes the following form :

$$\dot{x}(t) = \begin{bmatrix} A_{11} & & \\ A_{21} & A_{22} & \\ \vdots & & \diagdown \\ A_{s1} & \text{------} & A_{ss} \end{bmatrix} x(t) + \begin{bmatrix} B_1 \\ B_2 \\ \vdots \\ B_s \end{bmatrix} u(t)$$

$$y(t) = \begin{bmatrix} C_1 & C_2 & \text{----} C_s \end{bmatrix}$$

(1.3.4)

where $A_{ij} \in R^{ni \times nj}$, $B_i \in R^{ni \times m}$, $C_i \in R^{r \times ni}$, (i,j=1,...,s). Elaborated procedures for the determination of the strong components of a digraph can be found in (HAR-65), (KAU -68).

2.1. - Reachability condition

Using the above decomposition, the reachability condition of Theorem 1.9 is very easily expressed at the level of each subsystem in (1.3.4).

<u>Theorem 1.10</u> (TRA-86a). The system (1.2.1) (1.3.4) is input reachable if and only if :

$$(B_i \ A_{i1} \ A_{i2} \ \cdots \ A_{i,i-1}) \not\equiv 0 \qquad (i=1,\ldots,s) \qquad (1.3.5)$$

This result is straightforward from the definition of strong components and the hierarchically reordered digraph corresponding to (1.3.4). Therefore, the sole observation of the matrices in (1.3.4) allows to conclude on input reachability. The dual result for output reachability is obtained by replacing B_i by C'_{s-i+1} and A_{ij} by $A'_{s-j+1,s-i+1}$ in (1.3.5).

2.2. - Generic rank condition

Our main result is presented in this paragraph. Consider the system (1.3.4) and define the matrices A_i^* and B_i^*, (i=1,...,s) as follows :

$$A^*_1 = A_{11}, \ B^*_1 = B_1, \quad A^*_i = \begin{bmatrix} A^*_{i-1} & & & \vdots & 0 \\ \hline A_{i1} & \cdots & A_{i,i-1} & \vdots & A_{ii} \end{bmatrix} \quad B^*_i = \begin{bmatrix} B^*_{i-1} \\ \hline B_i \end{bmatrix} \qquad (1.3.6)$$

where $A_i^* \in R^{n_i^* \times n_i^*}$ and $B_i^* \in R^{n_i^* \times m}$ with $n_i^* = \sum_{j=1}^{s} n_j$.

<u>Theorem 1.11</u> (TRA-86a). A necessary and sufficient condition to have gr(A B)=n is that

$$gr \ (A_i^* \ B_i^*) \ = n_i^* \text{ for all } (i=1,\ldots,s) \qquad (1.3.7)$$

The problem consists, therefore, in finding the conditions to have $gr(A^*_i \ B_i^*)=n_i^*$ assuming that $gr(A^*_{i-1} \ B^*_{i-1}) = n^*_{i-1}$.

<u>Theorem 1.12</u> (TRA-86a). Given the rectangular pxq (p$\leqslant$q) matrix M and its associated digraph D=(V,E), assume that gr(M)=p and that $\{\Omega, T(\Omega)\}$ is the maximal G-dilation of order 0 in D. Then the necessary and sufficient condition to have

$$gr \begin{bmatrix} M & 0 \\ N_1 & N_2 \end{bmatrix} = gr\ (M_N) = p+p', \text{ where } N_1 \text{ is a } p'xq \text{ matrix and } N_2 \text{ is a } p'xp'$$

matrix, is that :

$$gr \begin{vmatrix} M^{\overline{T}(\Omega)} & 0 \\ N_1^{T(\Omega)} & N_2 \end{vmatrix} = p'+\text{card } \overline{\Omega} \tag{1.3.8}$$

where $\overline{\Omega} = V' - \Omega$ and M^E denotes the submatrix composed by the columns of M whose associated vertices do not belong to the set E.

This situation is clarified when the matrix M_N is rewrited with respect to $\{\Omega, T(\Omega)\}$ as follows :

$$V = T(\Omega)\ \cup\ \overline{T}(\Omega)\ \cup\ V_2$$
$$V' = \Omega \cup \overline{\Omega}\ \cup\ V_2$$

In order to apply Theorem 1.12 to our special problem, define :

$$M_{11} = [B_1\ A_{11}], \quad M_{ii} = \begin{bmatrix} M_{i-1,i-1} & 0 \\ N_{i,i-1} & N_{ii} \end{bmatrix} \tag{1.3.9}$$

$$N_{i,i-1} = [B_i\ A_{i1}\ A_{i2}\ \cdots\ A_{i,i-1}], \quad N_{ii} = [A_{ii}]$$

__Theorem 1.13__ (TRA-86a). The necessary and sufficient conditions to have gr(A B)=n where A and B are given in (1.3.4), is that :

$$1 - gr[M_{11}] = n_1$$

$$2 - gr \begin{bmatrix} M_{i-1,i-1}^{\overline{T}(\Omega_{i-1})} & 0 \\ N_{i,i-1}^{\overline{T}(\Omega_{i-1})} & N_{ii} \end{bmatrix} = gr[F_i] = n_i + \text{card}\,\overline{\Omega}_{i-1} \quad (i=2,\ldots,s) \tag{1.3.10}$$

where $\{\Omega_{i-1}, T(\Omega_{i-1})\}$ is the maximal G-dilation of order 0 of $M_{i-1, i-1}$.

Theorem 1.13 is straightforward from the successive application of Theorem 1.12 to the matrices M_{ii}, (i=1,...s), defined in (1.3.9). The above result is strengthened by the fact that $\{\Omega_i, T(\Omega_i)\}$ can also be determined in a sequential way.

Theorem 1.14 (TRA-86a). Assume that M_{ii} has full generic rank, then the maximal G-dilation of order 0 of M_{ii} is $\{\Omega_i, T(\Omega_i)\}$ with $\Omega_i = \Omega_{i-1} \cup \delta_i$ and $T(\Omega_i)=T(\Omega_{i-1})$ $\cup T(\delta_i)$ where $\{\delta_i, T(\delta_i)\}$ is the maximal G-dilation of order 0 of the matrix F_i defined in (1.3.10).

Dual results are easily obtained for structural observability by replacing B_i by C'_{s-i+1} and A_{ij} by $A'_{s-j+1,s-i+1}$ in (1.3.9).

The above results provide the basis for an iterative algorithm to conclude on structural controllability (observability) of large scale dynamical systems.

2.3. – Algorithm

Since the sequential reachability conditions (1.3.5) in Theorem 1.10 do not require any calculations, the subsequent algorithm refers to the generic rank conditions (1.3.10) in Theorem 1.13 only. At each step i, we need to verify whether or not F_i has a generic rank deficiency and, if not, to determine its maximal G-dilation of order 0. This can be performed by using a slightly modified version of the generic rank algorithm in (JOH-84). The first step is the same as in (JOH-84) and consists in a reordering of the matrix F_i.

Algorithm REORDER (M) :

Step 1   i=0, j=0, delete the null columns.

Step 2   Find the row with the minimal number of non-deleted entries (say α). If choice exists, select the one with minimal number of entries (deleted or non-deleted).

Step 3   Associate this row with index i and delete it, i=i+1.

Step 4   Associate the columns which have entries in this row with the indices j+1, j+2,..., j+α and delete them, j=j+α.

Step 5   If any row is left, go to step 2.
Otherwise, write M ordering the rows according to increasing indices from top to bottom and the columns according to increasing indices from left to right, stop.

This algorithm puts the matrix M in the following form :

$$
\begin{bmatrix}
\boxed{R_1} & & & \\
& \boxed{R_2} & & \\
& & \ddots & \\
& & & \boxed{R_K}
\end{bmatrix}
$$

Denote the number of rows and columns of the block R_i by p_i and q_i, (i=1,...,K).

In a second step, the matrix F_i is analyzed. If a dilation is detected, F_i has a generic rank deficiency. Otherwise, we determine its maximal G-dilation of order 0. The procedure is the same as in (JOH-84), with the modification that some rows and columns are memorized for the determination of the maximal G-dilation of order 0.

Algorithm ANALYZE (M) :

Step 1   i=1.

Step 2   If $p_i-q_i < 0$, go to step 8.
Otherwise, j=1, k=0.

Step 3   Tag the q_i+k first rows in R_i. Also tag all the rows in R_i which have entries only in the same columns as the entries of the tagged rows.

Step 4   Tag the columns which have nonzero entries in the tagged rows and calculate h_j^i = number of tagged rows−number of tagged columns.

Step 5   If $h_j^i > 0$, M contains a dilation, stop.

If $h_j^i = 0$, memorize the tagged rows and columns.

Step 6   If all the rows above R_i are tagged, go to step 7.
The rows above R_i with entries in the tagged columns are now scanned. Select the one with minimal number of entries in tagged columns. If choice exists, select the one with minimal number of entries and tag it, j=j+1, go to step 4.

Step 7   If all the rows in the block R_i are tagged, go to step 8.
Otherwise, delete all the tags, j=j+1, k=k+1.

Step 8   If i=K, stop, the maximal G-dilation of order 0 of M is given by :
  δ : memorized rows $T(\delta)$: memorized columns
Otherwise, i=i+1, go to step 2.

Assuming that (B A) has initially been put in form (1.3.4) by using any of the methods proposed in (SIL-78) (BOW-76), the global algorithm is the following :

Step 1   i=1, β =number of diagonal blocks, $M_{00}=0$.

Step 2   Successively delete the rows with only one entry and the corresponding columns.
If no row is left, gr(B A)=n, stop.

Step 3   If the row block i has been deleted or F_i has only one row left, go to step 4.
Otherwise, REORDER (F_i), ANALYZE (F_i).
If F_i contains a dilation, gr(A B)<n, stop.
Otherwise, $\{\delta_i, T(\delta_i)\}$ is returned by the algorithm ANALYZE.

Step 4   If i= β, gr(A B)=n, stop.
Otherwise, delete the columns corresponding to $T(\delta_i)$, delete the null rows. F_{i+1} is obtained by adding the row block (i+1).

Step 5   i=i+1, go to step 2.

It is a well known result that a rectangular pxq (p<q) structured matrix M has full generic rank if M includes a pxp permutation matrix (one nonzero entry by row and by column). The above algorithm is based on the fact that a G-dilation of order 0 $\{S^0, T(S^0)\}$ defines an independent permutation submatrix. Since the μ rows defined by S^0 have nonzero entries only in the μ columns defined by $T(S^0)$, the nonzero entries of the final permutation matrix (if any) are necessarily located in the square submatrix defined by $\{S^0, T(S^0)\}$. It is then clear that the remaining entries in the columns $T(S^0)$ are not available anymore (Step 4).

This remark has also been used to improve the algorithm by a preliminar reduction of the whole matrix at each iteration (Step 2).

2.4. - Example

To illustrate the algorithm, let us consider the system (1.2.1) (already put in form (1.3.4)) described by (B A) as follows :

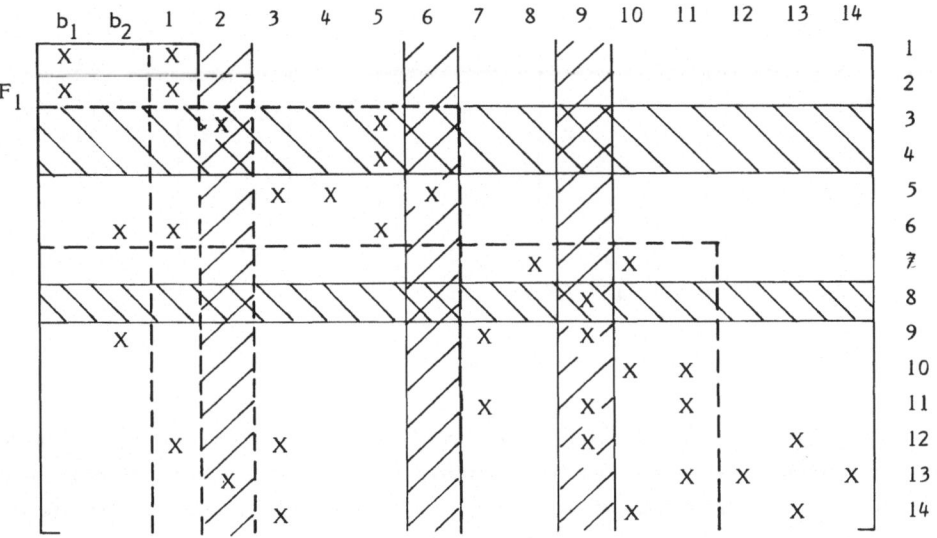

This system satisfies the reachability conditions (1.3.5) of Theorem 1.10. The generic rank algorithm is now applied.

* $\underline{i=1}$, $\beta = 5$.
Step 2 : Rows 3, 4, 8, and columns 2, 5, 9 are deleted.

* $\underline{i=2}$:
Step 3 : F_2 does not need to be reordered. ANALYZE F_2 --> $\begin{cases} \delta_1 = \{1,2\} \\ T(\delta_1) = \{b_1, 1\} \end{cases}$

Step 4 : These rows and columns are deleted. (B A) becomes :

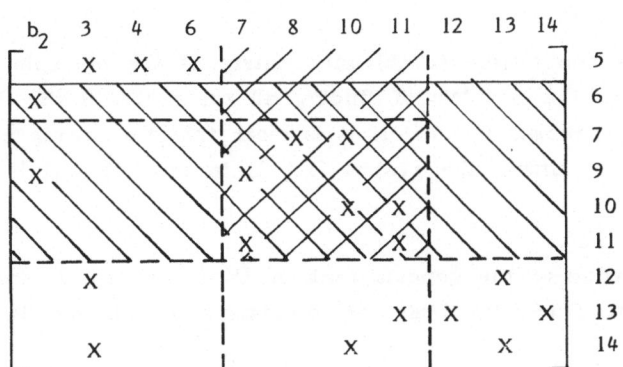

* $i=3$

Step 2 : Rows 6, 7, 9, 10, 11 and columns b_2, 7, 8, 10, 11 are deleted.

* $i=5$: ($i=4$ is jumped since the row block 4 has been deleted)
Step 3 :

$$F_5 = \begin{bmatrix} \overset{3}{X} & \overset{4}{X} & \overset{6}{X} & \vdots & & \\ X & & & \vdots & \overset{12}{X} & \\ & & & \vdots & X & \\ X & & & \vdots & & X \end{bmatrix} \begin{matrix} 5 \\ 12 \\ 13 \\ 14 \end{matrix} \xrightarrow{\text{REORDER}} \begin{bmatrix} \overset{3}{X} & \overset{13}{X} & \vdots & \overset{12}{} & \overset{14}{} & \vdots & \overset{4}{} & \overset{6}{} \\ X & X & \vdots & & & \vdots & & \\ & & \vdots & X & X & \vdots & & \\ X & & \vdots & & & \vdots & X & X \end{bmatrix} \begin{matrix} 12 \\ 14 \\ 13 \\ 5 \end{matrix} \begin{matrix} \text{ANALYSE}[F_5] \\ \delta_5 = \{ 12, 14 \} \\ T(\delta_5) = \{3, 13\} \end{matrix}$$

Step 4 : Since $i = \beta$, stop : gr (B A) = n.

Therefore, this system is structurally controllable.

Comments : It is clear that this algorithm does not present any advantage for strongly connected systems. Its applicability is therefore limited to large scale systems which can be decomposed into several hierarchically ordered subsystems and it is intuitively understandable that its efficiency increases with the number of subsystems. Some other aspects of the system like the location of the dilations or of the independent G-dilations of order 0 in (B A) must also be taken into consideration. It results that the efficiency of the algorithm may vary for each particular case so that it is difficult to carry out any meaningful quantitative comparison with other algorithms using a global approach. However, a qualitative discussion is possible.

The algorithm presented here must be considered as a whole for the structural controllability problem. If the generic rank alone is of importance, then applying the algorithm in (JOH-84) to the global matrix may be a good choice.

Solving the structural controllability problem without a decomposition scheme requires :

- The determination of the reachability matrix of the digraph of the system which is equivalent to the problem of finding all the paths in the digraph of the system. If this latter problem is solved, it is clear that the strong components are determined. Therefore, putting the system (1.2.1) in form (1.3.4) is an equivalent problem.

- the determination of the generic rank of (B A) by using, for example, the algorithm of (JOH-84). Then, the first step consists in reordering (B A) into a row echelon form (algorithm REORDER), which can be a heavy task for large scale systems. This step is avoided by our algorithm which uses REORDER at the subsystem level. Moreover, whereas the algorithm in (JOH-84) carries out the calculations on the whole reordered matrix, this algorithm proceeds to a reduction of (B A) at each step (Step 2), which can considerably reduce the task (see the example).

Over the scope of this section, the algorithm may be adapted for the sequential determination of minimal essential input (output) sets which are fundamental in the problem of determining the minimal feedback patterns avoiding structurally fixed modes (TRA-86b) (see Chapter V, Section 5.3.4b.3).

1.4. - CONCLUSION

This chapter provides an overview of the important concepts of controllability and observability that play a fundamental role in the problem of stabilization and pole assignment of linear time-invariant dynamic systems.

The solution of this problem is expressed in terms of these concepts. An analysis of the origins of uncontrollable and unobservable modes points out the interest for extending these concepts into a structural framework. Structural controllability and observability are thus introduced and appear as stronger properties than controllability and observability in the usual sense. Using these concepts, the above problem turns to which of generic stabilization and generic pole assignment. The structural properties are established from the structure of the system without any consideration of the parameters' values : they are in fact concerned with classes of systems. Therefore, this approach allows dealing with systems whose nonzero parameters are known with incertainties. It is worth noting that this situation occurs frequently in large scale systems.

The study of these properties can be achieved with a purely algebraic approach or with a graph-theoretic approach. Nevertheless, the easiest way to check structural controllability and observability is by combining the two approaches. Thus, the conditions to be examined involve both reachability of the graph associated with the system and generic rank determination.

In section 1.3.3.c, the algorithm of (TRA-86a) for checking structural controllability and observability is presented. It is specially suitable for application in large scale systems. Indeed, by using an appropriate decomposition of the system the global conditions are expressed in terms of several conditions refering to smaller dimensioned systems (TRA-86a). The checking can thus be carried out in a sequential way.

In the next chapter, the problem of stabilization and pole assignment will be investigated in the context of large scale systems whose characteristics generally require structural constraints on the control.

We will see that the fact of taking into account these new constraints yields to the introduction of new concepts which appear as an extension of those defined in the case of centralized control.

CHAPTER 2

STRUCTURALLY CONSTRAINED CONTROL :
STABILIZATION AND POLE ASSIGNMENT

2.1. - INTRODUCTION

In the control of large scale systems whose essential characteristic is their high dimensionality, conventional techniques fail to give reasonable solutions with reasonable computational efforts. The classical control theory generally stands on the assumption of a centralized information pattern ; i.e., all the information on the system is available at a given center, generally a geographical position, where all the calculations can be carried out.

For most large scale systems, this centralization assumption does not hold due to the geographical distribution of the information. This new constraint leads to economical and reliability problems related to the information transfer. This implies that the control system should be made of a number of local controllers that are only allowed to use part of the whole information in order to generate part of the whole control. In particular, the design of feedback controllers requires restrictions on the particular system output-input pairs that the controller can connect.

When no transfer of information between the different local stations is allowed, this yields to a decentralized scheme of control. When some but not all transfers (those of minimum cost for example) are allowed we obtain a nonstandard reduced information pattern. These situations are illustrated by Figure 2.1 :

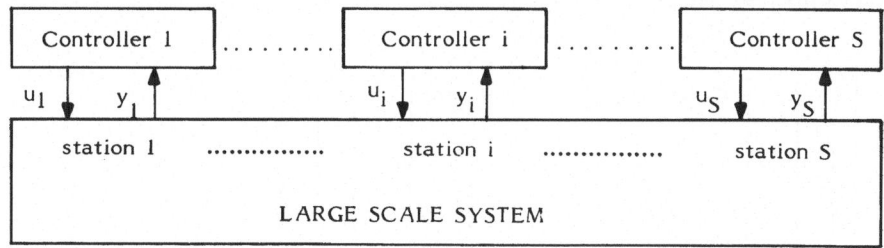

Figure 2.1.a - Decentralized structurally-constrained control

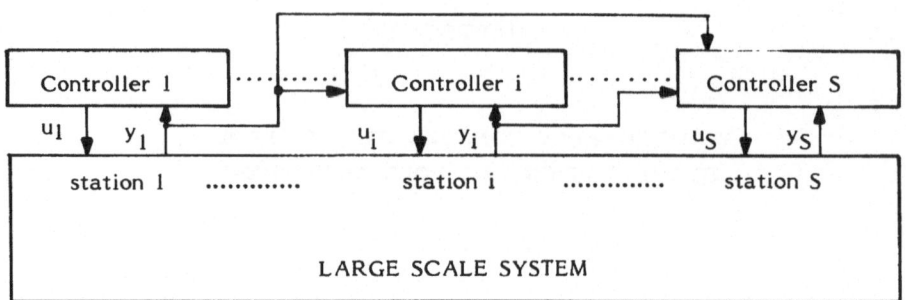

Figure 2.1.b. - Example of arbitrarily structurally-constrained control

It is clear that the decentralized control scheme is the most economically advantageous since no transfer of information from one geographical location to another is required : system inputs are assigned to a given set of local controllers (stations), which observe only local system outputs. This is the reason the first studies for the problem of stabilization and pole assignment were investigated within a decentralized control scheme. The results refering to this study are presented in the second paragraph of this chapter. These results were then extended to the more general case of arbitrarily structurally-constrained control which is considered in the third paragraph.

2.2. - DECENTRALIZED STRUCTURAL CONSTRAINTS

Consider a large scale system represented by a linear time-invariant dynamic model. As pointed out in the introduction, its geographical decomposition results in a partitioning of the control and the observation in several local control and observation stations. The models retained reflect this situation and appear in the following particular forms :

1-Frequency-domain model

$$
\begin{bmatrix} y_1 \\ \vdots \\ y_S \end{bmatrix} = \begin{bmatrix} W_{11}(p) & \cdots\cdots & W_{1S}(p) \\ \vdots & & \vdots \\ W_{S1}(p) & \cdots\cdots & W_{SS}(p) \end{bmatrix} \begin{bmatrix} u_1 \\ \vdots \\ u_S \end{bmatrix} \tag{2.2.1}
$$

2-Time-domain model

$$\begin{cases} \dot{x}(t) = A\,x(t) + \sum_{i=1}^{S} B_i\,u_i(t) \\ y_i(t) = C_i\,x(t) \quad (i=1,\ldots,S) \end{cases}$$

(2.2.2)

where $x(t) \in R^n$ is the state. The input and output vectors are partitioned : $u_i \in R^{m_i}$ and $y_i \in R^{r_i}$ are the input and the output vectors of the ith control and observation station, $(i=1,\ldots,S)$.

<u>Remark 2.1.</u> Note that the state vector is not partitioned, reflecting the fact that the system itself is taken as a whole ; i.e., it is not decomposed into several well-specified subsystems. The following results hold in this general framework. The results for "interconnected systems" will be presented as a particular case in a further paragraph.

Let us define :

$$m = \sum_{i=1}^{S} m_i \quad , \qquad r = \sum_{i=1}^{S} r_i$$

$$B = [B_1, \ldots, B_S] \qquad C' = [C_1', \ldots, C_S']$$

$$y' = [y_1', \ldots, y_S'] \qquad u' = [u_1', \ldots, u_S']$$

(2.2.3)

such that system (2.2.2) can be rewritten in the following global form :

$$\begin{cases} \dot{x}(t) = A\,x(t) + B\,u(t) \\ y(t) = C\,x(t) \end{cases}$$

(2.2.4)

2.2.1. - Problem formulation

If a decentralized control structure is desired, a local controller is associated with each station. Every local controller is described by the following general model :

$$\mathcal{C}_i \begin{cases} u_i(t) = Q_i\,z_i(t) + K_i\,y_i(t) + v_i(t) \\ \dot{z}_i(t) = S_i\,z_i(t) + R_i\,y_i(t) \end{cases}$$

(2.2.5)

where $z_i(t) \in R^{v_i}$ is the state of the ith controller and $v_i(t) \in R^{m_i}$ is the ith external input.

This control structure associated with system (2.2.2) (2.2.4) is represented by Figure 2.2 :

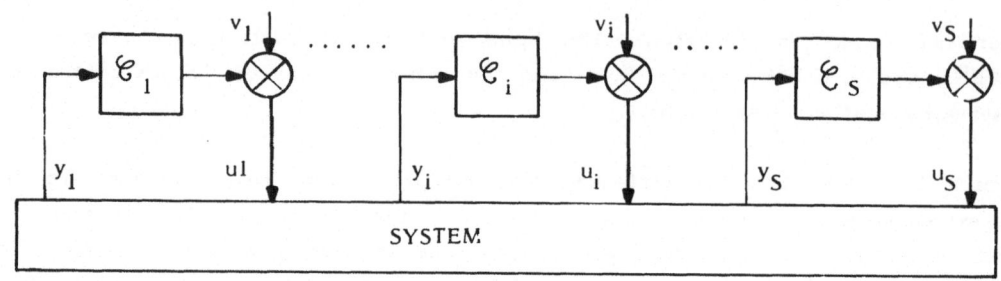

Figure 2.2

Let us define :

S = block–diag. $(S_1, \ldots, S_S)$
R = block–diag. $(R_1, \ldots, R_S)$
Q = block–diag. $(Q_1, \ldots, Q_S)$
K = block–diag. $(K_1, \ldots, K_S)$
$z'(t) = (z'_1, \ldots, z'_S)$
$v'(t) = (v'_1, \ldots, v'_S)$

then, the global controller is described by :

$$
\begin{cases}
u(t) = Q\ z(t) + K\ u(t) + v(t) \\
\\
\dot{z}(t) = S\ z(t) + R\ y(t)
\end{cases}
\tag{2.2.6}
$$

where the decentralization appears in the block-diagonal structure of the matrices S, R, Q and K.

We consider the problem of stabilizing the system (2.2.2) (2.2.4) (or assigning its poles) with a control law specified by (2.2.5) (2.2.6). This problem is equivalent to the problem of existence of a set of matrices S, R, Q and K such that the closed-loop system :

$$\begin{bmatrix} \dot{x}(t) \\ z(t) \end{bmatrix} = \begin{bmatrix} A + BKC & BQ \\ RC & S \end{bmatrix} \begin{bmatrix} x(t) \\ z(t) \end{bmatrix} + \begin{bmatrix} B \\ 0 \end{bmatrix} v(t)$$

is asymptotically stable (i.e., all its poles have a negative real part).

2.2.2. – Decentralized fixed modes

Even if the conditions for the stabilization (or pole assignment) of system (2.2.2) (2.2.4) hold with a centralized information pattern (see § 1.2.6), the system is not always stabilizable (pole assignable) with a decentralized control. Wang (WAN-78b) provided the example of a completely controllable interconnected system for which every subsystem (A_{ii}, B_i) was also controllable such that the conditions for its stabilization by means of centralized state feedback were satisfied. Nevertheless, he showed that the system could not be stabilized if decentralized constraints were imposed on the control.

This means that additional conditions are required. These conditions can be expressed in terms of the new concept of decentralized fixed modes. This concept was introduced by Wang and Davison in 1973 (WAN-73b).

Definition 2.1 (WAN-73b). Given the triple (C, A, B) associated with the system (2.2.4) with $A \in R^{nxn}$, $B \in R^{nxm}$ and $C \in R^{rxn}$ let us define Ω_d as the set of block-diagonal matrices :

$$\Omega_d = \{K / K = \text{block-diag.} \ (K_1, \ldots, K_S) \ , \ K_i \in R^{mxr}, \ (i=1,\ldots,S)\} \tag{2.2.7}$$

The decentralized fixed polynomial of system (2.2.4) is then defined as :

$$\psi \ (p, C, A, B \ \Omega_d) = \underset{K \in \Omega_d}{\text{g.c.d}} \ \{\det \ (pI-A-BKC)\} \tag{2.2.8}$$

Definition 2.2 (WAN-73b). The set of fixed modes of system (2.2.4) with respect to the control law (2.2.6) (decentralized fixed modes) is given by :

$$\Lambda \ (C, A, B \ \Omega_d) = \underset{K \subset \Omega_d}{\cap} \ \sigma(A + BKC) \tag{2.2.9}$$

where $\sigma \ (.)$ denotes the set of eigenvalues of $(.)$.

It is clear that the decentralized fixed modes are the zeros of the decentralized fixed polynomial and can thus be defined as :

$$\Lambda (C,A,B, \Omega_d) = \{ p \in \mathcal{C} \ / \quad \psi (p,C,A,B, \Omega_d) = 0 \} \tag{2.2.10}$$

Note that $\Lambda (C,A,B, \Omega_d) \subset \sigma (A)$ since K=0 is an element of Ω_d. The decentralized fixed modes of the system (2.2.4) are therefore the modes of the system that cannot be moved with a decentralized control (2.2.6), independently of the parameter values assigned to the controller (matrices S,Q,R and K).

The following figure illustrates the situation :

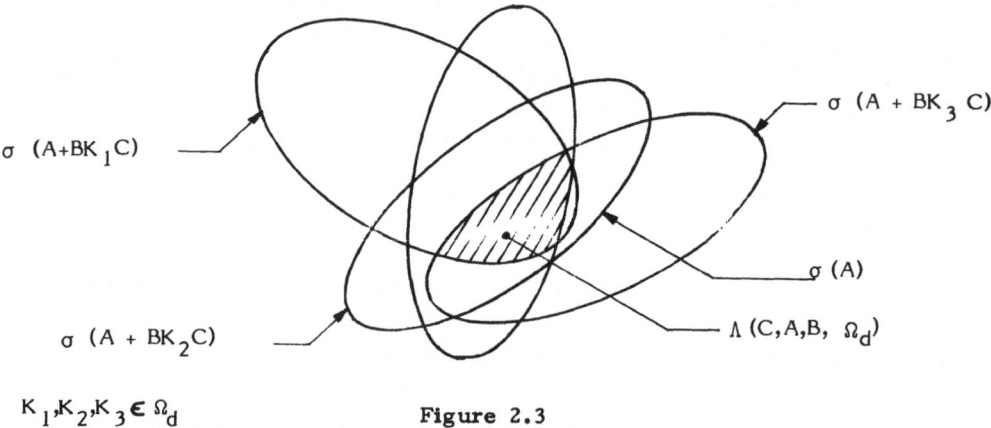

$\sigma (A+BK_1C)$

$\sigma (A + BK_3 C)$

$\sigma (A)$

$\sigma (A + BK_2C)$

$\Lambda (C,A,B, \Omega_d)$

$K_1,K_2,K_3 \in \Omega_d$

Figure 2.3

It is now interesting to note that the set of fixed modes does not depend on the dynamical or non-dynamical nature of the control. If a decentralized static feedback control is used instead of (2.2.6) ; i.e., the matrices S and R are identically zero, the set of decentralized fixed modes remains the same (WAN-73b). This set only depends on the particular output-input pattern specified by the structure of the matrix K.

From their definition, the decentralized fixed modes appear intuitively as an extension of the well-known uncontrollable and unobservable modes defined in Chapter 1. The relations between these concepts are clarified by further discussion.

Consider the triple (C,A,B) associated to the canonical form (1.2.4) of a general system (1.2.1). If the system is controlled by centralized output feedback of the form (1.2.9) with $K \in R^{mxr}$, Wang and Davison (WAN-73b) showed that :

$$\Lambda (C,A,B,R^{mxr}) = \sigma (A_{11}) \cup \sigma (A_{33}) \cup \sigma (A_{44}).$$

This result points out that the set of <u>centralized fixed modes</u> is the union of the uncontrollable and unobservable modes. The analogy is clear when we think that uncontrollable and unobservable modes are those that cannot be moved by the control : it was shown in Chapter 1 that the first ones are disconnected from the input and the second ones from the output. From another hand, if we denote the set of centralized fixed modes by Λ_c , the following definition holds :

$$\Lambda_c = \Lambda (C,A,B,R^{mxr}) = \bigcap_{K \in R^{mxr}} \sigma (A + BKC) \qquad (2.2.11)$$

where the analogy with the definition (2.2.9) of decentralized fixed modes is obvious.

Clearly, from the definitions (2.2.9) and (2.2.11), and since $\Omega_d \subset R^{mxr}$, the set of centralized fixed modes is included in the set of decentralized fixed modes :

$$\Lambda_c \subset \Lambda_d \qquad (2.2.12)$$

where $\Lambda_d = \Lambda (C,A,B, \Omega_d)$.

We understand intuitively that the modes that cannot be moved using a centralized control certainly cannot be moved using a decentralized control.

2.2.3. - Decentralized stabilization and pole assignment

Due to the great economical and reliability advantages provided by decentralization, the problem of stabilization and pole assignment by means of decentralized output feedback has been investigated by many authors (AOK-72) (COR-76a,b) (FES-79) (FES-80) (POT-79) (WAN-73a,b).

Only the most relevant results are presented here. The results of Wang and Davison (WAN-73b) that clearly point out the importance of decentralized fixed modes in this problem are presented first. Then, we present the study of Corfmat and Morse (COR-76), who had a different approach to the problem and provided a more complete solution.

2.2.3.a. - Wang and Davison results (WAN-73b)

The following theorem and corollary state the conditions for the existence of a solution in terms of decentralized output feedback in order to stabilize the system (or to assign its poles).

Theorem 2.1 (WAN-73b). Given the triple (C,A,B) associated with the system (2.2.4) and Ω_d the set of block-diagonal matrices defined in (2.2.7), a necessary and sufficient condition for the existence of a control law (2.2.6) such that the closed-loop system (2.2.7) is asymptotically stable is that :

$$\Lambda_d = \Lambda(C,A,B\ \Omega_d) \subset \mathscr{C}^- \tag{2.2.13}$$

where $\mathscr{C}^-$ denotes the left half part of the complex plan.

Corollary 2.1. Under the same assumptions as in theorem 2.1, a necessary and sufficient condition for the existence of a control law (2.2.6) such that all the poles of the closed-loop system are in $\mathscr{P}$ is that :

$$\Lambda_d = \Lambda\ (C,A,B\ \Omega_d) \subset \mathscr{P} \tag{2.2.14}$$

where $\mathscr{P}$ is an arbitrary prespecified symmetric region of the complex plan.

As a result of the above corollary, the poles of the system can be arbitrarily assigned if and only if :

$$\Lambda_d = \emptyset$$

The similarity between these results and those presented for centralized control in paragraphe 1.2.6 emphasizes the fact that decentralized fixed modes are the generalization of uncontrollable and unobservable modes to decentralized control.

Example 2.1. Consider the following 2-station linear time-invariant dynamic system :

$$\dot{x}(t) = \begin{bmatrix} 2 & 0 & 0 \\ 0 & a & 0 \\ 0 & 0 & -1 \end{bmatrix} x(t) + \begin{bmatrix} 1 \\ 0 \\ 0 \end{bmatrix} u_1(t) + \begin{bmatrix} 0 \\ 1 \\ 1 \end{bmatrix} u_2(t)$$

$$y_1(t) = \begin{bmatrix} 1 & 1 & 0 \end{bmatrix} x(t)$$

$$y_2(t) = \begin{bmatrix} 0 & 0 & 1 \end{bmatrix} x(t)$$

where a is an arbitrary real parameter.

The decentralized output feedback matrices K have the following structure :

$$K = \begin{bmatrix} k_1 & 0 \\ 0 & k_2 \end{bmatrix}$$

The poles of the system are the zeros of the characteristic polynomial : $\det(\lambda I-A)=(\lambda-2)(\lambda+1)(\lambda-a)$. Since $\lambda=2$ is a zero, the open-loop system is unstable independently of the sign of a.

The poles of the closed-loop system are the zeros of the closed-loop characteristic polynomial :

$$\det(\lambda I-A-BKC)=(\lambda-2-k_1)(\lambda+1-k2)(\lambda-a)$$

From Definition 2.1, the decentralized fixed polynomial is $(\lambda-a)$ and the system has a decentralized fixed mode at $\lambda=a$. Although the pole $\lambda=2$ can be arbitrarily assigned by an appropriate choice of k_1, the pole $\lambda=a$ cannot be moved. Nevertheless, if a<0, the system is stabilizable.

2.2.3.b. - Corfmat and Morse results (COR-76)

The above problem was considered by Corfmat and Morse in 1976 using a geometric approach (COR-76) (WON-74). The problem was approached by answering the following question :

"Is there a control law in the form of a <u>static</u> decentralized output feedback such that the system becomes controllable and observable by a single station ?"

If the answer is positive, the standard algorithms of centralized control can then be used for the design of an additional dynamic controller associated with this station such that the stabilization or pole assignment is achieved.

For the system (2.2.4), this approach leads to the following control laws and is illustrated by Figure 2.4 :

$$\begin{cases} u_i(t) = K_i\, y_i(t) + v_i(t) & (i=1,\ldots,S) \\ \begin{cases} \dot{z}_j(t) = S_j\, z_j(t) + R_j\, y_j(t) \\ v_j(t) = Q_j\, z_j(t) + N_j\, y_j(t) \end{cases} & j \in \{1,\ldots, S\} \quad (2.2.15) \end{cases}$$

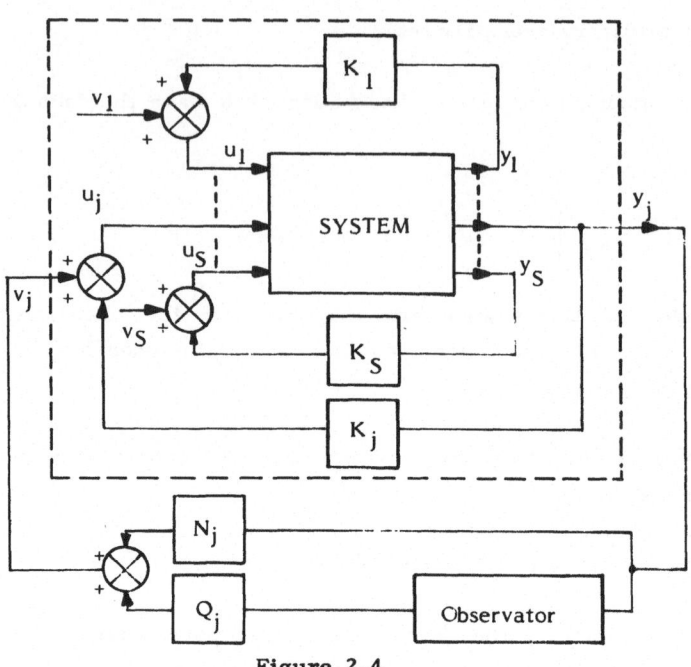

Figure 2.4

Of course, the initial system (2.2.4) is supposed to be globally controllable and observable.

Corfmat and Morse analysis is based on the association of a directed graph (digraph) to the system. A node is associated to every station and an edge exists from the node i to the node j if and only if :

$$W_{ij}(p) = C_j(pI-A)^{-1}B_i \neq 0$$

i.e., if the output at station j can be affected by the input at station i.

<u>Example 2.2.</u> Consider the following system described in the frequency-domain by :

$$W(p) = \begin{bmatrix} W_{11}(p) & 0 & 0 \\ W_{21}(p) & W_{22}(p) & W_{23}(p) \\ 0 & W_{32}(p) & 0 \end{bmatrix}$$

The associated graph is the following :

43

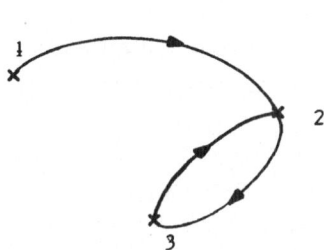

<u>Definition 2.3.</u> A system is said to be <u>strongly connected</u> if its associated digraph is strongly connected. (Note that this does not necessarily imply that all the transfer matrices $W_{ij}(p)$ are $\neq 0$).

A non-strongly-connected system can always be decomposed in several strongly connected subsystems constituted by the stations associated with the nodes defining the strong components of the digraph.

In the example 2.2, the system is not strongly connected and can be decomposed in two strongly connected subsystems, which are :

$$S_1 - y_1 = W_{11}(p) \ u_1$$
$$S_2 - \begin{bmatrix} y_2 \\ y_3 \end{bmatrix} = \begin{bmatrix} W_{22}(p) & W_{23}(p) \\ W_{32}(p) & 0 \end{bmatrix} \begin{bmatrix} u_2 \\ u_3 \end{bmatrix}$$

<u>Definition 2.4</u> (COR-76). A single station system given by the triple (C,A,B) is said to be <u>complete</u> if :

i - $C \ (pI-A)^{-1} \ B \neq 0$

ii – the <u>remnant polynomial</u> ρ (C,A,B) (COR-76) defined as the product of the n ($A \in R^{n \times n}$) first <u>invariant polynomials</u> (ROS-70) of the matrix $\begin{bmatrix} \lambda I-A & B \\ C & 0 \end{bmatrix}$ is equal to 1.

<u>Remark 2.2.</u> (COR-76). Condition ii is equivalent to :

$$\text{rank} \begin{bmatrix} \lambda I-A & B \\ C & 0 \end{bmatrix} > n \quad \text{for} \quad \forall \lambda \in \sigma(A)$$

i.e., no pole of the system is simultaneously uncontrollable and unobservable (see § 1.2.4).

Let us introduce now a fundamental definition in the study of multi-station systems : the concept of <u>complementary subsystems</u>.

<u>Definition 2.5.</u> Consider the S-station system (2.2.2). Define the sets $\pi = \{1,\ldots,S\}$ and the partition $\pi_\alpha = \{i_1,\ldots,i_k\}$, $\pi_\beta = \{i_{k+1},\ldots,i_S\}$. The system is thus partitioned in two aggregated stations α and β . Define the following matrices :

$$B_\alpha = \begin{bmatrix} B_{i_1} , \ldots, B_{i_k} \end{bmatrix} \qquad C_\alpha = \begin{bmatrix} C_{i_1} \\ C_{i_k} \end{bmatrix} \qquad C_\beta = \begin{bmatrix} C_{i_{k+1}} \\ C_{i_S} \end{bmatrix}$$

$$B_\beta = \begin{bmatrix} B_{i_{k+1}} , \ldots, B_{i_S} \end{bmatrix}$$

Hence, the set of <u>complementary subsystems</u> of system (2.2.2) is described by the triples (C_β, A, B_α), $(k=1,\ldots S-1)$.

The first result of Corfmat and Morse deals with strongly connected systems :

<u>Theorem 2.2</u> (COR-76b). Consider the globally controllable and observable system (2.2.1) (2.2.2). Then, using the control laws specified by (2.2.15), with arbitrary $j \in \{1,\ldots S\}$:

* the spectrum of the system can be assigned arbitrarily if and only if all its complementary subsystems are complete. This is equivalent to :

    i - $C_\beta (pI-A)^{-1} B_\alpha \neq 0$ $(k=1,\ldots,S-1)$
      $\Leftrightarrow$ (2.2.1) (2.2.2) is strongly connected.
    ii - $\rho(C_\beta, A, B_\alpha) = 1$ $(k=1,\ldots,S-1)$
      with $\rho (C_\beta, A, B_\alpha)$ denoting the remnant polynomial of the complementary subsystem (C_β, A, B_α).

* the system can be stabilized if and only if :
    - Condition (i) holds.
    iii- $\rho (C_\beta, A, B_\alpha)$ is a stable polynomial, $(k=1,\ldots,S-1)$.

Note that a necessary condition for the existence of a decentralized static output feeback that makes the system controllable and observable by a single station, is that the system is strongly connected.

<u>Example 2.3</u> (KAT-81). Consider the globally controllable and observable system described by the following matrices :

$$A = \begin{bmatrix} 0 & 1 & 1 & 0 \\ 0 & 0 & 0 & 1 \\ 0 & 0 & 1 & 0 \\ 0 & 0 & 0 & 2 \end{bmatrix} \quad B_1 = \begin{bmatrix} 1 \\ 0 \\ 1 \\ 0 \end{bmatrix} \quad B_2 = \begin{bmatrix} 0 \\ 0 \\ 0 \\ 1 \end{bmatrix}$$

$$C_1 = \begin{bmatrix} 1 & 0 & 0 & 0 \\ 0 & 1 & 0 & 0 \\ 0 & 0 & 1 & 0 \end{bmatrix}$$

$$C_2 = \begin{bmatrix} 0 & 0 & 1 & 0 \end{bmatrix}$$

The transfer matrix of the system is :

$$W(p) = \begin{bmatrix} W_{11}(p) & W_{12}(p) \\ W_{21}(p) & W_{22}(p) \end{bmatrix} = \frac{1}{p^2(p-1)(p-2)} \begin{bmatrix} p^2(p-2) & -(p-1) \\ 0 & p(p-1) \\ p^2(p-2) & 0 \\ p^2(p-2) & 0 \end{bmatrix}$$

Since $W_{12}(p) = C_1(pI-A)^{-1}B_2 \neq 0$ and $W_{21}(p) = C_2(pI-A)^{-1}B_1 \neq 0$, the system is strongly connected and condition (i) of theorem 2.2 holds.

The complementary subsystems are (C_2,A,B_1) and (C_1,A,B_2).

* <u>Calculation of $\rho\ (C_1,A,B_2)$</u> :

$$\begin{bmatrix} \lambda I - A & B_2 \\ C_1 & 0 \end{bmatrix} = \left[\begin{array}{cccc|c} \lambda & -1 & -1 & 0 & 0 \\ 0 & \lambda & 0 & -1 & 0 \\ 0 & 0 & \lambda-1 & 0 & 0 \\ 0 & 0 & 0 & \lambda-2 & 1 \\ \hline 1 & 0 & 0 & 0 & 0 \\ 0 & 1 & 0 & 0 & 0 \\ 0 & 0 & 1 & 0 & 0 \end{array} \right]$$

that can be put in Smith's form (ROS-70) to make apparent the invariant polynomials :

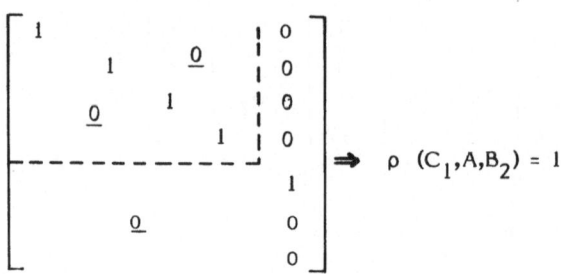

$$\Rightarrow \quad \rho \ (C_1, A, B_2) = 1$$

* Calculation of $\rho (C_2, A, B_1)$:

$$\begin{bmatrix} \lambda I - A & B_1 \\ C_2 & 0 \end{bmatrix} = \left[\begin{array}{cccc|c} \lambda & -1 & -1 & 0 & 1 \\ 0 & \lambda & 0 & -1 & 0 \\ 0 & 0 & \lambda-1 & 0 & 1 \\ 0 & 0 & 0 & \lambda-2 & 0 \\ \hline 0 & 0 & 1 & 0 & 0 \end{array} \right]$$

and Smith's form is the following :

$$\left[\begin{array}{ccc|c} 1 & & \underline{0} & \\ & 1 & & \underline{0} \\ \underline{0} & 1 & & \\ & & 1 & \\ \hline \underline{0} & & & \lambda^2(\lambda-2) \end{array} \right] \quad \Rightarrow \quad \rho \ (C_2, A, B_1) = 1$$

All the complementary subsystems are complete. According to Theorem 2.2, the spectrum of the system can be arbitrarily assigned by using first a static output feedback at station 2, which makes the system controllable and observable by station 1, and then applying an appropriate dynamic output feedback at station 1 for which standard design techniques can be used (BRA-70).

Example 2.4. Consider now the globally controllable and observable system described by the following matrices :

$$A = \begin{bmatrix} 0 & 1 & 0 & 0 \\ 0 & 1 & 0 & 0 \\ 0 & 0 & 1 & 0 \\ 0 & 0 & 0 & 1 \end{bmatrix} \qquad B_1 = \begin{bmatrix} 0 \\ 1 \\ 0 \\ 0 \end{bmatrix} \qquad B_2 = \begin{bmatrix} 0 & 0 \\ 0 & 0 \\ 1 & 0 \\ 0 & 1 \end{bmatrix}$$

$$C_1 = \begin{bmatrix} 1 & 0 & 0 & 0 \\ 0 & 1 & 0 & 0 \\ 0 & 0 & 0 & 1 \end{bmatrix}$$

$$C_2 = \begin{bmatrix} 0 & 1 & 1 & 0 \end{bmatrix}$$

which is also strongly connected.

* <u>Calculation of $\rho(C_1, A, B_2)$</u> :

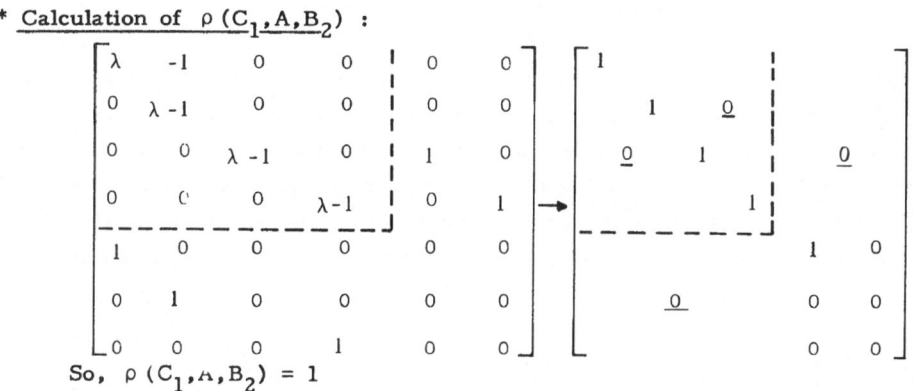

So, $\rho(C_1, A, B_2) = 1$

* <u>Calculation of $\rho(C_2, A, B_1)$</u> :

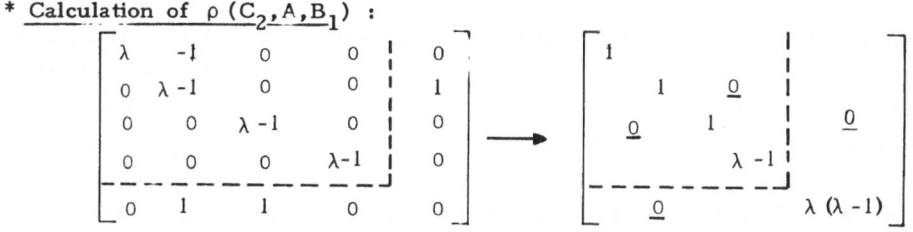

$\rho(C_2, A, B_1) = \lambda - 1$. Consequently, conditions (ii) and (iii) of Theorem 2.2 are not verified. This system cannot be stabilized by a control law of the kind (2.2.15) since the mode $\lambda_1 = 1$ cannot be moved by decentralized control.

In the case for which the conditions of Theorem 2.2 are satisfied, the procedure proposed by Corfmat and Morse leads to all memoriless controllers, save one of high complexity (order n-1). From a practical point of view, this situation presents some disadvantages. In particular, it generally requires high gains : even if all the modes of the system are controllable and observable at one station, some of them might have a weak degree of controllability or observability, that needs to be compensated by high gains. It would be better to spread the control complexity more equally among the local controllers, i.e., to endow each local controller with some dynamics. A result in this direction was obtained by Anderson and Linnemann (AND-84) for interconnected systems (the subsystems are interconnected through their inputs and outputs).

They showed that, for stabilization purpose, the order of the ith local controller can be taken at least equal to the number of stable zeros of the ith subsystem, while the total order of the decentralized controller (sum of the orders of the local controllers) does not exceed n-1.

Although Theorem 2.2 provides an interesting result, the whole problem is not solved yet. The conditions required by Theorem 2.2 guarantee the existence of a control law that uses only one dynamic controller at one station. They are therefore much too constraining for our problem which allows dynamic controllers at each station. In the second part of their work, Corfmat and Morse (COR-76b) relaxed the constraints of Theorem 2.2 and thereby provided the complete solution of the problem.

Theorem 2.3. Consider the globally controllable and observable system (2.2.4). Then, with a decentralized control such that (2.2.6) :

* its spectrum can be arbitrarily assigned if and only if the sum of the dimensions of the strongly connected subsystems is equal to the dimension of the whole system (2.2.4), and if the spectrum of every strongly connected subsystem can be arbitrarily assigned.

* the system (2.2.4) can be stabilized if and only if the complement of the spectra of the strongly connected subsystems with the spectrum of the system (2.2.4) is stable, and if every strongly connected subsystem can be stabilized.

Example 2.5. Consider the 2-station system described by the following transfer matrix :

$$W(p)= \begin{bmatrix} \dfrac{1}{p-1} & \dfrac{1}{p-1-\epsilon} \\ \\ 0 & \dfrac{1}{p} \end{bmatrix}$$

This system is not strongly connected and the two strongly connected subsystems are, in this case, reduced to single-station systems. Their spectra can therefore be arbitrarily assigned. Both are of dimension 1.

* Case 1 : $\epsilon = 0$. In this case, the dimension of the global system is 2 and it is equal to the sum of the dimensions of the strongly connected subsystems. Either stabilization or pole assignment are possible.

* Case 2 : $\epsilon \neq 0$. In this case, the dimension of the global system is 3. The pole $\lambda_1 = 1 + \epsilon$ cannot be moved by a decentralized control. Nevertheless, the stabilization of the system remains possible if $\lambda_1 = 1 + \epsilon$ is stable.

2.2.3.c. - Comments

The results of Corfmat and Morse were carried out using geometrical methods (WON-74) of system theory. With the same approach, Fessas (FES-79) (FES-80) obtained equivalent results using polynomial matrix methods (ROS-70) (WOL-74). Potter et al. (POT-79) also treated the same problem in the case of 2-station systems and provided equivalent conditions expressed in terms of a rank condition for the system matrix. Their result is stated in the following theorem :

Theorem 2.4.. Consider a globally controllable and observable 2-station system $(C_1\ C_2,\ A,\ B_1\ B_2)$ with $C_1(pI-A)^{-1}B_2 \neq 0$ and $C_2(pI-A)^{-1}B_1 \neq 0$ (i.e., the system is strongly connected). There exists an real-valued feedback matrix K_2 of appropriate dimension such that the system $(C_1, A + B_2K_2C_2, B_1)$ is controllable and observable if and only if :

$$\text{rank} \begin{bmatrix} \lambda I - A & B_1 \\ \\ C_2 & 0 \end{bmatrix} \geqslant n \qquad \forall\ \lambda \in \sigma(A) \tag{2.2.16a}$$

$$\text{rank} \begin{bmatrix} \lambda I - A & B_2 \\ C_1 & 0 \end{bmatrix} \geqslant n \qquad \forall \lambda \in \sigma (A) \qquad (2.2.16b)$$

As it will be seen in the next chapter, conditions (2.2.16) present the interest to be the same as the condition provided by Anderson and Clements (AND-81a) to characterize decentralized fixed modes. The next chapter deals with the different characterizations of fixed modes in the time-domain and in the frequency-domain. The equivalence between the results of Cormat and Morse and those of Wang and Davison expressed in terms of fixed modes will thus be clarified.

However, an immediate correlation can be established from the following theorem obtained by Fessas (FES-80).

Theorem 2.5. Consider the globally controllable and observable system (2.2.4) and suppose that it is strongly connected. This system can be made controllable and observable by a single station using a static decentralized control if and only if the system has no fixed modes.

2.3. - ARBITRARY STRUCTURAL CONSTRAINTS

As was pointed out in the introduction, the different local stations are sometimes allowed to share some information. This yields to a non-standard reduced information pattern reflecting a particular configuration of output-input pairs that the controller can connect. The analysis of the problem of control with arbitrary structural constraints does not require a model of the system in which the partitioning in specified stations is apparant.

The system is now described by the following general model :

$$\begin{cases} \dot{x}(t) = A \ x(t) + B \ u(t) \\ y(t) = C \ x(t) \end{cases} \qquad (2.3.1)$$

where $x \in R^n$, $u \in R^m$, $y \in R^r$ and :

$$B = (b_1, \ldots, b_m) \in R^{n \times m}$$
$$C' = (c'_1, \ldots, c'_r) \in R^{r \times n}$$

The particular information pattern can be adequately described by a binary matrix F of dimension mxr such that :

$f_{ij} = 1$ if a feedback is allowed from the output j to the input i.
$f_{ij} = 0$ otherwise.

For every input $u_i(t)$, (i=1,...,m), define the following indices set :

$$J_i = \{\, j \in \{1,\ldots,r\} \,/\, f_{ij} = 1 \,\} \qquad (2.3.2)$$

The dynamic controller corresponding to this information pattern is thus described as follows :

$$\begin{cases} \dot{z}_i(t) = S_i\, z_i(t) + \sum_{j \in J_i} r_{ij}\, y_j(t) \\[4mm] u_i(t) = q_i^!\, z_i(t) + \sum_{j \in J_i} k_{ij}\, y_j(t) + v_i(t) \end{cases} \qquad (i=1,\ldots,m) \qquad (2.3.3)$$

where $z_i(t) \in R^{\nu_i}$ is the state of the ith controller and $v_i(t)$ is the ith external input. S_i, q_i and r_{ij} and k_{ij} are constant matrices, vectors and scalars.

The concept of fixed modes introduced by Wang and Davison (WAN-73b) in the case of decentralized constraints on the control can now be generalized as follows :

Definition 2.6 (SEZ-81a). Define the set :

$$\Omega^* = \{K \in R^{mxr} \,/\, k_{ij} = 0 \text{ if } f_{ij} = 0\} \qquad (2.3.4)$$

then, the set of fixed modes of system (2.3.1) with respect to the control laws given by (2.3.3) is :

$$\Lambda(C,A,B,\ \Omega^*) = \bigcap_{K \in \Omega^*} \sigma\,(A + BKC) \qquad (2.3.5)$$

where $\sigma\,(.)$ denotes the set of eigenvalues of $(.)$.

This definition is the extension of Definition 2.2 to a set of matrices K having an arbitrary structure.

Note that if we consider the output feedback matrices :

$$K_1 \in \Omega_1^*,\ K_2 \in \Omega_2^*,\ \ldots,\ K_k \in \Omega_k^*$$

with the following inclusion relation :

$$\Omega_1^* \subset \Omega_2^* \subset \cdots \subset \Omega_k^*$$

the corresponding sets of fixed modes of system (2.3.1) are related by the inclusion relation :

$$\Lambda \, (C,A,B \, \Omega_k^*) \subset \ldots \subset \Lambda \, (C,A,B \, \Omega_1^*) \subset \sigma(A)$$

It is clear that if the system has uncontrollable or unobservable modes, they will be present in any of the preceding sets.

This situation is illustrated by the following figure :

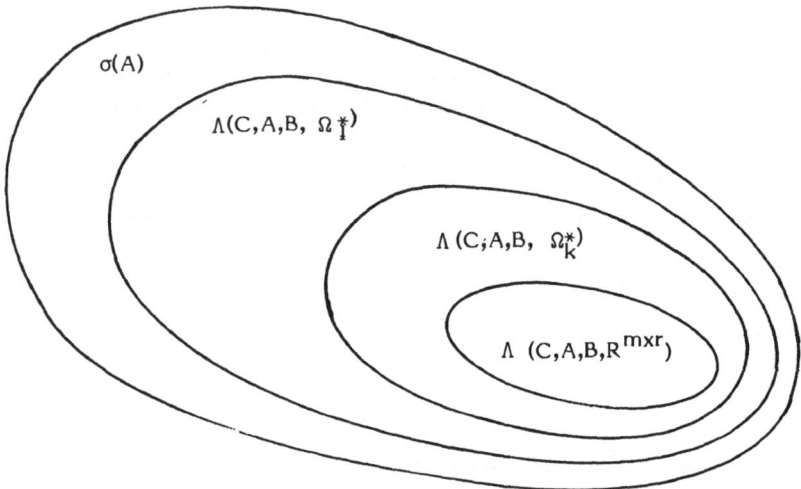

Figure 2.5

It was shown by Sezer and Siljak (SEZ-81a) (SIL-82b) that the conditions for the existence of a set of control laws (2.3.3) such that the system (2.3.1) can be stabilized (or have all its poles arbitrarily assigned) are the same as those in Theorem 2.1 and Corollary 2.2 but replacing the set $\Lambda(C,A,B, \Omega_d)$ by $\Lambda \, (C,A,B, \Omega^*)$.

Remark 2.3. Consider the particular case for which the geographical distribution of the system defines S specified control and observation stations such that the model of the system is given by (2.2.4). The information transfer is defined by the physical connections between stations and it can therefore be aggregated at the level of the stations $(u_i(t)$ and $y_i(t)$, $(i=1,\ldots,S)$, are no longer scalars but vectors of dimension m_i and $r_i)$.

The structurally constrained control is now represented by a binary <u>square</u> matrix F of dimensions SxS :

f_{ij} = 1 if a feedback is allowed from station j to station i
f_{ij} = 0 otherwise

and the sets that were associated to each scalar input in (2.3.2) are now associated to each control station :

$$J_i = \{ j \in \{1,\dots,S\} \; / \; f_{ij} = 1 \} \qquad (i=1,\dots,S)$$

The dynamic controller corresponding to this information pattern is described by :

$$\dot{z}_i(t) = S_i \, z_i(t) + \sum_{j \in J_i} Q_{ij} \, y_j(t) \qquad (i=1,\dots,S)$$

$$u_i(t) = H_i \, z_i(t) + \sum_{j \in J_i} K_{ij} \, y_j(t) + v_i(t)$$

where S_i, Q_{ij}, H_i, K_{ij} are matrices of dimensions (ν_i, ν_i), (ν_i, r_j), (m_i, ν_i), (m_i, r_j), respectively and v_i is the ith external input vector of dimension m_i.

Finally, the set Ω^* is defined as :

$$\Omega^* = \{ K \in R^{mxr} \; / \; K = block \, (K_{ij}), \; (i,j = 1,\dots,S) \text{ and } K_{ij} = 0 \text{ if } f_{ij} = 0 \}$$

2.4. - EVALUATION OF FIXED MODES

This paragraph presents two algorithms for determining the set of fixed modes of a system with respect to a specified control structure. The corresponding Fortran routines are provided in Appendices 2 and 3.

2.4.1. - By comparing the spectra of the open-loop and closed-loop dynamic matrix

This algorithm was proposed by Davison and Ozguner (DAV-76a) in the case of decentralized control. It is based on the definition of decentralized fixed modes (2.2.2) given by Wang and Davison (WAN.73b).

<u>Algorithm 2.1</u> : (DAV-76a)

The set of fixed modes of the system (2.2.4) with respect to a decentralized control whose structure is specified by the set of output feedback matrices Ω_d defined in (2.2.7) can be computed as follows :

1 - Compute the eigenvalues of matrix A : σ (A)

2 - Choose an arbitrary feedback matrix $K_d \in \Omega_d$ (by generation of pseudo-random numbers for example).

3 - Compute the eigenvalues of $(A + BK_dC)$: σ $(A + BK_dC)$.

4 - The fixed modes are the elements of the intersection of the two sets σ (A) and σ $(A + BK_dC)$ (with a probability 1).

It is clear that this algorithm can easily be extended to the case of arbitrary structural constraints by replacing the set Ω_d by Ω^* defined in (2.3.4).

<u>Example 2.6</u> (DAV-83). Consider the 3-station system given by the following triple :

$$\left\{ \begin{bmatrix} 1 & 0 & 0 & 0 \\ \hline 0 & 1 & 0 & 0 \\ \hline 0 & 0 & 1 & 0 \end{bmatrix} \begin{bmatrix} -1 & & & \\ & -1 & & \\ & & -2 & \\ & & & -3 \end{bmatrix} \begin{bmatrix} 0 & 1 & 0 \\ 1 & 1 & 1 \\ 0 & 0 & 1 \\ 0 & 0 & 1 \end{bmatrix} \right\}$$

Following the steps of algorithm 1, we have :

1 - $\sigma(A) = \{ -1(\text{order } 2), -2, -3\}$

2 - We choose arbitrarily : $K_d = \text{diag.}(0.13, -0.17, -0.1)$

3 - $\sigma (A + BK_dC) = \{ -0.01, -2.159, -2.1, -3\}$

4 - The set of fixed modes is given by : $\sigma(A) \cap \sigma (A + BK_dC) = \{ -3 \}$

Although our example considers a system of small dimension, this algorithm applies efficiently for large scale systems. For instance, it was used for a ship steam generator, which is a system with 119 state variables and 3 control stations (DAV-76a).

2.4.2. - By calculation of the system modes sensitivity (TAR-84) (TAR-85)

The preceding paragraphs showed that fixed modes are the modes of the system which remain invariant under structurally constrained feedback, independently of the numerical values of the nonzero feedback gains. We can take advantage of this property to define fixed modes in terms of the concept of eigenvalue sensitivity.

Definition 2.7 (TAR-84). The fixed modes of the system (2.3.1) with respect to a control whose structure is specified by the set of output feedback matrices Ω^* defined in (2.3.4) are the modes of the closed-loop system which are insensible (void of sensitivity) with respect to variations on the nonzero elements of the feedback matrix.

Consider an arbitrary real matrix D of dimension nxn and let us analyze the problem of determining the variation $d\lambda_r$ of any eigenvalue λ_r of D resulting from a variation dD on the matrix D. For the case in which λ_r is an eigenvalue of order 1, the quantities $d\lambda_r$ and dD are related by the following formula due to Faddeev and Faddeeva (FAD-63) :

$$d\lambda_r = w'_r \cdot dD \cdot v_r \qquad (2.4.1)$$

Where v_r and w_r are the normalized right and left eigenvectors of D corresponding to λ_r, respectively ; i.e. :

$$(D - \lambda_r I) \, v_r = 0 \qquad (2.4.2)$$
$$w'_r \, (D - \lambda_r I) = 0 \qquad (2.4.3)$$
$$w'_r \cdot v_r = 1 \qquad (2.4.4)$$

From (2.4.1), it can be shown (MOR-66) (ROS-65b) that :

$$d\lambda_r = \frac{Tr \, \{Q \, (\lambda_r) \cdot dD \, \}}{Tr \, \{ \, Q \, (\lambda_r) \}} \qquad (2.4.5)$$

where $Q(\lambda)$ is the adjoint matrix of $(\lambda I - D)$: $Q(p) = adj \, (\lambda I - D)$

Although Morgan (MOG-66) proposed an algorithm to evaluate $d\lambda_r$ directly from formula (2.4.5), it may be more convenient to use the formula given by Rosenbrock (ROS-65b) where $Q(\lambda_r)$ is replaced by its explicit form (ROS-65a) (GAN-79) :

$$d\lambda_r = \frac{\mathrm{Tr}\{[\underset{i(i\neq r)}{\pi} (D - \lambda_i I)].dD\}}{\underset{i(i\neq r)}{\pi} (\lambda_r - \lambda_i)} \qquad (2.4.6)$$

Another approach consists in determining the gradient of the eigenvalue λ_r with respect to the elements d_{ij} of the matrix D. Since Lancaster (LAN-64) showed that the right and left eigenvectors v_r and w_r associated with an eigenvalue λ_r of order 1 are continuous at d_{ij}, we derive (2.4.2) and we obtain :

$$(\frac{dD}{d(d_{ij})} = \frac{d\lambda_r}{d(d_{ij})} I) v_r + (D - \lambda_r I) \frac{dv_r}{d(d_{ij})} = 0$$

By multiplying on the left by w_r and using (2.4.3) and (2.4.4) we obtain :

$$\frac{d\lambda_r}{d(d_{ij})} = w'_r \frac{dD}{d(d_{ij})} v_r \qquad (2.4.7)$$

In our particular case, the matrix D is the dynamic closed-loop matrix of the system :

$$D = A + BKC = A + \underset{i,j}{\Sigma} b^i k_{ij} c_j \qquad (2.4.8)$$

where b^i and c_j are the ith column of B and the jth row of C, respectively. We are concerned by the variation of λ_r with respect to variations on the elements k_{ij} of K only. (2.4.7) becomes :

$$\frac{\partial\lambda_r}{\partial k_{ij}} = w'_r \frac{\partial D}{\partial k_{ij}} v_r$$

From (2.4.8), we have :

$$\frac{\partial D}{\partial k_{ij}} = b^i.c_j$$

and finally, we obtain :

$$\frac{\partial \lambda_r}{\partial k_{ij}} = w_r' \, b^i \, c_j \, v_r \qquad\qquad (2.4.9)$$

If the feedback matrix has some fixed zero elements, it is clear that $\frac{\partial \lambda_r}{\partial k_{ij}}$ is equal to zero for $k_{ij} = 0$.

Definition 2.8 (TAR-84). Consider a control structure defined by the set Ω^* in (2.3.4) and a feedback matrix $K \in \Omega^*$. The sensitivity matrix of a closed-loop matrix eigenvalue λ_r of order 1 with respect to this control is given by :

$$SK_r = (sk_{ij})$$
$$i=1,\ldots,m$$
$$j=1,\ldots,1$$

$$\qquad\qquad (2.4.10)$$

with

$$sk_{ij} = \begin{cases} w_r' \, b^i \, c_j \, v_r \\ 0 \qquad \text{if } k_{ij} = 0 \end{cases}$$

Theorem 2.6 (TAR-84). Given the system (2.3.1) and the set of feedback matrices Ω^*, λ_r is a fixed mode if and only if either of the two following equivalent conditions holds :

1 - the sensitivity matrix SK_r defined in (2.4.10) is identically zero

2 - $S_r = Tr \{ [\prod_{i(i \neq r)} (D - \lambda_i I) \ . \ dD \} = 0$ \qquad\qquad (2.4.11)

where $D = A + BKC$, $K \in \Omega^*$, $dD = BdKC$ and $dK \in \Omega^*$.

This result is straightforward from Definition (2.2.7) and the relations (2.4.6) and (2.4.10).

Using Theorem 2.6, the fixed modes of a system (not necessarily with distinct poles) can be computed by the following algorithm :

Algorithm 2.2 (TAR-84).

1 - Select an arbitrary feedback matrix $K \in \Omega^*$ such that the closed-loop system modes $\sigma(A+BKC)$ are distinct.

2 - Compute the sensibility matrix SK_r in (2.4.10) or S_r in (2.4.11) for $\forall \lambda_r \in \sigma(A+BKC)$

3 - The fixed modes are those for which SK_r is identically zero or S_r is zero.

Remark 2.4 : It is better to restrict the application of Algorithm 2 to systems with distinct poles. Indeed, the first step of this algorithm may not have a solution when the system has multiple fixed modes. Obviously, this situation is only possible for systems with multiple poles. This case will be discussed subsequently.

Example 2.7. Consider the globally controllable and observable 2-station system described by the following triple :

$$\left\{ \left[\begin{array}{ccc} 0 & 0 & 1 \\ \hline 1 & 0 & 0 \end{array} \right] \left[\begin{array}{ccc} 0 & 1 & 0 \\ 1 & 1 & 0 \\ 0 & 0 & 1 \end{array} \right] \left[\begin{array}{c|c} 1 & 0 \\ 0 & 0 \\ 0 & 1 \end{array} \right] \right\}$$

submitted to the decentralized control :

$$K = \text{diag.}(k_1, k_2) = \text{diag.}(1,5)$$

The results obtained by applying Algorithm 2 are the following :

Closed-loop eigenvalue : λ_r	$SK_r = \dfrac{\partial \lambda_r}{\partial K}$	S_r for $dK=K$
-2	$\left[\begin{array}{cc} -0.7645 & 0 \\ 0 & -0.1529 \end{array} \right]$	-29.9999
3	$\left[\begin{array}{cc} 1.088 & 0 \\ 0 & 0.2177 \end{array} \right]$	19.9999
1	$\left[\begin{array}{cc} -0.49 \ 10^{-15} & 0 \\ 0 & -0.32 \ 10^{-17} \end{array} \right]$	$0.666 \ 10^{-14}$

Therefore $\lambda_r = 1$ is a fixed mode with an accuracy $\geqslant 10^{-14}$.

This algorithm has been used for a ship steam generator model in (TAR-85).

The preceding analysis is restricted to simple fixed modes and the approach is specially adequate for systems with simple modes, which guarantee that the first step of the algorithm can be achieved.

The generalization of this approach to systems with multiple modes is not easy (TAR-85). Indeed, if λ_r is a multiple eigenvalue of order q of a real matrix D, its variations resulting from variations of the elements d_{ij} of D are given by the solutions of the following algebraic equation of order q (PAR-74) :

$$\frac{1}{q!} \, \{d^{q-1} \, [\text{Tr } Q \, (\lambda)]_{\lambda \, = \, \lambda_r} \} \cdot d\lambda_r = \text{Tr} \, \{ \sum_{k=1}^{q} \frac{1}{k!} \, [d^{k-1} \, Q(\lambda)]_{\lambda \, \neq \, \lambda_r} \cdot dD\} \qquad (2.4.12)$$

with $Q \, (\lambda) = \text{adj} \, (\lambda I - D)$.

Generally, a multiple eigenvalue λ_r of order q gives rise, after the perturbation dD, to q simple eigenvalues : $\lambda_r + (d\lambda_r)_1, \ldots, \lambda_r + (d\lambda_r)_i, \ldots, \lambda_r + (d\lambda_r)_q$, where $(d\lambda_r)_i$, (i=1,...,q), are the solutions of equation (2.4.12). From Definition (2.2.7), λ_r is a fixed mode of order q if $(d\lambda_r)_i = 0$, (i=1,...,q). So, from (2.4.12), the following condition must hold :

$$S_r = \text{Tr} \, \{ \sum_{k=1}^{q} \frac{1}{k!} \, [d^{k-1} \, Q(\lambda)]_{\lambda = \lambda_r} \cdot dD\} \, = 0 \qquad (2.4.13)$$

Remark 2.5.

A special case may occur for which not all the variations $(d\lambda_r)_i$, (i=1,...,q), are zero but only some of them, say q' $<$ q. Though condition (2.4.13) does not hold, λ_r is a fixed mode of order q'. Since D=A+BKC, this means that another matrix $K \in \Omega^*$ can be found whereby λ_r is an eigenvalue of order q' of D. In this situation, all the variations $(d\lambda_r)_i$, (i=1,...q'), are zero and we are brought back to the case for which condition (2.4.13) holds.

The evaluation of S_r requires the analytical calculation of the adjoint matrix of $(\lambda I - D)$ and of its variations from the order 1 to (q-1), in our case resulting from the variations of the elements k_{ij} of the feedback matrix K. Obviously, this is a very heavy task.

Like in the case of simple fixed modes, another approach consists in evaluating the gradients of λ_r with respect to the elements k_{ij} of K. We can thus use the following theorem due to Lancaster (LAN-64).

<u>Theorem 2.7</u> (LAN-64). Given λ_r a multiple eigenvalue of order q of a real matrix D(R), where R is a parameter, consider that the right and left eigenvector matrices V_q and W_q are chosen such that $W'_q Q_q = I$. Then, the q derivatives of λ_r with respect to R are the eigenvalues of the matrix W_q dD/dR V_q. (For the conditions of existence of W_q and V_q, the reader is refered to (LAN-64)).

In our case, D=A+BKC and the parameter R must be taken successively as each element k_{ij} of K. Moreover, as we have already seen :

$$\frac{\partial D}{\partial k_{ij}} = \frac{\partial (A + \sum\limits_{i,j} b^i k_{ij} c_j)}{\partial k_{ij}} = b^i c_j = 0$$

and we have the following result :

<u>Theorem 2.8</u> (TAR-85). Given the system (2.3.1) and the set of feedback matrices Ω^*, λ_r is a fixed mode of order q if and only if all the eigenvalues of every matrix $W_q b^i c_j V_q$, (i,j such that $k_{ij} \neq 0$), are zero. W'_q and V_q are the left and right eigenvector matrices corresponding to λ_r chosen such that $W'_q V_q = I$.

Note that the above theorem is not easy to use, either. This requires, first, the determination of the eigenvector matrices W_q and V_q and, then, the computation of the eigenvalues of μ matrices of dimension qxq (where μ is the number of nonzero elements in the feedback matrices $K \in \Omega^*$). The conclusion is that the approach based on the sensitivity properties of eigenvalues is not suitable to determine multiple fixed modes.

2.4.3 - Concluding remarks

The two preceding algorithms are easily implemented. It is interesting to note that they can also be used to compute the uncontrollable and unobservable modes of a system by replacing the set Ω^* by R^{mxr}. The only problem, which may induce a difficult interpretation of the results, is the computer accuracy. For the first algorithm, one has to decide in which limits two eigenvalues can be considered as equal and, in the second algorithm, the problem is to decide the accuracy of the computer zero. However, Davison et al. (DAV-78b) showed that if a closed-loop eigenvalue is 'very close' to an open-loop eigenvalue, the transfer matrix computed when considering this eigenvalue as a fixed mode is a good approximation of the transfer matrix of the system.

2.5. – CONCLUSION

This chapter deals with the problem of stabilization and pole assignment when a specified restricted information pattern is required, which constraints the feedback control structure.

The case of a decentralized scheme of control is presented first. The approaches of Wang and Davison (WAN-73b) and of Corfmat and Morse (COR-76) are both considered. Their results give, in different terms, the conditions for the existence of a solution to the above problem. The equivalence of these results will be discussed in the next chapter and we will see how the results of Corfmat and Morse can be expressed in terms of the concept of fixed modes introduced by Wang and Davison. It will also be pointed out how these results are related to the concept of controllability under decentralized information structure.

The case of arbitrary structural constraints on the control is then considered and the concept of fixed modes is extended to this general framework.

Finally, two algorithms for the evaluation of fixed modes are presented.

This chapter points out the fundamental importance of fixed modes. Indeed, the existence of unstable fixed modes indicates that stabilization is impossible, while the presence of any kind of fixed modes rules out arbitrary pole placement. Fixed modes appear thus as a generalization of uncontrollable and unobservable modes in the case of non-constrained control (centralized control). Many authors have been interested in this problem (for a survey, see (TRA-84a)) and their results provide different characterizations of fixed modes, which will be presented in the next chapter. The analysis of these characterizations, both in the time and in the frequency domain, will allow us to point out the situations that give rise to fixed modes, and to give an interpretation of fixed modes related to their origin.

CHAPTER 3

CHARACTERIZATION OF FIXED MODES

3.1. - INTRODUCTION

The last chapter pointed out the fundamental importance of the concept of fixed modes. Indeed, the existence of a solution for the stabilization or the pole placement problem of a linear invariant system requiring some structural constraints on the control depends critically on the properties of this finite set of numbers. The presence of unstable fixed modes indicates that stabilization is impossible while the presence of any sort of fixed modes rules out arbitrarily pole placement.

The first important problem of evaluating fixed modes for a given system under some prespecified structural constraints on the control has been treated in the last chapter and two efficient algorithms have been presented.

Our present objective is to obtain some insight into the behaviour of the fixed modes. For this purpose, the present chapter deals with the characterization of fixed modes and attempts to provide an understanding of their occurence.

We will successively consider the characterizations of fixed modes with a time-domain and a frequency-domain representation of the system. From the study and the comparison of the different characterizations, the reasons of the occurence of fixed modes and the conditions for their existence will be pointed out. It will appear that fixed modes may originate from two distinct sources : they may have a structural or a parametric origin. We will attempt to provide a clear classification of the different types of fixed modes and a whole paragraph will deal with the specific characterizations of fixed modes arising from structural particularities.

The last paragraph of this chapter is an overview of the graphical characterizations of fixed modes. Indeed, dynamic systems can be represented by directed graphs providing an efficient way to explore various structural properties by using the concepts of graph-theory. Numerous results have been obtained for the problem of characterizing fixed modes in this graph-theoretic framework.

3.2. - CHARACTERIZATION IN TERMS OF TRANSMISSION ZEROS

This characterization gives an interesting insight into the reasons of occurence of fixed modes and provides a simple explanation for their existence in terms of the familiar concept of system poles and zeros. It was provided independently by several authors (HUJ-84) (TAO-84) (DAV-85) (SER-82) (VID-83) whose results are proved and formulated using different methods. This characterization can equivalently be interpreted and used either in the frequency-domain context or in the time-domain context since transmission zeros are a concept as well defined by means of the system matrix as by means of the transfer function matrix (see Appendix 1).

Although the results are developped in a time-domain framework in (DAV-85a) and in a frequency-domain framework in (SER-82) (VID-83), their equivalence appears clearly in (HUJ-84) and (TAO-84).

In all the papers but (DAV-85), this characterization is obtained using an expression relating the closed-loop characteristic polynomial to the controller parameters and zero polynomials of certain constituant subsystems.

Consider the class of linear time-invariant systems described by the following state-space representation :

$$\begin{cases} \dot{x}(t) = A\, x(t) + B\, u(t) \\ y(t) = C\, x(t) \end{cases} \tag{3.2.1}$$

where $x(t) \in R^n$, $u(t) \in R^m$, $y(t) \in R^r$ are the state, input and output, respectively and A, B, C are real matrices of appropriate dimensions.

Many large scale systems appear to be geographically distributed or composed of an interconnection of subsystems. These characteristics can be taken into account in the model of the system by a partitioning of the input and output vectors resulting in several control and observation stations :

$$\begin{cases} \dot{x}(t) = A\, x(t) + \sum_{i=1}^{S} B_i\, u_i(t) \\ \\ y_i(t) = C_i\, x(t) \qquad (i=1,...,S) \end{cases} \tag{3.2.2}$$

which is related to (3.2.1) by :

$$B = (B_1,...,B_S)$$
$$C' = (C'_1,...,C'_S)$$

$$u'(t) = (u'_1(t),\ldots,u'_S(t))$$
$$y'(t) = (y'_1(t),\ldots,y'_S(t))$$

with $u_i(t) \in R^{m_i}$, $y_i(t) \in R^{r_i}$, $(i=1,\ldots,S)$,

and $m = \sum\limits_{i=1}^{S} m_i$, $r = \sum\limits_{i=1}^{S} r_i$, where S is the number of control and observation stations.

In order to point out the problems arising specially from the structural constraints on the control, we make the assumption that system (3.2.1) (3.2.2) is globally controllable and observable.

The following development is taken from (TAO-84) in which the chosen notations can be simply manipulated and do not lead to huge formula.

Consider the system (3.2.1) for which the transfer function matrix is given by :

$$W(p) = C (pI-A)^{-1} B = \frac{N(p)}{\phi (p)} \tag{3.2.3}$$

where $N(p)=C \text{ adj}(pI-A) B$ is the rxm numerator transfer function matrix and $\phi (p)$ is the characteristic polynomial of the system (3.2.1).

The set of i-input i-output subsystems of (3.2.1), $(i=1,\ldots,\min(m,r))$, are called i-dimensional subsystems and denoted by (C_q^i,A,B_s) or by their transfer function matrix $W_{q,s}^i(p)=C_q^i(pI-A)^{-1}B_s^i=W_j^i(p)$, where for notation convenience the subscripts q,s have been replaced by $j.C_q^i$, $(q=1,\ldots,q_c)$, are the set of submatrices formed from i rows of the matrix C and B_s^i, $(s=1,\ldots,s_b)$, are the set of submatrices formed from i columns of the matrix B, where :

$$q_c = \frac{r!}{(r-i)\; !i!} \qquad\qquad s_b = \frac{m!}{(m-i)\; !i!}$$

$W_j^i(p)$, $(j=1,\ldots,j_i)$, are the set of ixi submatrices of $W(p)$ and $j_i=q_c \cdot s_b$.

As was pointed out in Chapter 2, we know that the set of fixed modes is independent of the dynamic or static nature of the controller. Therefore, we will consider a static control law of the form :

$$u = K \; y + v \tag{3.2.4}$$

where v is the mx1 external input vector and the constant mxr matrix K can take any arbitary structure.

The closed-loop characteristic polynomial is thus obtained as :

$$\phi_c(p) = \left| pI-A-BKC \right| = \phi(p) \left| I - W(p) K \right| \qquad (3.2.5)$$

where $\left| . \right|$ stands for det (.).

The determinant in (3.2.5) can be expanded in terms of the principal minors of (-W(p) K) to give :

$$\phi_c(p) = \phi(p) \left(1 + \sum_{i=1}^{r} h^i(p) \right) \qquad (3.2.6)$$

where $h^i(p)$ is the sum of principal ith order minors of (-W(p) K). From Binet-Cauchy formula (GAN-79), these minors can be written in terms of determinants of submatrices of W(p) and K' as :

$$h^i(p) = \sum_{j=1}^{j_i} (-1)^i \left| K'^i_j \right| \left| W^i_j(p) \right| \qquad (3.2.7)$$

This result can equivalently be stated in a time-domain framework. From the definition of the zero polynomial of a system (KWA-72), the set of zero polynomials of the i-dimensional subsystems is given by :

$$Z^i_j(p) = \begin{vmatrix} pI-A & B^i_s \\ \\ C^i_q & 0 \end{vmatrix} = \left| pI-A \right| \left| C^i_q (pI-A)^{-1} B^i_s \right| = \phi(p) \left| W^i_j(p) \right| \qquad (3.2.8)$$

where the roots of $Z^i_j(p)$ are the set of transmission zeros of the i-dimensional subsystems.

From (3.2.6), (3.2.7) and (3.2.8), the two following results are equivalent :

- Frequency domain :

$$\phi_c(p) = \phi(p) + \sum_{i=1}^{r} \sum_{j=1}^{j_i} (-1)^i \left| K'^i_j \right| \phi(p) \left| W^i_j(p) \right| \qquad (3.2.9)$$

– Time domain :

$$\phi_c(p) = \phi(p) + \sum_{i=1}^{r} \sum_{j=1}^{j_i} (-1)^i |K'_j{}^i| \; z_j^i(p)$$

(3.2.10)

When i rows and i columns have been selected in K' to form $K'_j{}^i, C_q^i$ and B_s^i denote the submatrices of C and B obtained by selecting the same rows and columns. The subsystem of (C,A,B) or W(p) corresponding to the submatrix $K'_j{}^i$ is thus denoted by (C_q^i, A, B_s^i) or equivalently $W_j^i(p)$.

Now, consider that the controller (3.2.4) is constrained to have the following decentralized structure :

$$K = \text{block-diag.}(K_1, \ldots, K_S)$$

(3.2.11)

where $K_i \in R^{m_i \times r_i}$, (i=1,...,S).

The fixed modes are easily characterized using the above development.

3.2.1. – Tarock's results (TAO-84)

<u>Theorem 3.1.</u> The necessary and sufficient condition for a pole λ_0 of the system (C,A,B) or W(p) to be a fixed mode with respect to the controller K in (3.2.11) is that :

$$\text{rank} \begin{bmatrix} \lambda_0 I - A & B_s^i \\ \\ C_q^i & 0 \end{bmatrix} < n+i \; , \quad (i=1,\ldots,\min(m,r))$$

(3.2.12)

or equivalently :

$$\phi(\lambda_0) \; |W_j^i(\lambda_0)| = 0 \; , \quad (i=1,\ldots,\min(m,r))$$

(3.2.13)

where (C_q^i, A, B_s) or $W_j^i(p)$ are the subsystems of (C,A,B) corresponding to non-singular submatrices of K'.

The above theorem states that λ_0 is a fixed mode of (C,A,B) or $W(p)$ with respect to K in (3.2.11) if and only if all the subsystems (C_q^i, A, B_s) or $W_j^i(p)$ corresponding to non-singular submatrices $K_j^{'i}$ of K' have a common transmission zero corresponding to the system pole λ_0.

This result is straightforward from formulae (3.2.9) or (3.2.10).

3.2.2. - Hu and Jiang results (HUJ-84)

Hu and Jiang state their results using some preliminary definitions presented below.

From the integers m_i and r_i specifying the controller structure in (3.2.11), define :

$$m_0 = 0, \quad \overline{m}_i = \sum_{j=0}^{i} m_j, \quad m = \overline{m}_S$$
$$r_0 = 0, \quad \overline{r}_i = \sum_{j=0}^{i} r_j, \quad r = \overline{r}_S \qquad (i=1,\ldots,S)$$

such that the input and output vectors are partitioned in the following way :

$$u = \begin{bmatrix} u_1 \\ \vdots \\ \vdots \\ u_m \end{bmatrix} = \begin{bmatrix} u^1 \\ \hline \vdots \\ \hline u^S \end{bmatrix} \qquad u^i = \begin{bmatrix} u_{\overline{m}_{i-1}+1} \\ \vdots \\ u_{\overline{m}_i} \end{bmatrix}$$

$$y = \begin{bmatrix} y_1 \\ \vdots \\ y_r \end{bmatrix} \quad \begin{bmatrix} y^1 \\ \hline \vdots \\ \hline y^S \end{bmatrix} \qquad y^i = \begin{bmatrix} y_{\overline{r}_{i-1}+1} \\ \vdots \\ y_{\overline{r}_i} \end{bmatrix}$$

Let nonnegative integers s_i and q_i, $(i=1,\ldots,S)$, satisfy the condition :

$$0 \leqslant s_i \leqslant \overline{m}_i, \quad \sum_{i=1}^{S} s_i \, 0 \quad , \quad 0 \leqslant q_i \leqslant \overline{r}_i, \quad \sum_{i=1}^{S} q_i \, 0$$

and, for $s_i > 0$ and $q_i > 0$, define the positive integers :

$$f_{i,j}, \quad (j=1,\ldots,s_i)$$
$$g_{i,j}, \quad (j=1,\ldots,q_i) \tag{3.2.14}$$
$$\bar{m}_{i-1}+1 \leqslant f_{i,1} < \ldots < f_{i,si} \leqslant \bar{m}_i$$

$$\bar{r}_{i-1}+1 \leqslant g_{i,1} < \ldots < g_{i,qi} \leqslant \bar{r}_i$$

The following set of subsystems of (C,A,B) or $W(p)$ is now defined :

$$(\bar{C},A,\bar{B}) \quad \text{or} \quad \bar{W}(p) = \bar{C}(pI-A)^{-1}\bar{B} \tag{3.2.15}$$

$$B = \begin{bmatrix} \bar{B}^1,\ldots,\bar{B}^S \end{bmatrix} \qquad \bar{B}^i = \begin{bmatrix} b_{f_{i,1}} , \ldots, b_{f_i,s_i} \end{bmatrix}$$

$$\bar{C} = \begin{bmatrix} \bar{C}^1 \\ \vdots \\ \bar{C}^S \end{bmatrix} \qquad \bar{C}^i = \begin{bmatrix} c_{g_{i,1}} \\ \vdots \\ c_{g_{i,q_i}} \end{bmatrix}$$

with :

$$\bar{u} = \begin{bmatrix} \bar{u}^1 \\ \vdots \\ \bar{u}^S \end{bmatrix} \qquad \bar{u}^i = \begin{bmatrix} u_{f_{i,1}} \\ \vdots \\ u_{f_{i,s_i}} \end{bmatrix}$$

$$\bar{y} = \begin{bmatrix} \bar{y}^1 \\ \vdots \\ \bar{y}^S \end{bmatrix} \qquad \bar{y}^i = \begin{bmatrix} y_{g_{i,1}} \\ \vdots \\ y_{g_{i,q_i}} \end{bmatrix}$$

and

$$\bar{K} = \text{block-diag.}(\bar{K}_1,\ldots,\bar{K}_S), \quad \bar{K}_i \in R^{s_i \times q_i}, \quad (i=1,\ldots,S) \tag{3.2.16}$$

Note that when $s_i=0$, $\bar{B}^i$, $\bar{u}^i$ disappear and when $q_i=0$, $\bar{C}^i$, $\bar{y}^i$ disappear. In both cases, $\bar{K}_i$ disappears.

Definition 3.1. The subsystems $(\bar{C},A,\bar{B})$ or $\bar{W}(p)$ in (3.2.15) associated with the controller $\bar{K}$ in (3.2.16) are called the normal subsystems of (C,A,B) or $W(p)$. In particular, when $s_i=q_i$, they are called the nonsingularly normal subsystems.

The set of normal subsystems is denoted by N.sub.(C,A,B) and the set of nonsingularly normal subsystems by N.N.sub.(C,A,B).

With these definitions, Hu and Jiang state the following results :

<u>Theorem 3.2.</u> λ_0 is a fixed mode of (C,A,B) with respect to K in (3.2.11) if and only if λ_0 is a fixed mode of all the normal subsystems $(\overline{C},A,\overline{B})$ with respect to $\overline{K}$ in (3.2.16) ; i.e. :

$$\Lambda(C,A,B,K) = \bigcap_{(\overline{C},A,\overline{B}) \, \in \, N.sub.(C,A,B)} \Lambda(\overline{C},A,\overline{B},\overline{K})$$

<u>Theorem 3.3.</u> The necessary and sufficient condition for a pole λ_0 of the system (C,A,B) or W(p) to be a fixed mode with respect to K in (3.2.11) is that :

$$\text{rank} \begin{bmatrix} \lambda_0 I - A & \overline{B} \\ \overline{C} & 0 \end{bmatrix} < n + \sum_{i=1}^{S} s_i \qquad (3.2.18)$$

or equivalently :

$$\phi(\lambda_0) \ \det \overline{W}(\lambda_0) = 0 \qquad (3.2.19)$$

for all the nonsingularly normal subsystems of (C,A,B) or W(p).

This theorem states that λ_0 is a fixed mode of (C,A,B) or W(p) if and only if λ_0 is a common transmission zero of all the nonsingularly normal subsystems $(\overline{C},A,\overline{B})$ or $\overline{W}(p)$ which corresponds with a pole of the system.

The similarity of this result with the one of Tarock is obvious. Indeed, it is clear that $(\overline{C},A,\overline{B})$ or $\overline{W}(p)$ of dimension $\sum_{i=1}^{S} s_i$ correspond to the subsystems (C_q^i,A,B_s^i) or $W_j^i(p)$ of dimension i while the matrices $\overline{K}_i$ in (3.2.16) with $s_i = q_i$ correspond to the non singular matrices K_j^i. Formula (3.2.17) and (3.2.18) are therefore identical to formula (3.2.12) and (3.2.13) and Theorems 3.3 and 3.1 are equivalent. The advantage of Hu and Jiang formulation is that the subsystems to be considered are systematically defined by the set of integers in (3.2.14).

The same result was also provided previously by Seraji (SER-82) and Vidyasagar and Wiswanadham (VID-83) in a frequency-domain context. Though the contributions of these papers provide the same above established result using the same expansion of the closed-loop characteristic polynomial given in (3.2.9) and (3.2.10), they differ by the notations and the formulations which are used.

Their results are briefly presented in the two next paragraphs. The equivalent concepts are pointed out and the originalities which can be advantageously exploited are shown.

3.2.3. - Vidyasagar and Wiswanadham results (VID-83)

The main result of this paper is presented as a characterization of the fixed polynomial, whose zeros are the fixed modes of the system (3.2.1) with respect to the control structure specified by K in (3.2.11).

Define the sets $M = \{1,\ldots,m\}$ and $R = \{1,\ldots,r\}$ and denote by $*$ the number of elements of the set $(*)$. If $I = J$, then $W\begin{bmatrix} I \\ J \end{bmatrix}$ denotes the minor of $W(p)$ formed from the rows in the set I and the columns in the set J. $K\begin{bmatrix} I \\ J \end{bmatrix}$ is defined in the same way and $Z\begin{bmatrix} I \\ J \end{bmatrix} = (p) \; W\begin{bmatrix} I \\ J \end{bmatrix}$.

<u>Theorem 3.4.</u> The fixed polynomial $F(p)$ of system (3.2.1) with respect to K in (3.2.11) is given by :

$$F(p) = \text{g.c.d} \left\{ \phi(p), \quad Z \begin{bmatrix} I_{i_1} \cup I_{i_2} \cup \;\ldots\; \cup I_{i_S} \\ J_{i_1} \cup J_{i_2} \cup \;\ldots\; \cup J_{i_S} \end{bmatrix} \right\} \qquad (3.2.19)$$

with :

$$\begin{aligned} & I_{ij} \subset R_j = \{\bar{r}_{j-1} + 1, \ldots, \bar{r}_j\} \\ & J_{ij} \subset M_j = \{\bar{m}_{j-1} + 1, \ldots, \bar{m}_j\} \qquad (i=1,\ldots,S) \\ \text{and} \quad & \| I_{ij} \| = \| J_{ij} \| , \qquad (j=1,\ldots,S). \end{aligned} \qquad (3.2.20)$$

It is obvious that this result is equivalent to those of Tarock and Hu and Jiang stated in Theorems 3.1 and 3.3. Indeed,

$$Z \begin{bmatrix} I_{i1} \cup I_{i2} \cup \ldots \cup I_{iS} \\ J_{i1} \cup J_{i2} \cup \ldots \cup J_{iS} \end{bmatrix}$$

is a polynomial whose roots are the transmission zeros of some subsystem of dimension $\sum_{j=1}^{S} \| I_{ij} \|$ corresponding to some nonsingular matrix $K\begin{bmatrix} I \\ J \end{bmatrix}$. This polynomial corresponds to some $Z_j(p)$ in Tarock's notations.

The subsystems to be considered are specified in a similar way as in Hu an Jiang result by the sets of integers J_{ij} and I_{ij} in (3.2.20) which correspond to the sets of integers $\{f_{i,1}, \ldots, f_{i,s_i}\}$ and $\{g_{i,1}, \ldots, g_{i,q_i}\}$ in (3.2.14), respectively.

Note that this work appeared one year before the two other ones in the literature.

3.2.4. - Seraji's results (SER-82)

This result is even anterior to the one of Vidyasagar and Wiswanadham. The main result is given for diagonal scalar controllers for which the set of subsystems to be considered is very easily determinated. Then, the approach presents the originality to bring back the general case to this particular situation. Any arbitrary decentralized controller is treated by applying a transformation on the system such that the controller is brough back into a diagonal scalar structure in the new basis. The result can thus be applied on the transformed system.

Consider a controller whose structure is specified by the output feedback matrix in the following way :

$$K_s = \text{block-diag.}(k_1,\ldots,k_S) \qquad k_i \in R \tag{3.2.21}$$

Seraji gives the following result (equivalent to the three anterior ones) :

<u>Theorem 3.5.</u> The necessary and sufficient condition for λ_0 to be a fixed mode of system (3.2.1) with respect to K_s is that λ_0 is a common transmission zero of all the subsystems of dimension (i=1,...,S) formed selecting the same inputs and outputs of the system.

Now, if the controller structure is given by K as in (3.2.11) where $m_i \geqslant 1$, $r_i \geqslant 1$, Seraji applies a transformation on the system such that Theorem 3.5 can be used for the transformed system.

Consider the transformation matrix P :

$$
\begin{aligned}
K &= P \, K_s \\
W(p) &= W_t(p) \, P
\end{aligned} \tag{3.2.22}
$$

$$P = K \text{ diag.} \{ \frac{1}{k_1} , \cdots, \frac{1}{k_S} \}$$

The fixed modes of the system $W(p)$ with respect to K are then given by the fixed modes of system $W_t(p)$ with respect to K_s, which can be determined by using Theorem 3.5.

Despite of the originality of this approach, it is clear that the direct determination, from the original system, of the subsystems for which the transmission zeros have to be checked seems to be more convenient.

3.2.5. - Davison and Wang results (DAV-85)

Recently, the same characterization of fixed modes was established by Davison and Wang by using a more direct method. Their result is straightforward to the application of a lemma taken from (AND-82).

<u>Lemma 3.1</u> (AND-82). Given $M_0 \in R^{n \times n}$, det $M_0 \neq 0$, let $M_i = b_i c'_i$ where b_i, c_i $\in R^n$, and let q_i be a real scalar ; then

$$\det (M_0 + \sum_{i=1}^{s} q_i M_i) = 0 \qquad \forall q_i \in R, (i=1,...,s)$$

if and only if the following conditions are satisfied :

$$\det \begin{bmatrix} M_0 & b_{i_1} & b_{i_2} & \cdots & b_{i_t} \\ c'_{i_1} & 0 & 0 & \cdots & 0 \\ c'_{i_2} & 0 & 0 & \cdots & 0 \\ \vdots & \vdots & \vdots & & \vdots \\ c'_{i_t} & 0 & 0 & \cdots & 0 \end{bmatrix} = 0$$

$(i_1 = 1,...,s)$; $(i_2 = 1,...s)$; ... ; $(i_t = 1,...,s)$ for all disjoint values of i_1, $i_2,...,i_t$; (t=1,...,s).

Now, the system (3.2.1) can be rewritten as :

$$\begin{cases} \dot{x}(t) = A x(t) + (b_1, b_2,...,b_m) u(t) \\ y(t) = (c_1, c_2,...,c_r)' x(t) \end{cases}$$

and the controller information flow constraint represented by the output feedback matrix K in (3.2.11) can be represetend by the following pairs of integers :

$$\{ (i_1, j_1), (i_2, j_2),...,(i_s, j_s) \} \qquad\qquad (3.2.33)$$

where $i_k \in \{1,\ldots,m\}$, $j_k \in \{1,\ldots,r\}$, $(k=1,\ldots,s)$; i.e., each pair (i_k,j_k) represents a nonzero element in K.

The fixed modes of (3.2.1) with respect to K in (3.2.11) (3.2.23) have thus the properties :

$$\lambda_0 \in \sigma(A) \tag{3.2.24}$$

$$\det \{ A - \lambda_0 I + \sum_{k=1}^{s} q_k b_{i_k} c'_{j_k} \} = 0 \qquad \forall q_k \in R, (k=1,\ldots,s)$$

and from Lemma 3.1, the following result is straightforward :

Theorem 3.6 (DAV-85). The necessary and sufficient condition for $\lambda_0 \in \sigma(A)$ to be a fixed mode of the system (3.2.1) with respect to K in (3.2.11) (3.2.23) is that λ_0 is a transmission zero of all of the following subsystems :

$$\left(\begin{bmatrix} c'_{j_{k_1}} \\ c'_{j_{k_2}} \\ \vdots \\ c'_{j_{k_t}} \end{bmatrix} , \ A \ , \ (b_{i_{k_1}} , \ b_{i_{k_2}} , \ \ldots, \ b_{i_{k_t}}) \right) \tag{3.2.25}$$

where $(k_1=1,2,\ldots,s+1-t)$; $(k_2=k_1+1,k_1+2,\ldots,s+2-t)$; $(k_t=k_{t-1}+1,k_{t-1}+2,\ldots,s)$; for $(t=1,2,\ldots,\min(m,r))$ \hfill (3.2.26)

It is clear that the above result is equivalent to those stated before. The originality of this approach stands on the fact that it arises directly from the application of Lemma 3.1.

Example 3.1. Consider the linear dynamic system described in the frequency-domain by the following transfer matrix :

$$W(p) = \begin{bmatrix} \dfrac{1}{p(p-1)} & 0 & 0 \\[2mm] \dfrac{1}{p-1} & 0 & 0 \\[2mm] 0 & 0 & \dfrac{1}{p-1} \\[2mm] \dfrac{1}{p-1} & \dfrac{1}{p-1} & 0 \end{bmatrix}$$

for which $\phi(p) = p(p-1)^3$.

The controller structure is specified by the following output feedback matrix :

$$K = \begin{bmatrix} k_{11} & k_{12} & k_{13} & 0 \\ 0 & 0 & 0 & k_{24} \\ 0 & 0 & 0 & k_{34} \end{bmatrix} \quad \text{and} \quad K' = \begin{bmatrix} k_{11} & 0 & 0 \\ k12 & 0 & 0 \\ k13 & 0 & 0 \\ 0 & k_{24} & k_{34} \end{bmatrix}$$

Adopting the notations of Vidyasagar and Wiswanadham, the polynomial Z_J^I giving the transmission zeros of the subsystems corresponding to the non singular submatrices of K' are the following :

* One-dimensional subsystems :

$$Z\begin{bmatrix} 1 \\ 1 \end{bmatrix} = (p-1)^2 \qquad Z\begin{bmatrix} 3 \\ 1 \end{bmatrix} = 0 \qquad Z\begin{bmatrix} 4 \\ 3 \end{bmatrix} = 0$$

$$Z\begin{bmatrix} 2 \\ 1 \end{bmatrix} = p(p-1)^2 \qquad Z\begin{bmatrix} 4 \\ 2 \end{bmatrix} = p(p-1)^2$$

* Two-dimensional subsystems :

$$Z\begin{bmatrix} 1 & 4 \\ 1 & 2 \end{bmatrix} = p-1 \qquad Z\begin{bmatrix} 2 & 4 \\ 1 & 2 \end{bmatrix} = p(p-1) \qquad Z\begin{bmatrix} 3 & 4 \\ 1 & 2 \end{bmatrix} = 0$$

$$Z\begin{bmatrix} 1 & 4 \\ 1 & 3 \end{bmatrix} = 0 \qquad Z\begin{bmatrix} 2 & 4 \\ 1 & 3 \end{bmatrix} = 0 \qquad Z\begin{bmatrix} 3 & 4 \\ 1 & 3 \end{bmatrix} = p(p-1)$$

All these subsystems have a common transmission zero at $\lambda_0=1$ which corresponds to a pole of the system. Therefore, $\lambda_0=1$ is a fixed mode.

3.2.6. – Comments

From the above characterization of fixed modes, several interesting results can immediately be deduced. They are summarized in the following corollaries.

Corollary 3.1 (TAO-84). Conditions (i) and (ii) below are necessary for a system to have a fixed mode at λ_0 :

i – The entries of $W(p)$ corresponding to nonzero elements of K' must have pole-zero cancellations at $p = \lambda_0$.

ii – For the case $m = r$, $p = \lambda_0$ must be a transmission zero of the entire system.

Corollary 3.2 (TAO-84). If any of the subsystems corresponding to nonsingular submatrices of K' is minimum phase, the system has no unstable fixed modes. (All transmission zeros of minimum phase systems lie in the open left half plan).

It is interesting to notice that for single-input single-output systems a transmission zero is the frequence for which the transfer function numerator vanishes. Therefore, a fixed mode can be viewed as the frequence which cuts the information flow between the input and ouput of some subsystems. This can physically explain why a decentralized control may fail in the stabilization or pole assignment task.

3.3. ALGEBRAIC CHARACTERIZATIONS : TIME-DOMAIN

The following characterizations are stated by using a state-space representation of the system.

3.3.1. – Matrix rank test characterization

First, let us consider a partitioned system in the form (3.2.2) and the following decentralized control in accordance with the decomposition of the system :

$$
\begin{cases}
u_i(t) = Q_i\, z_i(t) + K_i\, y_i(t) + v_i(t) \\[2mm]
\dot{z}_i(t) = S_i\, z_i(t) + R_i\, y_i(t)
\end{cases}
\qquad (i=1,\ldots,S) \qquad (3.3.1)
$$

where $z_i(t)$ is the state of the ith controller and $v_i(t)$ is the ith external input.

In this context, the following interesting algebraic characterization of de-centralized fixed modes was provided by Anderson and Clements (AND-81a):

<u>Theorem 3.7.</u> Let be the set π $\{1,\ldots,S\}$ and define a partition of π into disjoint subsets $\pi_\alpha = \{i_1,\ldots,i_k\}$ and $\pi_\beta = \{i_{k+1},\ldots,i_S\}$. Define also the matrices :

$$B_\alpha = [B_{i_1},\ldots, B_{i_k}] \qquad\qquad B_\beta = [B_{i_{k+1}},\ldots,B_{i_S}]$$

$$C_\alpha = \begin{bmatrix} C_{i_1} \\ \vdots \\ C_{i_k} \end{bmatrix} \qquad\qquad C_\beta = \begin{bmatrix} C_{i_{k+1}} \\ \vdots \\ C_{i_S} \end{bmatrix}$$

Consider the system (3.2.2) and the decentralized control (3.3.1). Then, a necessary and sufficient condition for λ_0 σ (A) to be a decentralized fixed mode of (3.2.2) is that there exists at least one complementary subsystem (see Definition 2.6, § 2.2.3b, Chapter II) such that :

$$\text{rank} \begin{bmatrix} A-\lambda_0 I & B_\alpha \\ C_\beta & 0 \end{bmatrix} < n \tag{3.3.2}$$

<u>Remark 3.1.</u> Condition (3.3.2) is equivalent to $\rho(C_\beta, A, B_\alpha) \neq 1$ so that an immediate connection appears with the results of Corfmat and Morse (COR-76b) (see Theorem 2.2, Chapter II).

<u>Example 3.2.</u> Consider the following partitioned system ;

$$A = \begin{bmatrix} 0 & 1 & 0 & 0 \\ 0 & 1 & 0 & 0 \\ 0 & 0 & 1 & 0 \\ 0 & 0 & 0 & 1 \end{bmatrix} \qquad B_1 = \begin{bmatrix} 0 \\ 1 \\ 0 \\ 0 \end{bmatrix} \qquad B_2 = \begin{bmatrix} 0 \\ 0 \\ 1 \\ 0 \end{bmatrix} \qquad B_3 = \begin{bmatrix} 0 \\ 0 \\ 0 \\ 1 \end{bmatrix}$$

$$C_1 = \begin{bmatrix} 1 & 0 & 0 & 0 \end{bmatrix}$$

$$C_2 = \begin{bmatrix} 0 & 1 & 0 & 0 \end{bmatrix}$$

$$C_3 = \begin{bmatrix} 0 & 0 & 0 & 1 \\ 0 & 1 & 1 & 0 \end{bmatrix}$$

for which we concluded by using the definition of Wang and Davison (WAN-73) (see Example 2.1, § 2.2.3a, Chapter II) that it has a decentralized fixed mode at $\lambda_0 = 1$.

This result is verified by using the characterization stated in Theorem 3.1. Indeed, if we choose $\pi_\alpha = \{3\}$ and $\pi_\beta = \{1,2\}$, we have the following :

$$\text{rank} \begin{bmatrix} I-A & B_3 \\ C_1 & 0 \\ C_2 & 0 \end{bmatrix} = 3 < 4$$

which is sufficient to verify condition (3.3.2). Moreover, for this example, we have also :

$$\text{rank} \begin{bmatrix} I-A & B_1 & B_3 \\ C_2 & 0 & 0 \end{bmatrix} = 3 < 4 \quad \text{for} \quad \pi_\alpha = \{1,3\} \quad \text{and} \quad \pi_\beta = \{2\}$$

$$\text{rank} \begin{bmatrix} I-A & B_2 & B_3 \\ C_1 & 0 & 0 \end{bmatrix} = 3 < 4 \quad \text{for} \quad \pi_\alpha = \{2,3\} \quad \text{and} \quad \pi_\beta = \{1\}$$

A Fortran routine for detecting the real decentralized fixed modes of a system, which is based on Anderson and Clements' test, is provided in Appendix 4.

This result gives some insight into the reasons of occurence of fixed modes. Indeed, it provides an immediate interpretation : λ_0 is a decentralized fixed mode of system (3.2.2) if there exists a disjoint partition of the system in two aggregate stations α and β such that λ_0 is simultaneously uncontrollable by one of these stations and unobservable by the other one. This situation is illustrated by Figure 3.1 :

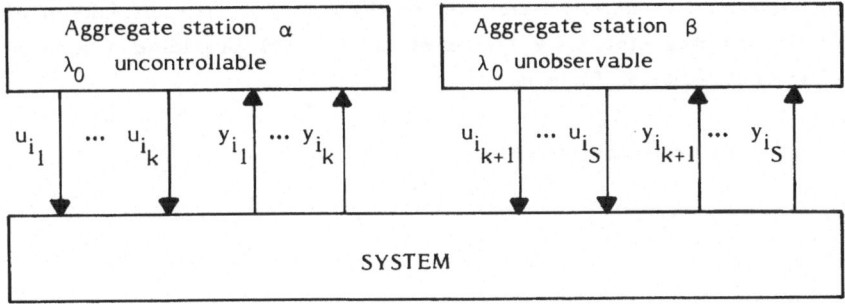

Figure 3.1

This interpretation leads to the following interesting conclusion : if $\lambda_0 \in \sigma(A)$ is simultaneously controllable and observable by one single station $i \in \{1,\ldots,S\}$, then λ_0 is not a decentralized fixed mode of the system.

Unfortunately, this characterization does not help particularly in the computation of fixed modes since it requires the evaluation of the rank of $2^S - 2$ matrices whose dimensions are superior to the order of the system. A method to overcome this desadvantage is presented by Petel and Misra (PET-84) in the case for which A has a superior Hesenberg form. They provide a condition which is numerically equivalent to condition (3.3.2) and in which fixed modes appear as the transmission zeros of certain transfer functions. They apply thus their algorithm for the evaluation of transmission zeros.

Consider now the general case of a non partitioned system (3.2.1) for which some arbitrary feedback structure constraints have been specified. For each input u_i, we define the index set :

$$J_i = \{j \in \{1,\ldots,r\} \text{ / the feedback is allowed from } y_j \text{ to } u_i\} \qquad (3.3.3)$$

As we already saw in paragraph 2.3 of Chapter II, the linear time-invariant dynamic controller associated with these feedback constraints is given by :

$$\begin{cases} \dot{z}_i(t) = S_i\, z_i(t) + \sum_{j \,\in\, J}\, r_{ij}\, y_j(t) \\[2mm] u_i(t) = q'_i\, z_i(t) + \sum_{j \,\in\, J}\, k_{ij}\, y_j(t) + v_i(t) \end{cases} \qquad (i=1,\ldots,m) \quad (3.3.4)$$

where $z_i(t)$ is the state of the controller and $v_i(t)$ is the ith external input. S_i, q_i and r_{ij} and k_{ij} are constant matrices, vectors and scalars of appropriate dimensions.

In the general context, an extention of the previous characterization of decentralizes fixed modes was stated by Pichai et al. (PIC-84) providing a general matrix rank test characterization of fixed modes.

Theorem 3.8. For an arbitrary subset :

$$I = \{i_1,\ldots,i_k\} \subset M = \{1,\ldots,m\}$$

define

$$J = \sum_{i \,\in\, M-I} J_i = \{j_1,\ldots,j_q\}$$

with J_i as defined in (3.3.3). Then, similarly to the case of decentralized control, the complementary subsystems of system (3.2.1) are given by : $(C_J,\ A,\ B_I)$, $(k=1,\ldots,m-1)$.

Then, a necessary and sufficient condition for λ_0 to be a fixed mode of (3.2.1) with respect to the control (3.3.4) is that there exists a complementary subsystem such that :

$$\text{rank} \begin{bmatrix} A-\lambda_0 I & B_I \\ C_J & 0 \end{bmatrix} < n \tag{3.3.5}$$

Example 3.3. Consider the following system with 3 inputs and 3 outputs :

$$A = \begin{bmatrix} -1 & & & \\ & -1 & & \\ & & -2 & \\ & & & -3 \end{bmatrix} \qquad B = \begin{bmatrix} 0 & 1 & 0 \\ 1 & 1 & 1 \\ 0 & 0 & 1 \\ 0 & 0 & 1 \end{bmatrix}$$

$$C = \begin{bmatrix} 0 & 0 & 1 & 1 \\ 1 & 0 & 0 & 0 \\ 0 & 1 & 0 & 1 \end{bmatrix}$$

for which the feedback structure constraints are given by :

$$J_1 = \{2\} \quad ; \quad J_2 = \{1,3\} \; ; \; J_3 = \{1\}$$

corresponding to the feedback matrix :

$$K = \begin{bmatrix} 0 & k_{12} & 0 \\ k_{21} & 0 & k23 \\ k_{31} & 0 & 0 \end{bmatrix}$$

Then, if we consider the subset $I = \{1,2\}$, we obtain $J = \{1\}$ and we have, for $\lambda_0 = -3$:

$$\text{rank} \begin{bmatrix} \lambda_0 I - A & b_1 \; b_2 \\ c_1 & 0 \; 0 \end{bmatrix} = 3 < 4$$

Therefore, $\lambda_0 = -3$ is a fixed mode.

In this general case, the interpretation does not appear so clearly as for the case of decentralized control. Indeed, we have to take into account that the partitioning of the system in two aggregated stations I and J is not disjoint. The interpretation can be given as follows : a necessary condition for $\lambda_0 \in \sigma(A)$ to be a fixed mode is that there exists a subset of inputs which cannot control λ_0 and that the subset of outputs involved in the feedback control of the complementary subset of inputs cannot observe λ_0. This situation is illustrated by Figure 3.2.

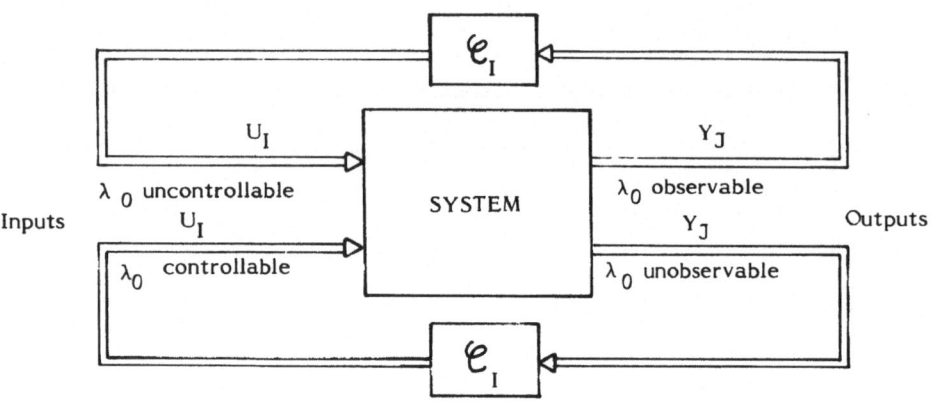

Figure 3.2

with $U_I \cup U_{\overline{I}} = \{u_1, \ldots, u_m\}$ and $Y_J \cup Y_{\overline{J}} = \{y_1, \ldots, y_r\}$
and $\mathcal{C}_I$: Controller of the subset of inputs U_I
$\quad \mathcal{C}_{\overline{I}}$: Controller of the subset of inputs $U_{\overline{I}}$

3.3.2. - Recursive characterization

Consider again a partitioned system in the form (3.2.2) and its associated decentralized feedback controller (3.3.1). In this context, the following characterization of fixed modes provided by Davison and Ozguner (DAV-83) presents the interest to show explicitely that the existence of fixed modes for a S-station system always reduces to the existence of fixed modes for a set of 2-station systems (by recursive application of Theorem 3.9 which brings back S to (S-1)).

Theorem 3.9.

I - Given the S-station system (3.2.2) with $S \geqslant 3$, then $\lambda_0 \in \sigma(A)$ is not a decentralized fixed mode of (3.2.2) if and only if λ_0 is not a decentralized fixed mode of any of the following (S-1)-station systems :

$$(1) \quad \left(\begin{bmatrix} \begin{bmatrix} C_1 \\ C_2 \end{bmatrix} \\ C_3 \\ \vdots \\ C_S \end{bmatrix} , \quad A , \quad \begin{bmatrix} [B_1, B_2], B_3, \ldots, B_S \end{bmatrix} \right)$$

$$(2) \quad \left(\begin{bmatrix} C_1 \\ \begin{bmatrix} C_2 \\ C_3 \end{bmatrix} \\ \vdots \\ C_S \end{bmatrix} , \quad A , \quad [B_1, [B_2, B_3], \ldots, B_S] \right)$$

$$(S\text{-}1) \quad \left(\begin{bmatrix} C_1 \\ \vdots \\ C_{S-2} \\ \begin{bmatrix} C_{S-1} \\ C_S \end{bmatrix} \end{bmatrix} , \quad A , \quad [B_1, \ldots, B_{S-2}, [B_{S-1}, B_S]] \right)$$

$$(S) \quad \left(\begin{bmatrix} C_1 \\ \vdots \\ \begin{bmatrix} C_{S-2} \\ C_S \end{bmatrix} \\ C_{S-1} \end{bmatrix} , \quad A, \quad [B_1, \ldots, B_{S-3}, [B_{S-2}, B_S], B_{S-1}] \right)$$

II – Given the S-station system (3.2.2) with S=2, then $\lambda_0 \in \sigma(A)$ is not a decentralized fixed mode of (3.2.2) if and only if the two following conditions both hold :

$$1 \quad \operatorname{rank} \begin{bmatrix} \lambda_0 I - A & B_1 \\ C_2 & 0 \end{bmatrix} \geqslant n$$

$$2 \quad \operatorname{rank} \begin{bmatrix} \lambda_0 I - A & B_2 \\ C_1 & 0 \end{bmatrix} \geqslant n$$

Now, it is of interest to analyse the meaning of the conditions (1) and (2) of part II of Theorem 3.9. These conditions correspond to the simultaneous requirements that :

- Station 1 (or 2) can control the mode λ_0
- Station 2 (or 1) can observe the mode λ_0
- λ_0 is not a transmission zero of certain subsystems of the system.

In fact, these conditions are equivalent to the condition (3.3.2) of Theorem 3.7 (see § 3.3.1) for a 2-station system. It was already obvious in condition (3.3.2) that the existence of decentralized fixed modes for a S-station system is reduced to the existence of fixed modes for a set of 2-station systems since this condition shows a partition of the system in 2 aggregated stations α and β. It stated that the system has no fixed modes if and only if the following condition holds for every possible partition :

$$
\text{rank} \begin{bmatrix} \lambda_0 I - A & B_\alpha \\ C_\beta & 0 \end{bmatrix} \geqslant n \qquad \forall \lambda_0 \in \sigma(A)
$$

This is equivalent to checking the existence of fixed modes for every 2-station system $(B_\alpha \, B_\beta, \, A, \, C_\alpha \, C_\beta)$, $\pi_\alpha \subset \pi$, $\pi_\alpha \cup \pi_\beta = \pi$.

Note that for a S-station system, the number of possible partitions is equal to $2^S - 2$. By using the procedure proposed by Davison and Ozguner and described in Theorem 3.9, we obtain the same tests. Nevertheless, the recursive characterization of the systems leads to some redundancies. Indeed, it results in $(S!/2)$ 2-station systems and it is clear that some are repeated.

Example 3.4. Consider a 4-station system $(C_1 \, C_2 \, C_3 \, C_4, \, A, \, B_1 \, B_2 \, B_3 \, B_4)$. Condition (3.3.2) must be tested for the 7 following 2-station systems :

(1) $(C_1 \, (C_2 \, C_3 \, C_4), \, A \, , \, B_1 \, (B_2 \, B_3 \, B_4))$
(2) $(C_2 \, (C_1 \, C_3 \, C_4), \, A \, , \, B_2 \, (B_1 \, B_3 \, B_4))$
(3) $(C_3 \, (C_1 \, C_2 \, C_4), \, A \, , \, B_3 \, (B_1 \, B_2 \, B_4))$
(4) $(C_4 \, (C_1 \, C_2 \, C_3), \, A \, , \, B_1 \, B_2 \, (B_3 \, B_4))$
(5) $(C_1 \, C_2 \, (C_3 \, C_4), \, A \, , \, B_1 \, B_3 \, (B_2 \, B_4))$
(6) $(C_1 \, C_3 \, (C_2 \, C_4), \, A \, , \, B_1 \, B_3 \, (B_2 \, B_4))$
(7) $(C_1 \, C_4 \, (C_2 \, C_3), \, A \, , \, B_1 \, B_4 \, (B_2 \, B_3))$

By using the procedure in Theorem 3.9, we obtain 12 2-station systems but 5 of them are redundant :

$$((C_1 \ C_2, \ C_3) \ C_4, \ A, \ (B_1 \ B_2 \ B_3) \ B_4 \quad) \quad (1)$$

$$((C_1 \ C_2) \ C_3 \ C_4, \ A, \ (B_1 \ B_2) \ B_3 \ B_4) \quad ((C_1 \ C_2) \ (C_3 \ C_4), \ A, \ (B_1 \ B_2) \ (B_3 \ B_4)) \quad (2)$$

$$((C_1 \ C_2 \ C_4) \ C_3, \ A, \ (B_1 \ B_2 \ B_4) \ B_3 \quad) \quad (3)$$

$$((C_1 \ C_2 \ C_3) \ C_4, \ A, \ (B_1 \ B_2 \ B_3) \ B_4) \quad (4)$$

$$(C_1 \ (C_2 \ C_3) \ C_4, \ A, \ B_1 \ (B_2 \ B_3)B_4) \quad (C_1 \ (C_2 \ C_3 \ C_4), \ A, \ B_1 \ (B_2 \ B_3 \ B_4)) \quad (5)$$

$$((C_1 \ C_4) \ (C_2 \ C_3), \ A, \ (B_1 \ B_4) \ (B_2 \ B_3)) \quad (6)$$

$$((C_1 \ C_2) \ (C_3 \ C_4), \ A, \ (B_1 \ B_2) \ (B_3 \ B_4)) \quad (7)$$

$$(C_1 \ C_2 \ (C_3 \ C_4), \ A, \ B_1 \ B_2 \ (B_3 \ B_4)) \quad (C_1 \ (C_2 \ C_3 \ C_4), \ A, \ B_1 \ (B_2 \ B_3 \ B_4)) \quad (8)$$

$$((C_1 \ C_3 \ C_4) \ C_2, \ A, \ (B_1 \ B_3 \ B_4) \ B_2) \quad (9)$$

$$((C_1 \ C_2 \ C_4) \ C_3, \ A, \ (B_1 \ B_2 \ B_4) \ B_3) \quad (10)$$

$$(C_1 \ (C_2 \ C_4) \ C_3, \ A, \ B_1 \ (B_2 \ B_4) \ B_3) \quad (C_1 \ (C_2 \ C_4 \ C_3), \ A, \ B_1 \ (B_2 \ B_4 \ B_3)) \quad (11)$$

$$((C_1 \ C_3) \ (C_2 \ C_4), \ A, \ (B_1 \ B_3) \ (B_2 \ B_4)) \quad (12)$$

Systems (1) and (4) ; (5), (8) and (11) ; (3) and (10) ; (2) and (7) are the same.

However, this recursive method presents the advantage to provide a systematic way to determine all the partitions.

3.3.3. - Particular cases

3.3.3.a. - Diagonal systems

Consider the following 2-station system in which the dynamic matrix A is diagonal :

$$\dot{x}(t) = \begin{bmatrix} \lambda_0 & & & \\ & \lambda_1 & & \\ & & \ddots & \\ & & & \lambda_{n-1} \end{bmatrix} x(t) + \begin{bmatrix} B_1^* \\ B_1 \end{bmatrix} u_1(t) + \begin{bmatrix} B_2^* \\ B_2 \end{bmatrix} u_2(t) \qquad (3.3.6)$$

$$y_1(t) = (C_1^*, C_1) \, x(t)$$
$$y_2(t) = (C_2^*, C_2) \, x(t)$$

where $u_1 \in R^{m_1}$, $u_2 \in R^{m_2}$, $y_1 \in R^{r_1}$, $y_2 \in R^{r_2}$ and $\lambda_i \in \mathscr{C}$, $(i=1,\ldots n)$. B_1^*, B_2^* are m_1, m_2 row vectors, respectively and C_1^*, C_2^*, are r_1, r_2 column vectors, respectively.

Let

$$B_1 = [b_1^1, \ldots, b_{m_1}^1] \qquad\qquad B_2 = [b_1^2, \ldots, b_{m_2}^2]$$

$$C_1 = \begin{bmatrix} c_1^1 \\ \\ c_{r_1}^1 \end{bmatrix} \qquad\qquad C_2 = \begin{bmatrix} c_1^2 \\ \\ c_{r_2}^2 \end{bmatrix}$$

and assume that λ_i, $(i=1,\ldots,n)$, are all distinct and occur in complex conjugate pairs.

Then, by applying Theorem 3.7 to this particular case and with some matrix manipulations, the following result is obtained :

Theorem 3.10 (DAV-83). λ_0 is not a decentralized fixed mode of system (3.3.6) if and only if the following conditions hold :

i - $(B_1^*, B_2^*) \neq 0$ and $\begin{bmatrix} C_1^* \\ C_2^* \end{bmatrix} \neq 0$

i.e., λ_0 is not a centralized fixed mode (it is controllable and observable).

ii - The following conditions do not simultaneously hold :

* $B_q^* = 0$
* $C_d^* = 0$
* λ_0 is a transmission zero of all the following single-input single-output subsystems :

$$c_i^d, \begin{bmatrix} \lambda_1 & 0 \\ 0 & \lambda_{n-1} \end{bmatrix} b_j^q \qquad \begin{array}{ll} \forall \, i \in \{1,2,\ldots,r_d\} & d=1,2 \\ & \\ \forall \, j \in \{1,2,\ldots,m_q\} & q=1,2 \end{array} \qquad q \neq d$$

Note that this theorem can easily be extended to the case for which the system has more than 2 stations by applying Theorem 3.9 (DAV-83) (see also (PET-84)).

3.3.3.b. - Interconnected systems

This paragraph deals with a particular class of systems of type (3.2.2) consisting of a number of subsystems interconnected together. These systems are represented in the state space by the following set of equations :

$$\begin{cases} \dot{x}_i(t) = A_{ii}\, x_i(t) + \sum_{\substack{j=1 \\ j \neq i}}^{S} A_{ij}\, x_j(t) + B_i\, u_i(t) \qquad (i=1,...,S) \\[4mm] y_i(t) = C_i\, x_i(t) \end{cases} \tag{3.3.7}$$

$$x_i \in R^{n_i} ,\ u_i \in R^{m_i} ,\ y_i \in R^{r_i}$$

$$A = \{A_{ij},\ (i=1,...,S),\ (j=1,...,S)\} \subset R^{n \times n}$$
$$B = \text{block.diag.}\ (B_1,...,B_S) \subset R^{n \times m}$$
$$C = \text{block.diag.}\ (C_1,...,C_S) \subset R^{r \times n}$$

A_{ii}, A_{ij}, B_i and C_i, $(i=1,...,S)$, $j \neq i$, are invariant matrices of appropriate dimension.

1 - Characterization with constrained interconnections

Consider the class of interconnections in the form :

$$A_{ij} = B_{ij}\, L_{ij}\, C_{ij} \quad (i,j=1,...,S) \quad j \neq i \tag{3.3.8}$$

where L_{ij} is the matrix of interconnection gains and B_{ij} and C_{ij} are arbitrary. Then, the following result was derived by Davison (DAV-83) :

Theorem 3.11. Given the system (3.3.7) with structure (3.3.8), if (C_i, A_{ii}, B_i) is controllable and observable for $\forall$ i=1,2,...,S then (3.3.7) (3.3.8) has no decentralized fixed modes for almost all interconnection gains L_{ij}, $(i=1,...,S)$, $(j=1,...,S)$, $i \neq j$, i.e. the class of nonzero gains L_{ij} for which (3.3.7) (3.3.8) has fixed modes is either empty or lies on a subset of a hypersurface in the parameter space of L_{ij}.

A more interesting result is provided if it is assumed that the system (3.3.7) is interconnected by the outputs ; i.e. :

$$A_{ij} = B_i \ L_{ij} \ C_j \quad (i=1,\ldots,S), \ (j=1,\ldots,S), \quad j\neq i \qquad (3.3.9)$$

Note that even if this class of systems seems to be very restrictive with respect to the class of general systems (3.2.2), a lot of physical systems have this particular structure. Indeed, the decentralized stabilizability study for this type of systems was the problem which motivated the extention to more general systems like (3.2.2) or (3.3.7).

Theorem 3.12. The system (3.3.7) with structure (3.3.9) has no decentralized fixed modes if and only if :

(C_i, A_{ii}, B_i) controllable and observable for all $(i=1,\ldots,S)$.

For this type of systems, the set of decentralized fixed modes is equal to the set of centralized fixed modes (uncontrollable or unobservable modes) which is itself equal to the union of the sets of centralized fixed modes of each disconnected system. This result was also derived by Saeks (SAE-79) and stated in the following way :

Theorem 3.12bis. The set of centralized fixed modes of (3.3.7) with structure (3.3.9) is given by :

$$\Lambda_d(C,A,B) = \Lambda_c(C,A,B) = \bigcup_{i=1}^{S} \Lambda_c(C_i, A_{ii}, B_i)$$

Therefore, for this class of systems, a decentralized control is equivalent to a centralized control as far as the pole assignment problem is concerned.

2 - Characterization using the property of block-diagonal dominance

Consider the system (3.3.7) and the following set of local controllers :

$$u_i = K_{ii} \ y_i \quad (i=1,\ldots,S) \text{ sucht that } K = \text{block.diag.}(K_{11},\ldots,K_{SS})$$

The dynamic matrix of the closed-loop system is :

$$A + BKC = \begin{bmatrix} \hat{A}_{11} & A_{12} & \cdots\cdots & A_{1S} \\ \vdots & \hat{A}_{22} & & \vdots \\ \vdots & & \ddots & \vdots \\ \vdots & & & \ddots \\ A_{S1} & \cdots\cdots\cdots & & \hat{A}_{SS} \end{bmatrix}$$

where $\hat{A}_{ii} = A_{ii} + B_i K_{ii} C_i$, $(i=1,\ldots,S)$.

If the diagonal submatrices $\hat{A}_{ii}$ are non singular and if :

$$\| \hat{A}_{ii}^{-1} \|^{-1} \;\leqslant\; \sum_{\substack{j=1 \\ j\neq i}}^{S} \| A_{ij} \| \qquad \text{for} \qquad \forall \; i=1,\ldots,S$$

then $(A+BKC)$ is strictly block-diagonal dominant. * denotes a norm of the matrix (*), for instance :

$$\| A \| = \max_i \;\; \sum_j | a_{ij} |$$

The following well-known result :

Theorem 3.13. If the matrix $(A + BKC)$ is strictly block-diagonally dominant, then $(A + BKC)$ is non singular.

leads to the subsequent characterization of fixed modes.

Corollary 3.3 (ARM-82). If $\lambda_0 \in \sigma(A)$ is a decentralized fixed mode of (3.3.7), then there exists $i \in \{1,\ldots,S\}$ such that :

$$\| (\hat{A}_{ii} - \lambda_0 \, I)^{-1} \|^{-1} \;\leqslant\; \sum_{\substack{j=1 \\ j\neq i}}^{S} \| A_{ij} \| \quad \text{for} \quad \forall \, K_{ii} \in R^{m_i x r_i} \tag{3.3.10}$$

The interest of this characterization will appear later (in Chapter 5) since it is used by Armentano and Singh to determine a control structure such that fixed modes are avoided.

3.3.4. - Comments

The characterizations presented in this paragraph are stated in a time-domain framework. It is clear that the most relevant is the matrix rank test characterization which was provided by Anderson and Clements (AND-81a) and, as it has been pointed out, all the other ones are equivalent.

This characterization allowed us to interprete the fixed modes in terms of the concepts of controllability and observability and the definition of complementary subsystems. We will find again this partitioning of the system in two aggregated stations in the frequency-domain characterizations which will give us the tools to interprete in a deeper way the reasons for the occurence of fixed modes.

Despite the theoretical interest of the matrix rank test characterization, it is clear that it does not seem to be very efficient from the computational point of view since it requires to test all the complementary subsystems.

An interesting result has been obtained for interconnected systems whose interconnections are made by the outputs since the fixed modes of these systems are just their uncontrollable and unobservable modes.

3.4. - ALGEBRAIC CHARACTERIZATIONS : FREQUENCY DOMAIN

This paragraph deals with the characterization of fixed modes from the input-output relations describing the system. The system is represented either by polynomial matrices ("matrix fraction description") or by a rational transfer function matrix ; i.e. :

$$U(p) = W(p) \ Y(p) \qquad (3.4.1)$$

where U and Y are the input and output vectors of dimension m and r respectively, and W(p) is the transfer function matrix of dimension mxr.

Let $S^{-1}(p)T(p)$ be a left coprime fraction description of $W(p)$, then the system can be described by :

$$S(p) \ Y(p) = T(p) \ U(p) \qquad (3.4.2)$$

where $S(p)$ and $T(p)$ are polynomial matrices with r and m columns, respectively.

3.4.1. - Necessary conditions on the transfer function matrix
for the existence of fixed modes

Before presenting the general frequency-domain characterizations of fixed modes, this paragraph provides some necessary conditions for their existence, which are interesting because they can be checked by the sole examination of the transfer matrix.

Consider the system described by (3.4.1) with :

$$W(p) = \frac{N(p)}{\phi(p)} \tag{3.4.3}$$

where $N(p)=Cadj(pI-A)B$ is a polynomial matrix and $\phi(p)$ is the characteristic polynomial of the system.

Now, if we consider the control law in (3.2.4) where the output feedback matrix K can take any arbitrary structure, the closed-loop transfer matrix is given by :

$$W_c(p,K) = \left[I-W(p)K\right]^{-1}W(p) = C(pI-A-BKC)^{-1}B \tag{3.4.4}$$

which can be rewritten as :

$$W_c(p,K) = \frac{N_c(p,K)}{\phi_c(p,K)} \tag{3.4.5}$$

where $N_c(p,K)=Cadj(pI-A-BKC)B$ is a polynomial matrix and $\phi_c(p,K)$ is the closed-loop characteristic polynomial.

If the system has fixed modes, it is clear that the fixed polynomial $F(p)$ divides the closed-loop characteristic polynomial such that we can write :

$$\phi_c(p,K) = F(p).P(p,K)$$

If we derive $\phi_c(p,K)$ with respect to K :

$$\frac{\partial \phi_c(p,K)}{\partial K} = F(p)\frac{\partial P(p,K)}{\partial K}$$

and it is clear that if $p=\lambda_0$ is a fixed mode, then :

$$\frac{\partial \phi_c(\lambda_0,K)}{\partial K} = 0 \tag{3.4.6}$$

Consider now the following theorem :

Theorem 3.14 (BIN-78) (BER-81). The Jacobian matrix of the closed-loop characteristic polynomial $\phi_c(p,K)$ with respect to K is given by :

$$\frac{\partial \phi_c(p,K)}{\partial K} = - N_c'(p,K) \tag{3.4.7}$$

where $N'_c(p,K)$ is the transpose of $N_c(p,K)$ in (3.4.5).

Theorem 3.14 and formula (3.4.6) lead to the following result :

<u>Theorem 3.15</u>. A necessary condition for λ_0 to be a fixed mode of (3.4.1) with respect to K in (3.2.4) is that the projection of the closed-loop transfer matrix numerator $N_c(p,K)$ on K' is zero for $p=\lambda_0$.

Since this result must hold independently of the parameter values in K, it must in particular hold for K=0. Therefore the following corollary directly follows :

<u>Corollary 3.4</u>. A necessary condition for λ_0 to be a fixed mode of (3.4.1) with respect to K in (3.2.4) is that the projection of the transfer matrix numerator $N(p)$ on K' is zero for $p=\lambda_0$.

It is interesting to notice that the condition in Corollary 3.4 is exactly the same as the condition (i) in Corollary 3.1 which was obtained from the characterization of fixed modes in terms of transmission zeros.

This result can also be expressed in the following form :

<u>Corollary 3.5</u>. A multiple mode of order q can be a fixed mode if it is not a pole of order superior to (q-1) for any subsystem formed from any local subsystem neither for the characteristic polynomial of the projection of the transfer matrix on K'.

<u>Remark 3.2</u>. As a consequence of Corollary 3.5, the fixed modes are poles of the complementary subsystems.

Note that Theorem 3.15, Corollaries 3.4 and 3.5 are valid for any arbitrary structure of the feedback matrix K (not necessarily block-diagonal as in (3.2.11)). For example, in the particular case of centralized control, we find again from Corollary 3.5 that a simple uncontrollable and/or unobservable mode does not belong to the characteristic polynomial (minimal realization), which is a well-known result.

<u>Example 3.5</u>. Consider the system described by the following transfer matrix :

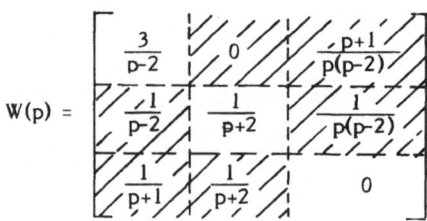

$$W(p) = \begin{bmatrix} \dfrac{3}{p-2} & 0 & \dfrac{p+1}{p(p-2)} \\ \dfrac{1}{p-2} & \dfrac{1}{p+2} & \dfrac{1}{p(p-2)} \\ \dfrac{1}{p+1} & \dfrac{1}{p+2} & 0 \end{bmatrix}$$

for which $\phi(p)=p(p-1)(p+2)(p-2)$.

If we consider a decentralized control in the form :

$$K=\text{block-diag.}(k_1,k_2,k_3)$$

the projection of $W(p)$ on K' is indicated by the non shaded blocks. Therefore, according to Corollary 3.5, $\lambda=2$ and $\lambda=-2$ are not fixed modes.

Note that the conditions derived in this paragraph can be easily used to determine a structure for the feedback matrix such that the system has no fixed modes.

3.4.2. – Transfer function matrix characterization for systems with distinct poles

Consider the system (3.4.1) partitioned in S control and observation stations such that its transfer function matrix appears in the following form :

$$W(p) = \begin{bmatrix} W_{11}(p) & \cdots\cdots & W_{1S}(p) \\ & & \\ W_{S1}(p) & \cdots\cdots & W_{SS}(p) \end{bmatrix} \qquad (3.4.8)$$

where $W_{ij}(p)$ is of dimension $r_i \times m_i$, $(i,j=1,\ldots,S)$, and assume that all its poles are distinct ; i.e. $W(p)$ can be factorized as :

$$W(p) = \frac{A_0}{p-\lambda_0} + \sum_{i=1}^{n-1} \frac{A_i}{p-\lambda_i} \qquad (3.4.9)$$

where $A_0 \neq 0$ and $\lambda_i \neq \lambda_0$ $(i=1,\ldots,n-1)$.

In the case of 2 stations, the time domain representation of (3.4.9) is given in (3.3.6) and $W(p)$ can be rewritten as follows :

$$W(p) = \frac{1}{p-\lambda_0} \begin{bmatrix} C_1^* B_1^* & C_1^* B_2^* \\ C_2^* B_1^* & C_2^* B_2^* \end{bmatrix} + \begin{bmatrix} C_1 \\ C_2 \end{bmatrix} \begin{bmatrix} \frac{1}{p-\lambda_1} & 0 \\ 0 & \frac{1}{p-\lambda_{n-1}} \end{bmatrix} [B_1 \quad B_2] \qquad (3.4.10)$$

Using (3.4.10), Davison and Ozguner (DAV-83) provided in the frequency domain the transposition of the result established by Theorem 3.10 in a time-domain framework. The case for which the system has more than 2 stations is considered by applying the recursive characterization of fixed modes given in Theorem 3.9.

<u>Theorem 3.16.</u> λ_0 is not a decentralized fixed mode of (3.4.1) (3.4.8) if and only if none of the following conditions occur with respect to the matrices :

$$A_0 \text{ and } \left[W(p) - \frac{A_0}{p-\lambda_0} \right]_{p=\lambda_0}$$

or their respective transposes :

<u>Case 1</u> : (S=2)

(i) $\quad A_0 = \begin{bmatrix} 0 & X \\ 0 & 0 \end{bmatrix}$ and $\left[W(p) - \frac{A_0}{p-\lambda_0} \right]_{p=\lambda_0} = \begin{bmatrix} X & X \\ 0 & X \end{bmatrix}$

<u>Case 2</u> : (S=3)

(i) $\quad A_0 = \begin{bmatrix} 0 & 0 & X \\ 0 & 0 & X \\ 0 & 0 & 0 \end{bmatrix}$ and $\left[W(p) - \frac{A_0}{p-\lambda_0} \right]_{p=\lambda_0} = \begin{bmatrix} X & X & X \\ X & X & X \\ 0 & 0 & X \end{bmatrix}$

(ii) $\quad A_0 = \begin{bmatrix} 0 & X & X \\ 0 & 0 & 0 \\ 0 & 0 & 0 \end{bmatrix}$ and $\left[W(p) - \frac{A_0}{p-\lambda_0} \right]_{p=\lambda_0} = \begin{bmatrix} X & X & X \\ 0 & X & X \\ 0 & X & X \end{bmatrix}$

(iii) $\quad A_0 = \begin{bmatrix} 0 & X & 0 \\ 0 & 0 & 0 \\ 0 & X & 0 \end{bmatrix}$ and $\left[W(p) - \frac{A_0}{p-\lambda_0} \right]_{p=\lambda_0} = \begin{bmatrix} X & X & X \\ 0 & X & 0 \\ X & X & X \end{bmatrix}$

Case 3 :

(i)
$$A_0 = \begin{bmatrix} 0 & 0 & 0 & X \\ 0 & 0 & 0 & X \\ 0 & 0 & 0 & X \\ 0 & 0 & 0 & 0 \end{bmatrix} \text{ and } \left[W(p) - \frac{A_0}{p - \lambda_0} \right]_{p = \lambda_0} = \begin{bmatrix} X & X & X & X \\ X & X & X & X \\ X & X & X & X \\ 0 & 0 & 0 & X \end{bmatrix}$$

(ii)
$$A_0 = \begin{bmatrix} 0 & 0 & X & X \\ 0 & 0 & X & X \\ 0 & 0 & 0 & 0 \\ 0 & 0 & 0 & 0 \end{bmatrix} \text{ and } \left[W(p) - \frac{A_0}{p - \lambda_0} \right]_{p = \lambda_0} = \begin{bmatrix} X & X & X & X \\ X & X & X & X \\ 0 & 0 & X & X \\ 0 & 0 & X & X \end{bmatrix}$$

(iii)
$$A_0 = \begin{bmatrix} 0 & 0 & X & 0 \\ 0 & 0 & X & 0 \\ 0 & 0 & 0 & 0 \\ 0 & 0 & X & 0 \end{bmatrix} \text{ and } \left[W(p) - \frac{A_0}{p - \lambda_0} \right]_{p = \lambda_0} = \begin{bmatrix} X & X & X & X \\ X & X & X & X \\ 0 & 0 & X & 0 \\ X & X & X & X \end{bmatrix}$$

(iv)
$$A_0 = \begin{bmatrix} 0 & X & X & X \\ 0 & 0 & 0 & 0 \\ 0 & 0 & 0 & 0 \\ 0 & 0 & 0 & 0 \end{bmatrix} \text{ and } \left[W(p) - \frac{A_0}{p - \lambda_0} \right]_{p = \lambda_0} = \begin{bmatrix} X & X & X & X \\ 0 & X & X & X \\ 0 & X & X & X \\ 0 & X & X & X \end{bmatrix}$$

(v)
$$A_0 = \begin{bmatrix} 0 & X & X & 0 \\ 0 & 0 & 0 & 0 \\ 0 & 0 & 0 & 0 \\ 0 & X & X & 0 \end{bmatrix} \text{ and } \left[W(p) - \frac{A_0}{p - \lambda_0} \right]_{p = \lambda_0} = \begin{bmatrix} X & X & X & X \\ 0 & X & X & 0 \\ 0 & X & X & 0 \\ X & X & X & X \end{bmatrix}$$

(vi)
$$A_0 = \begin{bmatrix} 0 & X & 0 & 0 \\ 0 & 0 & 0 & 0 \\ 0 & X & 0 & 0 \\ 0 & X & 0 & 0 \end{bmatrix} \text{ and } \left[W(p) - \frac{A_0}{p - \lambda_0} \right]_{p = \lambda_0} = \begin{bmatrix} X & X & X & X \\ 0 & X & 0 & 0 \\ X & X & X & X \\ X & X & X & X \end{bmatrix}$$

(vii)
$$A_0 = \begin{bmatrix} 0 & X & 0 & X \\ 0 & 0 & 0 & 0 \\ 0 & X & 0 & X \\ 0 & 0 & 0 & 0 \end{bmatrix} \text{ and } \left[W(p) - \frac{A_0}{p - \lambda_0} \right]_{p = \lambda_0} = \begin{bmatrix} X & X & X & X \\ 0 & X & 0 & X \\ X & X & X & X \\ 0 & X & 0 & X \end{bmatrix}$$

where X denotes elements whose values are not necessarily zero.

In the case for which S=2, the interpretation of this result follows. $W_{11}(p)$, $W_{21}(p)$, $W_{22}(p)$ have no element with pole at λ_0 and consequently, λ_0 is not conttrollable and observable by a single station. All the elements of $W_{21}(p)$ have a zero at λ_0 and those of $W_{12}(p)$ have a pole at λ_0, what produces a parametric cancellation when one local feedback is applied. This will be explain in details in Paragraph 3.4.4.

It is clear that this condition is strictly equivalent to the conditions in Theorem 3.10.

A similar interpration can be given for the case $S>2$.

Example 3.6. Consider again the system in Example 3.5 and assume that we want to check if $\lambda_0 =-1$ is a fixed mode with respect to $K = \text{block-diag}(k_1,k_2,k_3)W(p)$ can be written in the form (3.4.9) :

$$W(p) = \begin{bmatrix} 0 & 0 & 0 \\ 0 & 0 & 0 \\ \dfrac{1}{p+1} & 0 & 0 \end{bmatrix} + \begin{bmatrix} \dfrac{3}{p-2} & 0 & \dfrac{p+1}{p(p-2)} \\ \dfrac{1}{p-2} & \dfrac{1}{p-2} & \dfrac{1}{p(p-2)} \\ 0 & \dfrac{1}{p+2} & 0 \end{bmatrix}$$

and we have :

$$A_0' = \begin{bmatrix} 0 & 0 & X \\ 0 & 0 & 0 \\ 0 & 0 & 0 \end{bmatrix} \qquad \left[W(p) - \dfrac{A_0}{p+1} \right]'_{p=-1} = \begin{bmatrix} X & X & 0 \\ 0 & X & X \\ 0 & X & 0 \end{bmatrix}$$

which satisfies condition (ii) for the case 2 (S=3). Therefore $\lambda_0=-1$ is a decentralized fixed mode.

3.4.3. – Polynomial matrix rank test characterization

Consider a linear time-invariant system partitionned in S control and observation stations such that its matrix fraction description in (3.4.2) takes the form :

$$(S_1(p),\ldots,S_S(p))\, U(p) = (T_1(p),\ldots,_S(p))\, Y(p) \tag{3.4.11}$$

where $S_i(p)$ and $T_i(p)$, $(i=1,\ldots,S)$, have r_i and m_i columns, respectively, corresponding to the number of outputs and inputs of the ith station.

In this framework, Anderson (AND-81a) gives the following characterization of decentralized fixed modes :

Theorem 3.17. λ_0 is a decentralized fixed mode of system (3.4.2) (3.4.11) with respect to K in (3.2.11) if and only if there exists some nonempty subset $\{i_1,\ldots i_k\}$ of $\{1,\ldots,S\}$ such that :

$$\text{rank} \quad [S_{i_1}(\lambda_0), \ldots, S_{i_k}(\lambda_0), T_{i_1}(\lambda_0), \ldots, T_{i_k}(\lambda_0)] \quad \sum_{j=1}^{k} r_{i_j} \qquad (3.4.12)$$

It we set $\sum_{j=i}^{k} r_{ij} = \beta$, then the degree of the fixed mode λ_0 is given by the largest positive integer d such that all $\beta \times \beta$ minors of the matrix in the left hand side of (3.4.12) have a zero at λ_0 of order at least d.

Note that this result was recently proved in a different way by Zheng (ZHE-84).

3.4.4. – General transfer function matrix characterization

This general characterization of fixed modes is the most complete one in the frequency domain. It was derived by Anderson in 1982 (AND-82) using the result that he obtained together with Clement in 1981 (AND-81a) and which was presented in the preceding paragraph.

3.4.4.a. – Particular case : 2x2 transfer function matrix with simple pole at λ_0

Consider the system (3.4.1) where $W(p)$ is given by :

$$W(p) = \begin{bmatrix} W_{11}(p) & W_{12}(p) \\ W_{21}(p) & W_{22}(p) \end{bmatrix} \qquad (3.4.13)$$

Suppose that $W(p) = S^{-1}(p)T(p)$ with :

$$S(p) = \begin{bmatrix} s_{11}(p) & s_{12}(p) \\ s_{21}(p) & s_{22}(p) \end{bmatrix}$$

$$T(p) = \begin{bmatrix} t_{11}(p) & t_{12}(p) \\ t_{21}(p) & t_{22}(p) \end{bmatrix} \qquad (3.4.14)$$

and $S(p)$ and $T(p)$ are left coprime.

If we apply Theorem 3.17 to this particular case, there is a fixed mode at λ_0 if and only if, by reordering the inputs and outputs if necessary, the following condition holds :

$$\text{rank} \begin{bmatrix} s_{11}(\lambda_0) & t_{11}(\lambda_0) \\ s_{21}(\lambda_0) & t_{21}(\lambda_0) \end{bmatrix} < 1$$

i.e., $s_{11}(\lambda_0)=s_{21}(\lambda_0)=t_{11}(\lambda_0)=t_{21}(\lambda_0)=0$.

$W_{ij}(p)$ in (3.4.13) can easily be expressed in terms of $s_{ij}(p)$ and $t_{ij}(p)$ for $i,j=1,2$ and the following result follows :

<u>Theorem 3.18</u> (AND-82). λ_0 is a decentralized fixed mode of system (3.4.13) if and only if $W(p)$ or $W'(p)$ has the following form :

$$\begin{bmatrix} \text{entry with no pole at } \lambda_0 & \text{entry with pole at } \lambda_0 \\ \text{entry with zero at } \lambda_0 & \text{entry with no pole at } \lambda_0 \end{bmatrix} \qquad (3.4.15)$$

This result can easily be interpreted. Indeed, consider system (3.4.13) as illustrated by Figure 3.4a and suppose that we set $u_2 = k_2 \, y_2$, producing thus a system with input u_1 and output y_1 as illustrated by Figure 3.4b.

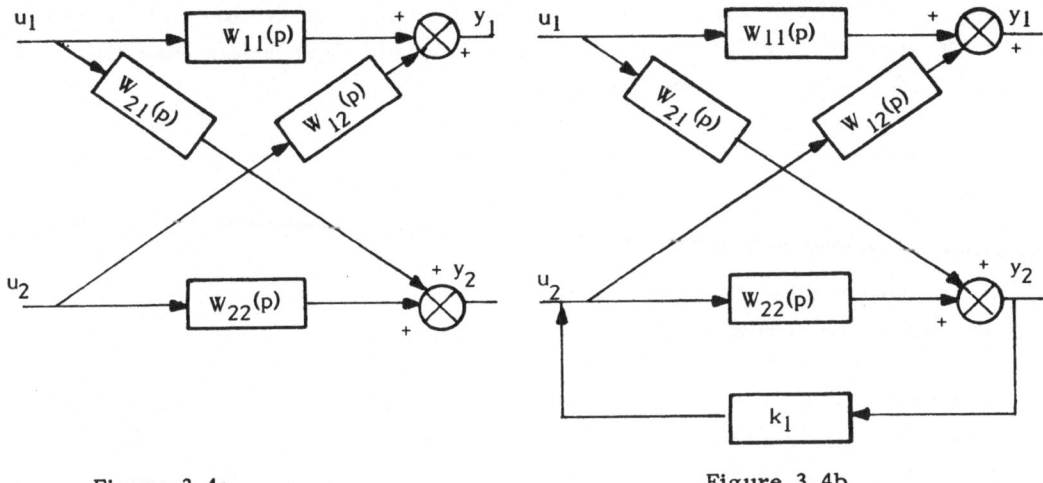

Figure 3.4a   Figure 3.4b

The transfer function of the new system is given by :

$$\frac{y_1}{u_1} = W_{11} + W_{12} \quad \frac{k_2\,W_{21}}{1 - W_{22}\,k_2} \qquad (3.4.16)$$

If condition (3.4.15) holds for $W(p)$, the pole-zero cancellation at λ_0 in the product $W_{12} \cdot W_{21}$, associated with the fact that W_{11} and W_{22} have no pole at λ_0 (λ_0 is not simultaneously controllable and observable by a single station), makes that (3.4.16) is uncontrollable. If condition (3.4.15) holds for $W'(p)$, then system (3.4.16) is unobservable. Of course, the application of feedback at station 1 ($u_1 = k_1\,y_1$) do not change this situation and, in both cases, λ_0 is a fixed mode of system (3.4.13).

Note that Theorem 3.18 agrees with the result of Davison and Ozguner (DAV-81) given in Theorem 3.16 for the case 1 and also with the necessary condition expressed in Theorem 3.15. Moreover, this particular case is a perfect illustration of Theorem 2.5 (Chapter 2) (FES-80) that shows a first relation between the results of Corfmat and Morse (COR-76a,b) and the concept of fixed modes.

In the case for which λ_0 is not a simple pole, the above result can be extended as follows :

<u>Theorem 3.19</u> (AND-82). λ_0 is a decentralized fixed mode of system (3.4.13) if and only if the order of λ_0 as a pole in $W_{11}(p)$, $W_{22}(p)$ and $W(p)$ is less than the order it has as a pole in $W_{12}(p)$ or $W_{21}(p)$.

In this case, a pole-zero cancellation also occurs in the transfer function (3.4.16).

3.4.4.b. – General case : systems with more than 2 multi-input multi-output stations

Consider now a general system (3.4.1) described by a transfer function matrix $W(p)$ as in (3.4.8) or, equivalently, by (3.4.11), where $S^{-1}(p)T(p)$ is a left coprime matrix fraction description of $W(p)$, and the decentralized control in (3.2.11).

For a matrix M, $M\begin{bmatrix} i_1 \cdots i_k \\ j_1 \cdots j_k \end{bmatrix}$ denotes the minor of M formed by the rows $(i_1,...,i_k)$ and the columns $(j_1,...,j_k)$ of M. By reordering of inputs and outputs, the system is put in the following form :

$$W(p) = \begin{bmatrix} W_{\alpha\alpha}\ (p) & W_{\alpha\beta}\ (p) \\ W_{\beta\alpha}\ (p) & W_{\beta\beta}\ (p) \end{bmatrix} \qquad (3.4.17)$$

$$W(p) = [S_\alpha\ (p)\ S_\beta\ (p)\ ]^1\ [T_\alpha\ (p)\ T_\beta\ (p)\ ] \qquad (3.4.18)$$

and we assume control structures in the form :

$$U_i = K_i\ Y_i \qquad i = \alpha\ \beta \qquad (3.4.19)$$

Assume that r_α (r_β) is the number of rows and m_α (m_β) the number of columns of $W_{\alpha\alpha}\ (p)$ $(W_{\beta\beta}\ (p))$.

The number of zeros at λ_0 of $W\begin{bmatrix} i_1 \cdots i_k \\ j_1 \cdots j_k \end{bmatrix}$ is denoted by with : $\#\begin{bmatrix} i_1 \cdots i_k \\ j_1 \cdots j_k \end{bmatrix}$

$\# = 0$ λ_0 is neither a pole nor a zero

$\# < 0$ λ_0 is a pole of order $- \#$

$\# = \infty$ the minor is identically zero

We have $m = m_\alpha + m_\beta$ and $r = r_\alpha + r_\beta$. Among the first r_α rows, let us define the number of rows which do not belong to the considered minor by :

$$\Theta_r \begin{bmatrix} i_1 & \cdots & i_k \\ j_1 & \cdots & j_k \end{bmatrix} = \left\| \{ i'_1, \ldots, i'_{i-k} \} \cap \{ 1, \ldots, r_\alpha \} \right\|$$

where : $\{i_1, \ldots, i_k\} \cup \{i_1, \ldots, i_{r-k}\} = \{1, \ldots, m\}$

and among the m_α first columns, the number of columns which belong to the considered minor by :

$$\Theta_c \begin{bmatrix} i_1 & \cdots & i_k \\ j_1 & \cdots & j_k \end{bmatrix} = \left\| \{ j_1, \ldots, j_k \} \cap \{ 1, \ldots, m_\alpha \} \right\|$$

* denotes the number of elements of the set (*).

In this framework, Anderson (AND-82) showed the following result :

Theorem 3.20. The following two conditions are equivalent :

1 - rank $(S_\alpha (\lambda_0) \ T_\alpha(\lambda_0)) < r_\alpha$ $\qquad\qquad$ (3.4.20)

and the fixed mode $_0$ has degree d (see Theorem 3.17)

2 - there exists d such that $0 < d < D$ and such that, whenever $\Theta_r + \Theta_c > r_\alpha$

$$\# \begin{bmatrix} i_1 & \cdots & i_k \\ j_1 & \cdots & j_k \end{bmatrix} \geqslant (d-D) + (\Theta_r + \Theta_c - r_\alpha) \qquad\qquad (3.4.21)$$

for all minors of $W(p)$. D is the degree of λ_0 in the characteristic polynomial $\phi(p)$.

Remark 3.3. The preceding theorem states that λ_0 is a fixed mode of degree d if certain minors have λ_0 as a zero of certain minimum order or as a pole of limited multiplicity, while λ_0 is at the same time a pole of $W(p)$.

We recall that Θ_r indicates the number of rows among the first r_α rows which are not in the minor under scrutiny and Θ_c indicates the number of columns among the first m_α columns which are in this same minor. Therefore, the quantity $\Theta_r + \Theta_c - r_\alpha$ is associated with the position of the minor in $W(p)$. When $\Theta_r + \Theta_c - r_\alpha$ is nonnegative, the minor is constrained by the inequality (3.4.2). In the case for which this quantity is negative, the minor is not constrained.

Example 3.7. Consider the 2-station system described by the following transfer function matrix :

$$W(p) = \begin{bmatrix} \dfrac{1}{p(p-1)} & \vdots & 0 & 0 \\[2ex] \dfrac{1}{p-1} & \vdots & 0 & 0 \\[2ex] 0 & \vdots & 0 & \dfrac{1}{p-1} \\[1ex] \hdashline \\[-1ex] \dfrac{1}{p-1} & \vdots & \dfrac{1}{p-1} & 0 \end{bmatrix}$$

for which $\phi(p)=p(p-1)^3$.

The dotted lines denote the partition in 2 stations and the associated decentralized control has the following structure :

$$\begin{bmatrix} U_\alpha \\[2ex] U_\beta \end{bmatrix} = \begin{bmatrix} k_{11} & k_{12} & k_{13} & 0 \\[1ex] 0 & 0 & 0 & k^{24} \\[1ex] 0 & 0 & 0 & k^{34} \end{bmatrix} \begin{bmatrix} Y_\alpha \\[2ex] Y_\beta \end{bmatrix}$$

We have : $r_\alpha = 3$, $r_\beta = 1$, $m_\alpha = 1$, $m_\beta = 2$ and D=3. The minors verifiying $\theta_r + \theta_c > 3$ are the following :

$$W\begin{bmatrix} 1 \\ 1 \end{bmatrix} \quad W\begin{bmatrix} 2 \\ 1 \end{bmatrix} \quad W\begin{bmatrix} 3 \\ 1 \end{bmatrix} \quad W\begin{bmatrix} 4 \\ 1 \end{bmatrix} \quad W\begin{bmatrix} 4 \\ 2 \end{bmatrix} \quad W\begin{bmatrix} 4 \\ 3 \end{bmatrix} \quad W\begin{bmatrix} 1 & 4 \\ 1 & 2 \end{bmatrix} \quad W\begin{bmatrix} 1 & 4 \\ 1 & 3 \end{bmatrix}$$

$$W\begin{bmatrix} 2 & 4 \\ 1 & 2 \end{bmatrix} \quad W\begin{bmatrix} 2 & 4 \\ 1 & 3 \end{bmatrix} \quad W\begin{bmatrix} 3 & 4 \\ 1 & 2 \end{bmatrix} \quad W\begin{bmatrix} 3 & 4 \\ 1 & 3 \end{bmatrix}$$

From theorem 3.20, $\lambda_0 = 1$ is a fixed mode of degree d if and only if there exists d such that :

$$\#\begin{bmatrix} 1 \\ 1 \end{bmatrix} = -1 \geqslant (d-3) \qquad \#\begin{bmatrix} 4 \\ 1 \end{bmatrix} = -1 \geqslant (d-3)+1$$

$$\# \begin{bmatrix} 2 \\ 1 \end{bmatrix} = -1 \geqslant (d-3) \qquad \# \begin{bmatrix} 4 \\ 2 \end{bmatrix} = -1 \geqslant (d-3)+1$$

$$\# \begin{bmatrix} 3 \\ 1 \end{bmatrix} = \infty \geqslant (d-3) \qquad \# \begin{bmatrix} 4 \\ 3 \end{bmatrix} = \infty \geqslant (d-3)+1$$

$$\# \begin{bmatrix} 1 & 4 \\ 1 & 2 \end{bmatrix} = -2 \geqslant (d-3) \qquad \# \begin{bmatrix} 2 & 4 \\ 1 & 3 \end{bmatrix} = \infty \geqslant (d-3)$$

$$\# \begin{bmatrix} 1 & 4 \\ 1 & 3 \end{bmatrix} = \infty \geqslant (d-3) \qquad \# \begin{bmatrix} 3 & 4 \\ 1 & 2 \end{bmatrix} = \infty \geqslant (d-3)$$

$$\# \begin{bmatrix} 2 & 4 \\ 1 & 2 \end{bmatrix} = -2 \geqslant (d-3) \qquad \# \begin{bmatrix} 3 & 4 \\ 1 & 3 \end{bmatrix} = -2 \geqslant (d-3)$$

These twelve inequalities are satisfied for d=1. Therefore, $\lambda_0 = 1$ is a decentralized fixed mode of degree 1.

Theorem 3.20 in its general form is rather complex. Therefore, it is interesting to note that it takes a particular simple form if we consider the case for which λ_0 is a simple pole of the system. The following result is easily obtained from condition 2 of theorem 3.20 when d=D=1.

Theorem 3.21 (AND-82). Under the same assumptions as in Theorem 3.20, assume that λ_0 is a simple zero of $\phi(p)$. Then, λ_0 is a decentralized fixed mode of system (3.4.1) (3.4.8) if and only if there exists a partition of the system as in (3.4.17) such that $W(p)$ or $W'(p)$ appears in the following form :

$$
\begin{array}{c}
 \\
r_\alpha \\
\\
r_\beta
\end{array}
\begin{bmatrix}
\begin{array}{c} m_\alpha \\ \text{no entry has a pole} \\ \text{at } \lambda_0 \end{array} & \begin{array}{c} m_\beta \\ \lambda_0 \text{ is a simple zero of the} \\ \text{characteristic polynomial} \\ \text{of this block} \end{array} \\
\hline
\begin{array}{c} \text{every entry has a zero} \\ \text{at } \lambda_0 \end{array} & \begin{array}{c} \text{no entry has a pole} \\ \text{at } \lambda_0 \end{array}
\end{bmatrix}
\qquad (3.4.22)
$$

Example 3.8. Consider the same system as in Example 3.5 and 3.6. If the system is put in form (3.4.17) taking station 1 as the first aggregated station and stations 2 and 3 as the second, $W(p)$ is partitioned as follows :

$$W(p) = \begin{bmatrix} \dfrac{3}{p-2} & 0 & \dfrac{p+1}{p(p-2)} \\ \hline \dfrac{1}{p-2} & \dfrac{1}{p+2} & \dfrac{1}{p(p-2)} \\ \dfrac{1}{p+1} & \dfrac{1}{p+2} & 0 \end{bmatrix}$$

and it is clear that $W'(p)$ has the form in Theorem 3.21 with $\lambda_0 = -1$. Therefore, we find again as in Example 3.6 that $\lambda_0 = -1$ is a fixed mode.

Theorem 3.21 is strictly equivalent to Theorem 3.16 which was derived by Davison and Ozguner (DAV-83) (see § 3.4.2) for the case of simple poles. The only difference consists in the fact that in Theorem 3.16, all the possible partitionings of the system which lead to form (3.4.22) in Theorem 3.21 are listed without reordering of the stations in 2 aggregated stations. The two aggregated stations can easily be deduced from the structures of the matrices A_0 or

$$W(p) - \frac{A_0}{p - \lambda_0} \quad \text{as follows :}$$

* rows r_α (r_β) : rows of A_0 $\left(W(p) - \dfrac{A_0}{p-\lambda_0} \Big|_{p=\lambda_0} \right)$ where there is a block specified

at X (0) or rows of $W(p) - \dfrac{A_0}{p-\lambda_0}\Big|_{p=\lambda_0}$ (A_0) where there is no block specified at 0 (X).

* columns m_α (m_β) : columns of A_0 $\left(W(p) - \dfrac{A_0}{p-\lambda_0}\Big|_{p=\lambda_0} \right)$ where there is no block

specified at X (0) or columns of $W(p) - \dfrac{A_0}{p-\lambda_0}\Big|_{p=\lambda_0}$ (A_0) where there is a block specified at 0 (X).

Note also that the two diagonal blocks for which no entry has a pole at λ_0 agree with the necessary condition for λ_0 to be a fixed mode in Theorem 3.15 (see § 3.4.1).

3.4.5. - Interpretation

It is interesting to note that in the results obtained for the general transfer function matrix characterization of fixed modes (Theorems 3.20 and 3.21), we find again the partitioning of the system in two aggregated stations which was already present in the state-space characterization of Anderson and Clements (see § 3.3.1, Theorem 3.7).

This situation can be intuitively interpreted as follows. When a system is controlled in a decentralized way, a station i can transmit some information to a station j through the transfer channel $C_j(pI-A)^{-1}B_i = W_{ij}(p)$. A similar scheme holds for a set of stations to another.

The lack of information at one station or at one set of stations, which results from the decentralization constraint (only local variables are available), can therefore be compensated by the information transfer through the transfer functions of the interconnection terms.

Hence, for a S-station system, if all the subsets of stations $\{i_1, \ldots, i_k\} \subset \{1, \ldots, S\}$ can transmit some information to their complementary subsets of stations $\{i_{k+1}, \ldots, i_S\}$ and vice versa, the information is spreaded among all the stations of the system and a decentralized control will succeed either for the stabilization task or for the pole assignment task.

As we already saw, the information transfer can be prevented by special parametric configurations, which result in the existence of fixed modes. It is also clear that this transfer is impossible if the transfer function blocks corresponding to the interconnections :

$$C_\alpha (pI-1)^{-1} B_\beta = W_{\alpha\beta}(p) \text{ and } C_\beta (pI-A)^{-1} B_\alpha = W_{\beta\alpha}(p)$$

are identically zero.

In order to illustrate this remark, consider again the case of a system with two single-input single-ouput stations. We saw in Paragraph 3.4.4a that if the feedback $u_2 = k_2 y_2$ is applied at station 2, the transfer function of the resulting system is given by (3.4.16) :

$$\frac{y_1}{u_1} = W_{11} + W_{12} \frac{k_2 W_{21}}{1 - W_{22} k_2}$$

Assume that the pole of the system λ_0 is not simultaneously controllable and observable by a single station ($W_{11}(p)$ and $W_{22}(p)$ have no pole at λ_0). However, λ_0 is controllable and observable by the global system ($W_{12}(p)$ or $W_{21}(p)$ have a pole at λ_0, say $W_{12}(p)$).

We saw in Paragraph 3.4.4.a that if λ_0 is a zero of $W_{21}(p)$, the pole-zero cancellation in the product $W_{12} \cdot W_{21}$ leads to the uncontrollability of (3.4.16) : λ_0 is a fixed mode. Now, if $W_{21}(p) = 0$, it is obvious that system (3.4.16) is also uncontrollable and λ_0 is also a fixed mode.

In the last case ($W_{21}(p) = 0$), λ_0 is a fixed mode independently of the parameter values : this type of fixed modes is called a <u>structurally fixed mode</u>.

In the first case ($W_{21}(p) \neq 0$), λ_0 is a fixed mode depending of the parameter values : a <u>non structurally fixed mode</u>.

If we consider now the special case for which $\lambda_0 = 0$, the condition for λ_0 to be a fixed mode is also met for generic values of the parameters even when $W_{21}(p) \neq 0$. The fixed modes at the origine may thus be a particular case of structurally fixed modes.

The extention of the preceding discussion to the case of multi-station systems is easily derived. λ_0 is a structurally fixed mode if $W_{\alpha\beta}(p)$ or $W_{\beta\alpha}(p)$ is identically zero and the conditions of Theorem 3.20 (or Theorem 3.21 if λ_0 is a simple pole of the system) hold. If $\lambda_0 = 0$, the conditions in Theorem 3.20 and 3.21 may also hold for generic values of the parameters and λ_0 is also a structurally fixed mode in this case.

The analogy with the case of centralized control becomes now more precise. As we saw in Chapter II, fixed modes appear as an extention of the concepts of uncontrollable and unobservable modes and it is clear that structurally fixed modes establish the relation with structurally uncontrollable and unobservable modes (see § 1.3, Chapter I). Both are defined for generic values of the parameters of the system and are therefore structural properties. A similar study to the one made for structural controllability and structural observability in chapter I can thus be carried out for structurally fixed modes. The next paragraph presents the results which have been obtained on this purpose.

3.5. - STRUCTURALLY FIXED MODES

3.5.1. - Preliminaries

It has been pointed out in the last paragraph that a fixed mode may originate from two distinct sources :

- a perfect matching in system parameters (in this case, a perturbation of the nonzero parameters can eliminate the fixed mode) : it is then called a <u>non structurally fixed mode</u>,

- the special structure within the system (in this case the fixed mode remains however the nonzero parameters values are perturbed) : it is then called a <u>structurally fixed mode</u>.

It was also pointed out that structurally fixed modes themselves may have two different origines, though both are of a structural nature. They may result from :

- the lack of interconnections between the different stations,
- the special location of a mode at the origin.

The following example illustrates the above situations :

Example 3.9. Consider the following 2-station system :

$$
A = \begin{bmatrix} 0 & 1 & 0 \\ 1 & 1 & 0 \\ 0 & 0 & 1 \end{bmatrix} \qquad B_1 = \begin{bmatrix} 1 \\ 0 \\ 0 \end{bmatrix} \qquad B_2 = \begin{bmatrix} 0 \\ 0 \\ 1 \end{bmatrix}
$$

$$
C_1 = \begin{bmatrix} 0 & 0 & 1 \end{bmatrix}
$$

$$
C_2 = \begin{bmatrix} 1 & 0 & 0 \end{bmatrix}
$$

which is globally controllable and observable. If a decentralized control law of the form :

$$
u = K y
$$

$$
K = \begin{bmatrix} k_1 & 0 \\ 0 & k_2 \end{bmatrix}
$$

is applied to this system, the resulting closed-loop system has the following dynamic matrix :

$$A_K = A + B K C = \begin{bmatrix} 0 & 1 & k_1 \\ 1 & 1 & 0 \\ k_2 & 0 & 1 \end{bmatrix}$$

and $\det(pI-A_K) = (p-1)(p^2-p-1-k_1 k_2)$ where it appears that the eigenvalue at $\lambda_0=1$ is independent of k_1 and k_2. Therefore, $\lambda_0=1$ is a decentralized fixed mode.

However, if one of the nonzero entries in A is slightly perturbed as follows :

$$A = \begin{bmatrix} 0 & 1 & 0 \\ 1 & 1 & 0 \\ 0 & 0 & 1+\epsilon \end{bmatrix}$$

where ϵ is an arbitrary small constant, we have :

$$\det (pI-A_K) = p(p-1)(p-1-\epsilon)-(p-1-\epsilon)-k_1 k_2 (p-1)$$

and the fixed mode does not exist any more : the fixed mode at $\lambda_0=1$ was therefore a non structurally fixed mode.

Consider now that the dynamic matrix of the system is given by :

$$A = \begin{bmatrix} 0 & 1 & 0 \\ 1 & 0 & 0 \\ 0 & 0 & 0 \end{bmatrix}$$

such that : $\det (pI-A_K) = p(p^2-1-k_1 k_2)$.

In this case, the system has a fixed mode at $\lambda_0=0$ which will remain regardless of any perturbation of the non zero elements of the matrices A, B_1, B_2, C_1 and C_2.

Consider also the following system :

$$A^* = \begin{bmatrix} 1 & 0 & 0 \\ 0 & 0 & 0 \\ 0 & 0 & -1 \end{bmatrix} \qquad B_1^* = \begin{bmatrix} 1 \\ 1 \\ 0 \end{bmatrix} \qquad B_2^* = \begin{bmatrix} 0 \\ 0 \\ 1 \end{bmatrix}$$

$$C_1^* = \begin{bmatrix} 1 & 0 & 0 \end{bmatrix} \qquad C_2^* = \begin{bmatrix} 0 & 1 & 1 \end{bmatrix}$$

We have : det $(pI-A_K^*) = p(p-1-k_1)(p+1-k_2)$ and the system $(B_1^*,B_2^*,A^*,C_1^*,C_2^*)$ has also a fixed mode at $\lambda_0=0$ which will remain however the nonzero elements of the matrices are perturbed. Nevertheless, in the two above cases, the origin of the structurally fixed mode at $\lambda_0=0$ is not the same. This appears clearly in the associated transfer function matrices.

Case 1 :

$$\text{System } (B_1^*,B_2^*,A^*,C_1^*,C_2^*) \qquad W_1(p) = \left[\begin{array}{c|c} \dfrac{1}{p-1} & 0 \\ \hline \dfrac{1}{p} & \dfrac{1}{p+1} \end{array}\right]$$

It is now clear that the fixed mode at $\lambda_0=0$ arises from the lack of the interconnection term between stations 1 and 2 which is shown by $C_1^*(pI-A^*)^{-1}B_2^* = 0$ in $W_1(p)$. In this case, $\lambda_0=0$ will be called a structurally fixed mode of type (i).

Case 2 :

$$\text{System } (B_1,B_2,A,C_1,C_2) \qquad W_2(p) = \left[\begin{array}{cc} 0 & \dfrac{1}{p} \\ \dfrac{p}{(p+1)(p-1)} & 0 \end{array}\right]$$

In $W_2(p)$, none of the interconnection terms between the stations are missing. The fixed mode at $\lambda_0=0$ arises from its special location at the origin that results in a pole-zero cancellation at $\lambda_0=0$ between the interconnection terms.

This fixed mode is called a structurally fixe mode of type (ii).

In both cases, it is clear that no perturbation of the nonzero entries in the matrices can change the situation. Nevertheless, the different origins of the fixed mode in case 1 and 2 can be significantly shown by affecting a value to a zero parameter, which results in shifting the pole $\lambda_0=0$ without affecting the interconnection structure of the systems :

$$\text{Case 1 :} \begin{bmatrix} 1 & 0 & 0 \\ 0 & 2 & 0 \\ 0 & 0 & -1 \end{bmatrix} \qquad \text{Case 2 :} \begin{bmatrix} 0 & 1 & 0 \\ 1 & 0 & 0 \\ 0 & 0 & 2 \end{bmatrix}$$

The pole at the origin is now replaced by a pole at $\lambda_1 = 2$. In case 1, we have now :

$$W_1(p) = \begin{bmatrix} \dfrac{1}{p-1} & 0 \\[2ex] \dfrac{1}{(p-2)} & \dfrac{1}{p+1} \end{bmatrix}$$

where nothing is changed with respect to the zero interconnection term. The structurally fixed mode at $\lambda_0 = 0$ is replaced by a structurally fixed mode at $\lambda_1 = 2$, reflecting that the origin of the structurally fixed mode is a lack of interconnections.

On the contrary, in case 2, we have :

$$W_2(p) = \begin{bmatrix} 0 & \dfrac{1}{p-2} \\[2ex] \dfrac{p}{(p-2)(p-1)} & 0 \end{bmatrix}$$

where the pole-zero cancellation at $\lambda_0 = 0$ is no longer possible ; i.e. the system has no fixed mode. Therefore, in case 2, the origin of the fixed mode was its special location at the origin.

The next paragraphe will show that these two types of structurally fixed modes can be characterized in different ways.

Refering to the above discussion, we would like to point out that one should not understand that all fixed modes at the origin are of a structural nature. Indeed, consider the following system :

$$A = \begin{bmatrix} 0 & 0 & 0 \\ 0 & a_{22} & 0 \\ 0 & 0 & 2 \end{bmatrix} \qquad B_1 = \begin{bmatrix} 1 \\ 0 \\ 0 \end{bmatrix} \qquad B_2 = \begin{bmatrix} 0 \\ -1 \\ 2 \end{bmatrix}$$

$$C_1 = \begin{bmatrix} 0 & 1 & 1 \end{bmatrix}$$

$$C_2 = \begin{bmatrix} 1 & 0 & 0 \end{bmatrix}$$

where a_{22} is an arbitrary real paramater.

The transfer function matrix is given by :

$$W(p) = \begin{bmatrix} 0 & \dfrac{p+2(1-a_{22})}{(p-a_{22})\,(p-2)} \\[3mm] \dfrac{1}{p} & 0 \end{bmatrix}$$

where it is clear that the system has a fixed mode at $\lambda_0=0$ if and only if $a_{22} = 1$; therefore, in this case $\lambda_0=0$ is not a structurally fixed mode.

Note also that the lack of interconnection terms in the transfer function matrix is necessary for a fixed mode to be structural (of type (i)) but not sufficient. Indeed, consider the following system :

$$A = \begin{bmatrix} 0 & 1 & 0 \\ 1 & 0 & 0 \\ 0 & 0 & 2 \end{bmatrix} \qquad B_1 = \begin{bmatrix} 1 \\ 1 \\ 1 \end{bmatrix} \qquad B_2 = \begin{bmatrix} 1 \\ -1 \\ 0 \end{bmatrix}$$

$$C_1 = \begin{bmatrix} 1 & 1 & 0 \end{bmatrix}$$

$$C_2 = \begin{bmatrix} 1 & -1 & 1 \end{bmatrix}$$

for which $\det(pI-A_K) = (p-2)(p-2k_1-1)(p-2k_2+1)$. $\lambda_1=2$ is therefore a fixed mode. The transfer function matrix of this system is given by :

$$W(p) = \begin{bmatrix} \dfrac{2}{p-1} & 0 \\[2ex] \dfrac{1}{p-2} & \dfrac{2}{p+1} \end{bmatrix}$$

which exibits a zero block. One could thus think that $\lambda_1 = 2$ is a structurally fixed mode of type (i). However, it must be notify that the zero block results from a cancellation. Indeed, if one replaces, for example, the first entry in B_2 by 2, the transfer function matrix becomes :

$$W(p) = \begin{bmatrix} \dfrac{2}{p-1} & \dfrac{1}{p-1} \\[2ex] \dfrac{1}{p-2} & \dfrac{3}{p+1} \end{bmatrix}$$

which has no more zero block. $\lambda_1 = 2$ is no longer a fixed mode since there is no possible cancellation in the product :

$$\frac{1}{p-2} \quad \times \quad \frac{1}{p-1}$$

The necessary and sufficient conditions for a pole to be a structurally fixed mode will be presented in the subsequent paragraphs.

As was pointed out in Paragraph 3.4.5, fixed modes are the extention of the concepts of uncontrollable and unobservable modes existing in centralized control and structurally fixed modes establish the relation with structurally uncontrollable and unobservable modes. The properties of controllability, observability and their structural counterparts, structural controllability and structural observability, were defined in Chapter I. In the decentralized control case, these properties were stated and studied in (KOB-78) (KOB-82) and (MOM-83) which introduced the concepts of controllability and observability under decentralized information structure and their structural counterparts.

These properties are presented in the next paragraph and their relations with fixed modes appear clearly. The analogy with the centralized control case is then even more obvious.

Some of the definitions used in the following were already introduced in Chapter I and the reader will be refered to it whenever it will be necessary.

3.5.2. - Controllability and observability under decentralized information structure

Consider the S-station system (3.2.2) and a decentralized control feedback in the form (3.2.4) (3.2.11).

The following definition was given by Momen and Evans (MOM-83).

Definition 3.4. The system (3.2.2) is said to be <u>controllable under decentralized information structure</u> if there exists a decentralized output feedback of the form (3.2.4) (3.2.11), which transfers any initial state x(0) to the origine in a finite interval of time T, i.e., x(T)=0.

Definition 3.5. The system (3.2.2) is said to be <u>structurally controllable under decentralized control structure</u> if there exists a structurally equivalent system (see § 1.3) which is controllable under decentralized structure.

Kobayashi et al (KOB-78) (KOB-82) provided also a definition for a system controllable under decentralized information structure. Nevertheless, the controllability under decentralized information structure in the sense of Kobayashi et al appears to be different from which defined in Definition 3.4 and even from which defined in Definition 3.5. Though the differences do not seem to be of a major importance at first glance, we will see that their interpretation is strongly connected with the different types of fixed modes.

Definition 3.6. The system (3.2.2) is said to be controllable under decentralized information structure (in the sense of Kobayahsi et al) if there exists a decentralized control of the form :

$$u_i(t) = F_i \, I_i(t) \qquad (i=1,\dots,S) \qquad\qquad (3.5.1)$$

where $I_i(t)$ is the set of available data at station i at time t :

$$I_i(t) = \{y_i(p), \, u_i(q), p \in (0,t), \, q \in (0,t) \}$$

which transfers any initial state x(0) to the origine in a finite interval of time T ; i.e., x (T)=0.

The difference between Definitions 3.6 and 3.4 can be found in the different types of decentralized control laws which are used. Undeed, the control law in (3.2.4) (3.2.11) has not the degree of freedom of the one in (3.5.1). Controllability under decentralized information structure in the sense of Kobayashi et al is, in fact, more similar to structural controllability under decentralized information structure defined in Definition 3.5.

We will see that the required conditions for a system to have this property are also of a structural nature. However, a slight difference still remains, which is based on the existence of two different types of structurally fixed modes, as was pointed out in Paragraph 3.4.5 and Example 3.9.

3.5.2.a. - Kobayashi et al. approach (KOB-78) (KOB-82)

The approach of Kobayashi et al. is based on the observation that, in decentralized control systems, a control station can transmit necessary information to other control stations through the state space by using signaling strategies. That is, station j can transmit information to station i if (and only if) $C_i(pI-A)^{-1}B_j = W_{ij}(p)$ $\neq 0$. In this case, station j can enlarge its controllable subspace by sending information to station i and station i can enlarge its observable subspace by receiving information from station j. Kobayahsi et al. represent this situation by defining the matrices C_i^* and B_i^*, (i=1,...,S), which are interpreted as the observation and the driving matrices of station i under the cooperation of the other stations :

$$B_i^* = [B_i, B_{j_l}, ..., B_{j_k}] \qquad \text{if } C_j(pI-A)^{-1}B_i \neq 0 \qquad \forall j \in \{j_l, ..., j_k\}$$

$$C_i^* = \begin{bmatrix} C_i \\ C_{q_l} \\ \vdots \\ C_{q_v} \end{bmatrix} \qquad \text{if} \qquad C_i(pI-A)^{-1}B_j \neq 0 \qquad \forall j \in \{q_l,...,q_v\}$$

Then they develop, their study by associating with the system the same directed graph as Corfmat and Morse (COR-76a,b), which was defined in Paragraph 2.2.3b of Chaper II. Like Corfmat and Morse, they decompose the system in strongly connected subsystems.

For a system (3.2.2) with N strongly connected subsystems, the reordering of the stations put the system in the following form :

$$\dot{x}(t) = A\ x(t) + \tilde{B}_1\ \tilde{u}_1(t) + \dots M\ \tilde{B}_N\ \tilde{u}_N(t)$$
$$\tilde{y}_i(t) = \tilde{C}_i\ x(t) \qquad (i=1,\dots,N)$$

with $\qquad \tilde{B}_i = [B_{j_1}, \dots, B_{j_i}] \in R^{n \times \tilde{m}_i} \qquad\qquad \tilde{C}_i = \begin{bmatrix} C_{j_1} \\ \\ C_{j_i} \end{bmatrix} \in R^{r_i \times n}$

where $i_1, \dots, i_k$ are the stations belonging to the strong component i. Note that the matrices $\tilde{B}_i$ and $\tilde{C}_i$ are composed of the driving and observation matrices of the station within the ith strong component. $\tilde{B}_i$ and $\tilde{C}_i$ are also the driving and observation matrices of every station with in the strong component i under the cooperation of the remaining stations within this same strong component.

The reordered transfer function matrix of the system appears thus in a block-triangular form :

$$\tilde{W}(p) = \begin{bmatrix} \tilde{W}_{11} & & & \\ \tilde{W}_{21} & \tilde{W}_{22} & & \\ & & & \\ \tilde{W}_{N1} & \dots\dots\dots & & \tilde{W}_{NN} \end{bmatrix}$$

where $\tilde{W}_{ij}$ is the transfer function matrix from the strongly connected subsystem i to the strongly connected subsystem j.

The system put in the above form is called the <u>quotient system</u>. Kobayashi et al give then the following theorem :

<u>Theorem 3.22</u> (KOB-82). The system (3.2.2) is controllable under decentralized information structure (in the sense of Kobayashi et al., see Definition 3.6) if and only if the quotient system has no decentralized fixed modes with respect to the decentralized control law :

$$U = \tilde{K}\ Y \qquad\qquad\qquad (3.5.2)$$

$\tilde{K} = \text{block-diag.}\ (\tilde{K}_1, \dots, \tilde{K}_N) \qquad K_i \in R^{\tilde{m}_i \times \tilde{r}_i} \qquad (i=1,\dots,N)$

The fixed modes of the quotient system correspond to the structurally fixed modes of type (i) of the system (3.2.2). Indeed, none partition of the quotient system can be found such that the interconnection term between the first aggregated station and the second is not identically zero. As we saw in Paragraph 3.4.5, if Theorem 3.20 is satisfied in this case, the fixed modes are structural and arise from the lack of interconnections. The non structurally fixed modes and the structurally fixed modes of type (ii) of the system (3.2.2) (if any) are set aside by the fact that in (3.5.2) the control law associated to each strongly connected station (which may contain several stations of the system (3.2.2)) is centralized.

This leads to the following corollary :

<u>Corollary 3.6.</u> The system (3.2.2) is controllable under decentralized information structure (in the sense of Kobayashi et al.) if and only if it has no structurally fixed modes of type (i).

An interesting consequence is that a strongly connected system is always controllable under decentralized information structure (in the sense of Kobayashi et al) and <u>cannot have structurally fixed modes of type (i).</u>

If we establish the connection with the results of Corfmat and Morse (COR-76a,b) presented in Paragraph 2.2.3b of Chapter II, the fixed modes can be characterized in terms of the remnant polynomials of the complementary subsystems (see Definition 2.6, Chaper II) and the results of Corfmat and Morse can be restated in terms of fixed modes. Theorem 2.2 leads to the following corollary :

<u>Corollary 3.7.</u> The system (3.2.2) with S stations, controllable, observable and strongly connected can be stabilized by a control law of the form :

$$
\begin{cases}
u_i(t) = K_i \, y_i(t) + v_i(t) & (i=1,\ldots,S) \\
\dot{z}_j(t) = S_j \, z_j(t) + R_j \, y_j(t) \\
\hspace{3cm} j \in \{1,\ldots,S\} \\
v_j(t) = Q_j \, z_j(t) + N_j \, y_j(t)
\end{cases}
$$

if and only if the set of fixed modes (which cannot be structurally fixed modes of type (i)) is stable. The arbitrary pole placement is possible if and only if the set of fixed modes is empty. (Note that the control law uses a dynamic controller only at one station since strongly connected systems without fixed modes can be made controllable and observable by a single station using static decentralized feedback).

For this system, the set of decentralized fixed modes (including non structurally fixed modes and structurally fixed modes of type (ii)) is characterized by the fixed polynomial :

$$F(p) = \text{l.c.m.} \quad \rho\,(C_\alpha\,,\ A,B_\beta\,)$$
$$k=1,\ldots,S-1$$

$$\pi_\alpha = \{\,i_1,\ldots,i_k\,\} \subset \quad \pi = \{\,1,\ldots,S\,\}$$

If we consider now the general case of non strongly connected systems, Theorem 2.3 can be restated as follows :

Corollary 3.8. The controllable and observable system (3.2.2) can be stabilized using a dynamic decentralized control of the form (2.2.6) if and only if the set of fixed modes (structurally and non structurally) is stable. The arbitrary pole placement is possible if and only if the set of fixed modes is empty.

For this system, the set of structurally fixed modes of type (i) is given by the set of poles of the system which are not pole of any strongly connected subsystem (fixed modes of the quotient system). The remaining fixed modes (including non structurally fixed modes and structurally fixed modes of type (ii)) are given by the union of the sets of fixed modes of every strongly connected subsystem. For each strongly connected subsystem, they can be characterized as follows :

$$F_i(p) \quad = \text{l.c.m} \quad \rho\,(\tilde{C}_{i\alpha}\,,\ A\ ,\ \ \tilde{B}_{i\beta}) - \ \Lambda\,(\tilde{C}_i,A,\tilde{B}_i,R^{\tilde{m}_i x \tilde{r}_i}\,)$$
$$\pi_{i\,\alpha} \subset \pi_i$$

$$\pi_i = \{\,j_1,\ \ldots\ j_i\,\}$$

The equivalence between the results of Wang and Davison (WAN-73) (see § 2.2.3.a, Chapter II) and those of Corfmat and Morse is now clear. Note that the results of Corfmat and Morse are more complete since they include the differentiation between structurally and non structurally fixed modes.

3.5.2.b. - Momen and Evans approach (MOM-83)

The approach of Momen and Evans is purely structural. The principle remains the same as in Kobayashi et al. approach since their results are also based on the determination of the enlarged structurally controllable and observable subspaces of every station under the collaboration of the other ones.

Let L_i, (i=1,...,S) be the structurally controllable and observable subspace of station i and L_i^* the enlarged structurally controllable and observable subspace of station i under the cooperation of the other stations. Then, we have the following result :

<u>Theorem 3.23</u> (MOM-83). The system (3.2.2) is structurally controllable under the decentralized information structure (see Definition 3.5) if and only if :

$$\overset{S}{\underset{i=1}{u}} \; L_i^* = R^n \tag{3.5.3}$$

The condition (3.5.3) means that the union of the enlarged structurally controllable and observable subspaces over all the stations is equal to the state space of the system. Note that if condition (3.5.3) holds, the system is also structurally observable under decentralized information structure. The connection with fixed modes is then made in the following theorem :

<u>Theorem 3.24</u> (MOM-83). The system (3.2.2) is structurally controllable (and observable) under decentralized information structure if and only if it has no structurally fixed modes.

Therefore, Momen and Evans (MOM-83) define a controllable system under decentralized structure (see Definition 3.4) as a system without decentralized fixed modes of any type and extend this result in a structural way in Definition 3.5 and Theorem 3.24 (As was pointed out in the examples given in Paragraph 3.5.1, for a system with structurally fixed modes of type (i) or (ii), none structurally equivalent system without fixed modes can be found).

The difference between structural controllability under decentralized information structure and controllability under decentralized information structure in the sense of Kobayashi et al. can be found in the fact that, in Theorem 3.24, structurally fixed modes include both structurally fixed modes of type (i) and of type (ii). Indeed, the determination of the enlarged structurally controllable and observable subspaces is based on the conditions for structural controllability and observability (see Theorems 1.6 and 1.7, Chapter I). These conditions involve both reachability conditions and generic rank conditions. The reachability conditions turn into connectivity conditions in the case of decentralized control ; i.e., the system must be strongly connected and, therefore, it has no structurally fixed modes of type (i). The generic rank conditions take into account the particular case of a pole at the origin which may also lead to structural uncontrollability or unobservability. By this way, in decentralized control, the system is also required not to have structurally fixed modes of type (ii).

Remember that in Kobayashi et al, the enlarged controllable and observable subspaces were determined using connectivity conditions only : this is why their controllability under decentralized information structure only requires the system not to have structurally fixed modes of type (i).

It is clear now that decentralized fixed modes can be defined as the modes that are uncontrollable under the decentralized information structure using a dynamic decentralized control in the form (2.2.6). However, an interesting remark can be made. From Theorem 3.24 and Corollary 3.6 and the Definitions 3.5 and 3.6, we can deduce the following result :

Corollary 3.9. Non structurally fixed modes and structurally fixed modes of type (ii) are controllable under the decentralized information structure using a dynamic time-varying decentralized controller in the form (3.5.1).

Structurally fixed modes of type (i) are not controllable under decentralized information structure even if the time-invariance constraint on the decentralized control is relaxed.

This point will be discussed again in Chapter IV and the physical interpretation of this result will be pointed out. It is interesting to mention that Corollary 3.9 expresses a major difference with respect to the case of centralized control since uncontrollable and unobservable modes remain uncontrollable and unobservable even if a dynamic time-varying controller is used.

Although the above study was carried out in a decentralized information structure framework, it is clear that the results can be extended to the general case of arbitrary constrained information structure.

3.5.2.c. - Potential pole assignment using decentralized static output feedback

The issue of controllability and structural controllability under decentralized information structure has been discussed in the preceding section and related to the concepts of fixed modes. It has been pointed out that structural controllability (absence of structurally fixed modes) is the necessary and sufficient condition for potential pole assignment using a decentralized dynamic controller. The present paragraph gives the necessary and sufficient condition for potential pole assignment using a decentralized static controller. Of course, this condition is sufficient to assure that the system has no structurally fixed modes but it is not necessary. The approach of Evans and Kruse (EVA-84) is original in the sense that it does not refer explicitly to the concepts of controllability, observability or fixed modes, although they are implied in the final result.

The result is carried out by a graph-theoretic method using the work of Mason (MAS-56) which establishes the relation between the structure of the digraph associated to a square matrix and its characteristic polynomial.

Consider a nxn matrix $M=(m_{ij})_{i,j=1,\ldots n}$. One defines the digraph $D=(V,E)$ such that $V= v_1,\ldots,v_n$ is a set of n vertices and E is a set of edges from V to $V.(v_i,v_j) \in E$ if and only if $m_{ji} \neq 0$ in M. The weight m_{ji} is associated with the edge (v_i,v_j). The characteristic polynomial of M is given by :

$$\phi(p) = \det(pI-M) = p^n + a_{n-1}p^{n-1} + \ldots + a_1 p + a_0$$

then, the coefficients $a_0,\ldots,a_{n-1}$ are given by :

$$a_{n-1} = \Sigma (\lambda)$$

$$a_{n-2} = \Sigma [\pi_2 (\lambda)]$$

$$a_1 = \Sigma [\pi_{n-1} (\lambda)]$$

$$a_0 = \Sigma [\pi_n (\lambda)]$$

in which $\Sigma[\pi_j (\lambda)]$ denotes the sum over the cycles of length j (or product of disjoint cycles whose sum of lengths is j) of the products of the weights associated to the edges involved in each cycle. The complete assignment of all roots of $\phi(p)$ to some set of arbitrary values can only be achieved if none of the coefficients $a_0,\ldots,a_{n-1}$ is zero.

Consider the system (3.2.1) and the control law :

$$U = K Y$$

where K is the output feedback matrix with an arbitrary specified structure. Then, from the above result, potential pole assignment can be achieved if :

1-All the coefficients in the characteristic polynomial of (A+BKC) can be assigned independently by the existing cycle gains of the associated graph.

2-Sufficient cycle gains can be assigned independently by the feedback gains in K.

Using the above discussion, Evans and Kruser (EVA-84) provided the following result :

Let $\{k_1,\ldots,k_q\}$ be the set of feedback gains in K and $\{L_1,\ldots,L_c\}$ be the set of existing cycles in the graph associated with (A+BKC). Define :

$$M_{La} = \begin{array}{c} L_1 \\ \vdots \\ L_c \end{array} \begin{array}{c} a_{n-1} \cdots a_0 \\ \left[\right] \end{array}$$

such that $m_{La}(i,j)=1$ if and only if L_i is of length j or appear in a product of disjoint cycles whose sum of lengths is j, and $m_{La}(i,j)=0$ otherwise. Define also :

$$M_{KL} = \begin{array}{c} k_1 \\ \vdots \\ k_q \end{array} \begin{array}{c} L_1 \cdots L_c \\ \left[\right] \end{array}$$

where $m_{KL}(i,j)=1$ if and only if k_i appears in the gain of the cycle L_j, and $m_{KL}(i,j) \neq 0$ otherwise.

Then, the necessary and sufficient condition for potential pole assignment is that :

$$gr \left[M_{KL} * M_{La} \right] = n$$

where * denotes the boolean product of two binary matrices.

This result provides an interesting test, in the form of a simple generic rank test determination, for potential pole assignment using decentralized static output feedback. For the practical implementation of the test, Evans and Kruser (EVA-84) provided an APL routine which gives the list of existing cycles in a digraph.

3.5.3. – Characterization of structurally fixed modes

Similarly to the case of centralized control, the characterization of structurally fixed modes (and therefore of structural controllability under constrained information structure) can be carried out either within a pure algebraic framework or within a graph-theoretic framework.

3.5.3.a. - Algebraic approach

Using the results obtained for structural controllability and observability (LIN-74) (SHI-76) (see § 1.3, Chapter I), Sezer and Siljak derived the following characterization of structurally fixed modes :

<u>Theorem 3.25</u> (SEZ-81a). The system (3.2.2) with S stations has structurally fixed modes if and only if one of the following conditions holds :

i - There exists a $\pi_\alpha \subset \pi = \{1,\ldots,S\}$ ($\pi_\alpha \cup \pi_\beta = \pi$) and a permutation matrix P such that :

$$P'AP = \begin{bmatrix} A_{11} & 0 & 0 \\ A_{21} & A_{22} & 0 \\ A_{31} & A_{32} & A_{33} \end{bmatrix} \quad P'B = \begin{bmatrix} B_\alpha^1 \\ B_\alpha^2 \\ B_\alpha^3 \end{bmatrix} \quad P'B = \begin{bmatrix} 0 \\ 0 \\ B_\beta^3 \end{bmatrix}$$

$$C\,P = \begin{bmatrix} C_\alpha^1 & 0 & 0 \end{bmatrix}$$

$$C\,P = \begin{bmatrix} C_\beta^1 & C_\beta^2 & C_\beta^3 \end{bmatrix}$$

(3.5.4)

ii - There exists a $\pi_\alpha \subset \pi = \{1,\ldots,S\}$ such that :

$$\text{gr} \begin{bmatrix} A & B_\beta \\ C_\alpha & 0 \end{bmatrix} \quad n$$

Structurally fixed modes of type (i) are characterized by condition (i) which exibits an obvious lack of connectivity of the system (of course, a strongly connected system can never be put in the form (3.5.4)). They are the eigenvalues of the matrix A_{22}. This situation is illustrated by Figure 3.5. The state vector is splited in 3 subvectors $x=(x^1,x^2,x^3)$ where x^2 are the states associated with the structurally fixed modes of type (i). It appears clearly in the figure that none information transfer is possible from the aggregated station α to the aggregated station β because of the block-triangular structure of the matrix A.

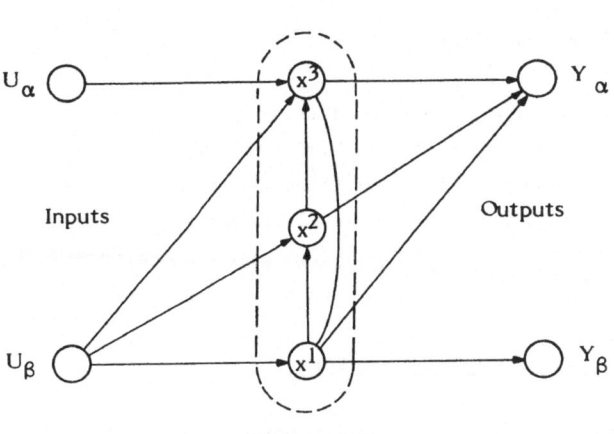

Figure 3.5

Structurally fixed modes of type (ii) are characterized by condition (ii) which is the structural counterpart of condition (3.3.2) (Characterization of fixed modes of Anderson and Clement. see § 3.3.1, Chapter II) for a pole at the origine.

Note that a structurally fixed mode of type (i) at the origin satisfies also condition (3.5.5) but it arises from the lack of connectivity of the system and remains uncontrollable under dynamic time-varying decentralized control. Clearly, the inverse is not true ; i.e. if $\lambda_0=0$ satisfies (3.5.5), condition (3.5.4) is not necessarily satisfied.

The similarity with structurally uncontrollable and unobservable modes appears clearly since, as we saw in Chapter I, they may arise from the lack of reachability or from their special location at the origin. The connectivity condition (3.5.4) corresponds to the reachability conditions (Conditions 1 and 2 of Theorems 1.6 and 1.7 in Chapter I) and the generic rank condition (3.5.5) corresponds to the generic rank conditions (Conditions 3 and 4 of Theorems 1.6 and 1.7 in Chapter I). Of course, in the case of decentralized control, we have only the concept of fixed mode (instead of the two dual concepts of uncontrollable and unobservable mode in centralized control) since the controllability of a mode is achieved through the cooperation of several stations by using the feedback from some output stations in which the observability concept is already involved.

Note that Theorem 3.25 can be generalized to arbitrarily constrained control : in this case, the sets π_α and π_β are replaced by the sets I and J (see Definition 3.3, § 3.3.1).

3.5.3.b. - Graph-Theoretic approach

Similarly to the case of centralized control, the conditions for structural controllability under decentralized information structure can be derived in a graph-theoretic framework. Several characterizations exist, depending on the graph that one associates to the system. The one which is presented first is the exact graphical counterpart of the algrebraic characterization given in Paragraph 3.5.3a.

1-Using the graph of the closed-loop system

Given the system (3.2.1), one associates a digraph $\Gamma = (U \cup X \cup Y, E)$, where $U = \{u_1, \ldots, u_m\}$ is the set of input vertices, $X = \{x_1, \ldots, x_n\}$ is the set of state vertices and $Y = \{y_1, \ldots, y_r\}$ the set of output vertices. E is the set of edges such that (u_i, x_j) E if and only if $b_{ji} \neq 0$ in B, $(x_i, x_j) \in E$ if and only if $a_{ji} \neq 0$ in A and $(x_i, y_j) \in E$ if and only if $c_{ji} \neq 0$ in C. Note that this digraph is the same as the one defined in Paragraph 1.3.1.b (Chapter I) for the study of structural controllability and observability in a graphical framework. Nevertheless, in our present study, we add a set of edges E_K from Y to U which represents the particular structure of the control. $(y_i, u_j) \in E_K$ if and only if $k_{ji} \neq 0$ in the feedback matrix K. The digraph associated to the system and the control structure is therefore $\Gamma_K = (U \cup X \cup Y, E \cup E_K)$.

In this graphical framework, the following graphical characterization of structurally fixed modes was provided independently by Linnemann (LIE-83) and by Pichai et al. (PIC-84) :

Theorem 3.26. The system (3.2.1) has no structurally fixed modes with respect to the arbitrarily structurally constrained control (3.3.4) if and only if both of the following two conditions hold :

i - each state vertex $x_k \subset X$ is involved in a strong component of Γ_K which includes an edge from E_K.

ii - there exists a set of disjoint cycles $\mathcal{C}_k = (V_k, E_k)$, (k=1,...,c) in Γ_K such that :

$$X \subset \bigcup_{k=1}^{c} V_k$$

(i.e., the whole state vertices set is spanned by a disjoint union of cycles).

Condition (i) corresponds to the structurally fixed modes of type (i) and condition (ii) to thoses of type (ii). Obviously, condition (i) in Theorem 3.26 is equivalent to condition (i) in Theorem 3.25 and Figure 3.5 shows that the states corresponding to x_2 are not contained in a strong component including feedback edges when feedback is allowed from Y_α to U_α and from Y_β and U_β. The equivalence of condition (ii) in Theorem 3.26 and condition (ii) in Theorem 3.25 is not so obvious. The proof can be found in (PIC-84).

Example 3.10. Consider the two following cases already treated in Example 3.9 :

Case 1 :

$$A = \begin{bmatrix} 1 & 0 & 0 \\ 0 & 2 & 0 \\ 0 & 0 & -1 \end{bmatrix} \qquad B_1 = \begin{bmatrix} 1 \\ 1 \\ 0 \end{bmatrix} \qquad B_2 = \begin{bmatrix} 0 \\ 0 \\ 1 \end{bmatrix}$$

$$C_1 = \begin{bmatrix} 1 & 0 & 0 \end{bmatrix}$$

$$C_2 = \begin{bmatrix} 0 & 1 & 1 \end{bmatrix}$$

This system already appears in the form (3.5.4) with $B_\alpha = B_1$, $B_\beta = B_2$, $C_\alpha = C_1$ and $C_\beta = C_2$. So, condition (i) of Theorem 3.25 is satisfied. Therefore, $\lambda_0 = 2$ is a structurally fixed mode of type (i). It is easy to verify that condition (ii) of Theorem 3.25 does not hold ; i.e., the system has no structurally fixed modes of type (ii). The same conclusions can be stated by considering the digraph associated to this system :

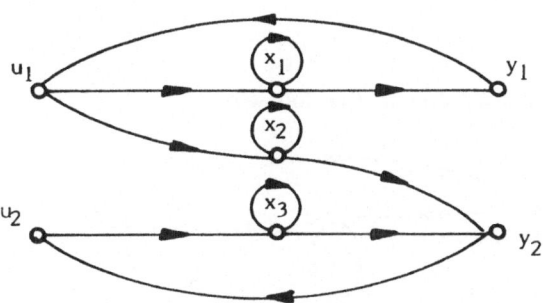

where it appears clearly that the vertex x_2 is not contained in a strong component including a feedback edge. However, the state vertices set is spanned by a disjoint union of cycles as it is shown below :

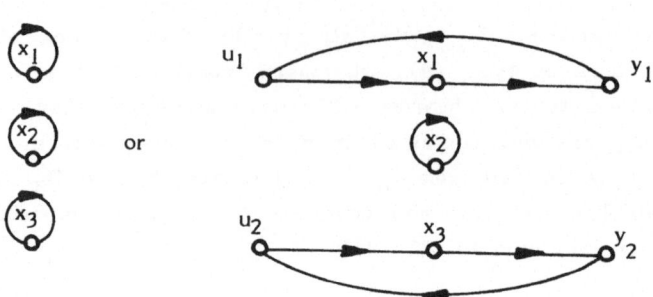

Therefore, condition (i) of Theorem 3.26 is satisfied but not condition (ii).

Case 2 :

$$A = \begin{bmatrix} 0 & 1 & 0 \\ 1 & 0 & 0 \\ 0 & 0 & 0 \end{bmatrix} \qquad B_1 = \begin{bmatrix} 1 \\ 0 \\ 0 \end{bmatrix} B_2 = \begin{bmatrix} 0 \\ 0 \\ 1 \end{bmatrix}$$

$$C_1 = \begin{bmatrix} 0 & 0 & 1 \end{bmatrix}$$

$$C_2 = \begin{bmatrix} 1 & 0 & 0 \end{bmatrix}$$

No permutation matrix can be found to put the system in the form (3.5.4) ; i.e., the system has no structurally fixed modes of type (i). Nevertheless, condition (ii) of Theorem 3.25 holds since :

$$\text{gr} \begin{bmatrix} A & B_1 \\ \\ C_2 & 0 \end{bmatrix} = 2 < 3$$

The digraph of this system is the following :

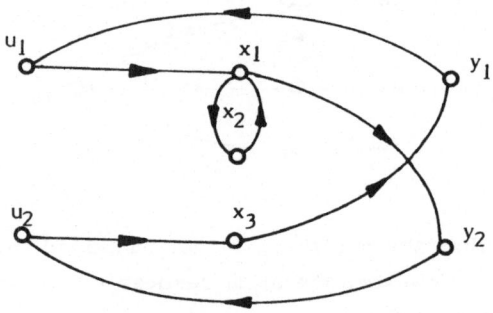

We find that condition (ii) of Theorem 3.26 does not hold since the graph contains no disjoint union of cycles such that all the state vertices are involved. Therefore, this system has a structurally fixed mode of type (ii) at the origin.

2-Using the graph of the open-loop system

This characterization was also provided by Pichai et al (PIC-83) (the same authors as those of the preceding characterization) but it was presented one year before. It will appear clearly that, from the computational point of view, it is much less interesting than the preceding one. Indeed, it consists of a test that has to be applied to all the 2^S-2 complementary subsystems of the system while the characterization given before is stated in terms of the whole system. Nevertheless, it presents the interest of characterizing both structurally fixed modes of type (i) and of type (ii) in a single test.

The digraph associated to the system (3.2.1) described by the triple (C,A,B) is the same as in the preceding paragraph : $\Gamma=(U \cup X \cup Y, E)$.

For a given digraph, the following special subgraphs are first defined :

Definition 3.7 (PIC-83a)

Input Stem : A path from an input vertex (the root) to a state vertex (the tip), or a single input vertex.

Output Stem : A path from a state vertex (the root) to an output vertex (the tip), or a single output vertex.

Input-Output Stem : A path from an input vertex to an output vertex.

State Stem : A path between two state vertices, or a single state vertex.

Cycle : Already defined (see § 1.3.1b, Chapter I).

Input cactus : An input stem with at least one state vertex. The root and the tip of the input stem are also the root and the tip of the input cactus. An input cactus connected to a cycle from any point other than the tip is also an input cactus.

Output Cactus : Similar definition to an input cactus.

Chain : A group of disjoint cycles connected to each other in sequence, or a single cycle.

Link : An input stem connected to the first cycle of a chain (from any point other than the tip), the last cycle of which is connected to an output stem (from any point other than the root).

These special subgraphs are illustrated in Figure 3.6 :

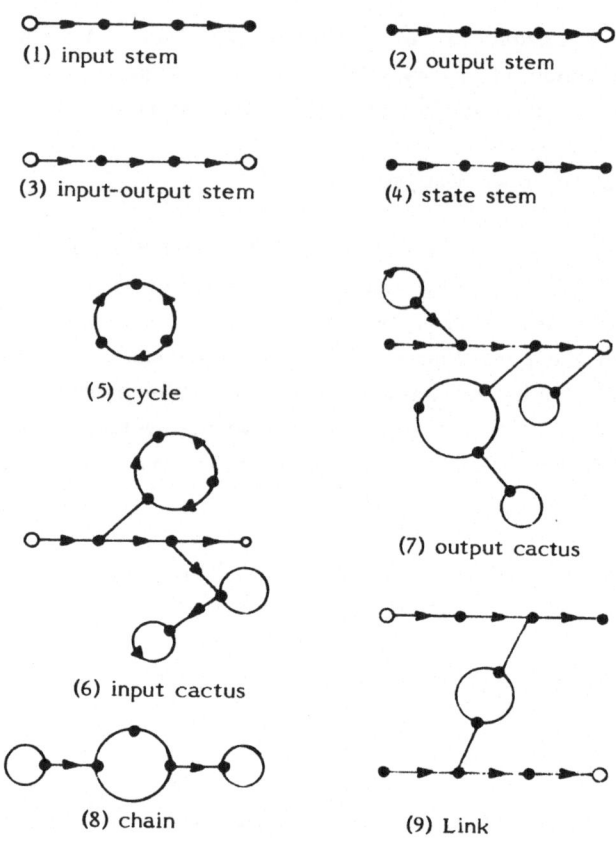

(1) input stem

(2) output stem

(3) input-output stem

(4) state stem

(5) cycle

(7) output cactus

(6) input cactus

(8) chain

(9) Link

Figure 3.6

A _generalized cactus_ is then defined as follows :

<u>Definition 3.8</u> (PIC-83a). Each of the following digraphs is called a generalized cactus :

    - Disjoint union of one or more input-output stems plus the same number of state stems plus some (or none) cycles, input stems, output stems.

    - Disjoint union of a link with some (or none) cycles, input stems, output stems.

    - Disjoint union of input cacti and output cacti.

    Consider that system (3.2.1) is controlled by an arbitrarily constrained control in the form (3.3.4). The complementary subsystems (C_J, A, B_I) of (3.2.1) with respect to (3.3.4) were defined in Definition 3.3 (Paragraph 3.3.1).

Pichai et al. (PIC-83) stated the following characterization of structurally fixed modes :

<u>Theorem 3.27.</u> The system (3.2.1) has no structurally fixed modes with respect to the control law (3.3.4) if and only if each digraph associated with a complementary system is spanned by a generalized cactus.

Pichai et al. (PIC-83a) provided an algorithm to check whether a digraph is spanned by a generalized cactus, which involves only binary computations. The test for the existence of structurally fixed modes requires to apply this algorithm to each "complementary subgraph", i.e. 2^S-2 times. This is due to the fact that the structure of the control is not involved in the graph associated with the system (contrarily to the case of the preceding characterization). Consequently, this characterization requires an enormous computational task.

3-Using the generic rank of the product of several real matrices

Recently, Papadimitriou and Tsitsiklis (PAP-84) provided an interesting graphical result to evaluate the generic rank of the product of several real matrices. Its application to the determination of structurally fixed modes of type (ii) leads to a nice computational algorithm.

Consider the sequence of structured matrices (see § 1.3, Chapter I), M_i, (i=1,...,k), associated to the sequence of real matrices $M_i \in R^{p_i \times q_i}$. We define the graph G=(V,E) as follows :

$V = \{ v_i^j : (i=1,...,k), (j=1,...,p_i) ; v_{k+1}^j : (j=1,...,q_{k+1}) \}$ is a set of vertices. One vertex is associated to every row of every matrix M_i, (i=1,...,k). The last q_{k+1} vertices are associated to the columns of the last matrix M_k of the sequence. There is an edge from the vertex v_i^p to the vertex v_{i+1}^q if the (p,q)th entry of the matrix M_i is non zero.

An information path is defined as a path from a vertex with subscript 1 to one with subscript k+1. Two information paths are said to be independent if they are vertex-disjoint.

Then, we have the following result :

<u>Theorem 3.27b(PAP-84).</u> $gr\left[\prod_{i=1}^{k} M_i\right]$ is equal to the maximum number of independent information paths in G.

In order to apply this result to the characterization of structurally fixed modes of type (ii), note that condition (ii) in Theorem 3.25 is equivalent to gr $(A+BKC) <$ n, where A, B, C are the structured matrices of system (3.2.1) and K the structured matrix specifying the feedback information pattern. Note moreover that $(A+BKC)$ can be rewritten as the product of three matrices :

with :

$$(A+BKC) = M_1 \cdot M_2 \cdot M_3$$
$$M_1 = [\, I \quad B \,]$$
$$M_2 = \begin{bmatrix} A & 0 \\ 0 & K \end{bmatrix} \qquad (3.5.6)$$
$$M_3 = \begin{bmatrix} I \\ C \end{bmatrix}$$

I denotes the identity structured matrix and 0 the null matrix of appropriate dimensions.

The following characterization of structurally fixed modes of type (ii) results :

Theorem 3.28 (PAP-84). System (3.2.1) has no structurally fixed modes of type (ii) with respect to the feedback pattern specified by K if and only if the number of information paths in the graph associated to the sequence of structured matrices M_1, M_2, M_3 defined in (3.5.6) is equal to n.

Example 3.11. Consider again the same system as in the preceding examples :

$$A = \begin{bmatrix} 0 & 1 & 0 \\ 1 & 0 & 0 \\ 0 & 0 & 0 \end{bmatrix} \qquad B_1 = \begin{bmatrix} 1 \\ 0 \\ 0 \end{bmatrix} \quad B_2 = \begin{bmatrix} 0 \\ 0 \\ 1 \end{bmatrix}$$

$$C_1 = \begin{bmatrix} 0 & 0 & 1 \end{bmatrix}$$

$$C_2 = \begin{bmatrix} 1 & 0 & 0 \end{bmatrix}$$

and

$$K = \begin{bmatrix} k_1 & 0 \\ 0 & k_2 \end{bmatrix}$$

The structured matrix corresponding to A+BKC is :

$$\begin{bmatrix} 0 & X & X \\ X & 0 & 0 \\ X & 0 & 0 \end{bmatrix}$$

and can be written as the product $M_1 \cdot M_2 \cdot M_3$ where :

$$M_1 = \begin{bmatrix} X & 0 & 0 & X & 0 \\ 0 & X & 0 & 0 & 0 \\ 0 & 0 & 0 & 0 & X \end{bmatrix} \quad M_2 = \begin{bmatrix} 0 & X & 0 & 0 & 0 \\ X & 0 & 0 & 0 & 0 \\ 0 & 0 & 0 & 0 & 0 \\ 0 & 0 & 0 & X & 0 \\ 0 & 0 & 0 & 0 & X \end{bmatrix}$$

$$M_3 = \begin{bmatrix} X & 0 & 0 \\ 0 & X & 0 \\ 0 & 0 & X \\ 0 & 0 & X \\ X & 0 & 0 \end{bmatrix}$$

The graph associated with the sequence of matrices M_1, M_2, M_3 is the following :

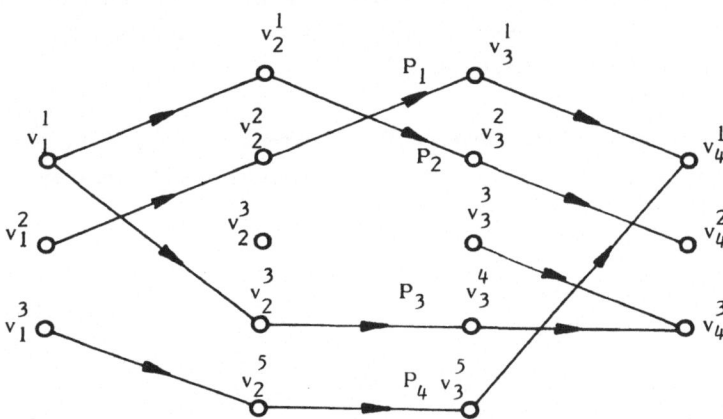

where there are only pairs of independent information paths : (P_1, P_2), (P_1, P_3), (P_2, P_4), (P_3, P_4). Therefore, the system has a structurally fixed mode of type (ii) at the origin.

The interest of this characterization compared to the one given by condition (ii) in Theorem 3.26 is that the available algorithms to determine the vertex-disjoint information paths of a graph are more efficient than those to determine the vertex-disjoint cycles. Indeed, it is proved in (PAP-84) that the test can be carried out in time $0 (n^{5/2})$.

3.5.4. - Evaluation of structurally fixed modes by calculation
of the structural sensitivity of the modes of the system

In Paragraph 2.4.2 (Chapter II), the fixed modes were characterized using the fact that such modes are the modes of the closed-loop system which are insensible to variations of the elements of the feedback matrix. Although this condition is verified either by structurally or non structurally fixed modes, this approach can be used to characterize structurally fixed modes if one adds the following supplementary conditions :

1-Structurally fixed modes of type (i) are only sensible to variations of some parameters of the system : these parameters are those of the matrix A_{22} defined in (3.5.4) in Theorem 3.25.

2-Structurally fixed modes of type (ii) are insensible to variations of any parameter of the system.

These conditions refer to the fact that, though the existence of a structurally fixed mode is insensible to variations of the parameters of the system, the value of a structurally fixed mode of type (i) is defined by the eigenvalues of A_{22} and the location of a structurally fixed mode of type (ii) is always at the origin. This situation was already pointed out in Example 3.9.

Consider an arbitrary real matrix D of dimension nxn. The gradient of a simple eigenvalue λ_r of D with respect to the elements of D is given by (2.4.7) (see § 2.4.2, Chapter II) :

$$\frac{d\lambda_r}{d(d_{ij})} = w'_r \frac{dD}{d(dj_{ij})} v_r \tag{2.4.7}$$

where w_r and v_r are the left and right eigenvectors of D corresponding to λ_r. Assume that the elements d_{ij} are functions of q parameters and let P be one of them. Then, we have :

$$\frac{\partial D}{\partial P}_{ij} = \frac{\delta_{ij}(P) \quad I^i.I_j}{0 \quad \text{if } d_{ij} = 0} \qquad \text{with} \quad \delta_{ij}(P) = \frac{\partial d_{ij}(P)}{\partial P} \tag{3.5.7}$$

where I^i and I_j are the ith column and jth row of the nxn identity matrix respectively. So, (2.4.7) becomes :

$$\left[\frac{\partial\lambda_r}{\partial P}\right]_{ij} = \begin{cases} \delta_{ij}(P) \, w'_r \cdot I^i \cdot I_j \cdot v_r \\ 0 \quad \text{if } d_{ij}=0 \end{cases} = \begin{cases} \delta_{ij}(P) \, w_i \cdot v_j \\ 0 \quad \text{if } d_{ij}=0 \end{cases}$$

where w_i and v_j are the ith and jth entries of w_r and v_r respectively. If we consider now the q parameters $(P_1,\ldots,P_q)$, the gradients of an eigenvalue of D with respect to each of them are :

$$\left[\frac{\partial\lambda_r}{\partial P_t}\right]_{ij} = \begin{cases} \delta_{ij}(P_t) \, w_i \, v_j & i,j=1,\ldots,n \\ 0 \quad \text{if } d_{ij}=0 & t=1,\ldots,q \end{cases} \qquad (3.5.8)$$

Since at least one element d_{ij} is function of each parameter P_t, $(t=1,\ldots,q)$, all the derivatives $\delta_{ij}(P_t)$ cannot be simultaneously zero. Therefore, λ_r is insensible to the variations of P_t if and only if $w_i \cdot v_j=0$, $(i,j=1,\ldots,n)$. This result is stated in the following lemma :

Lemma 3.1. The necessary and sufficient condition for a simple eigenvalue λ_r of a real matrix $D \in R^{nxn}$ to be insensible to the variations of the elements of D is that :

$$w_i \cdot v_j=0 \quad (i,j=1,\ldots,n)$$

where w_i and v_j are the ith and jth entries of the left and right eigenvectors of D corresponding to λ_r.

Definition 3.9 (TAR-84). The structural sensitivity matrix SS_r of a simple eigenvalue λ_r is defined as follows :

$$SS_r = (ss_{ij}) \quad \begin{aligned} i&=1,\ldots,n \\ j&=1,\ldots,n \end{aligned}$$

with :

$$ss_{ij} = \begin{cases} 1 & \text{if } w_i \cdot v_j \neq 0 \\ 0 & \text{if } d_{ij}=0 \text{ or } w_i \cdot v_j=0 \end{cases} \qquad (3.5.9)$$

From Definition 3.9 and the above discussion, the following characterization of structurally fixed modes follows :

<u>Theorem 3.29</u> (TAR-84b). Consider the system (3.2.1) and an arbitrary constrained feedback pattern specified by K. A simple eigenvalue λ_r of the closed-loop feedback matrix (A+BKC) is a structurally fixed mode of the system if and only if one of the two following conditions holds :

i - The structural sensitivity matrix SS_r (with respect to (A+BKC)) of λ_r is equal to one of the structural sensitivity matrices SS_r of λ_r^*, (r=1,...,n), where $\{\lambda_1^*,...,\lambda_n^*\}$ is the set of eigenvalues of D' and D' is structurally equivalent to (A+BKC).

ii - The structural sensitivity matrix SS_r (with respect tot (A+BKC)) of λ_r is identically zero.

From the remarks stated at the begining of the paragraph, it is clear that condition (i) characterizes the structurally fixed modes of type (i) and condition (ii) those of type (ii).

The fixed modes of system (3.2.1) (with distinct modes) with respect to an arbitrary feedback pattern specified by K can be determined with distinction of their different types by the following algorithm :

<u>Algorithm 3.1</u> (TAR-84)

1-Choose a numerical value for the nonzero entries of the feedback matrix K such that the eigenvalues of (A+BKC) are distinct.

2-For all $\lambda_r \subset \sigma(A+BKC)$, evaluate the sensitivity matrices SK_r (see Definition 2.8, Chapter II). If $SK_r = 0$, then λ_r is a fixed mode ; i.e. $\lambda_r \in \Lambda$.

3-If $\Lambda = \emptyset$, STOP. The system has no fixed modes.

4-For all $\lambda_r \subset \Lambda$, calculate the structural sensitivity matrix SS_r. If $SS_r = 0$, then λ_r is a structurally fixed mode of type (ii) ; i.e. $\lambda_r \in \Lambda_{S2}$.

5- $\Lambda_1 = \Lambda - \Lambda_{S2}$.

If $\Lambda_1 = \emptyset$, go to 10.

6-Choose a matrix D* structurally equivalent to (A+BKC) and such that its eigenvalues are distinct.

7-Evaluate the set of fixed modes Λ^* of the new system by replacing (A+BKC) by D' in step 2. If $\Lambda^* = \emptyset$, $\Lambda_{NS} = \Lambda_1$. Go to 10.

8-For all $\lambda_q^* \in \Lambda^*$, evaluate the structural sensitivity matrices SS_q^*.

9-For all $\lambda_r \in \Lambda_1$, compare SS_r with all the SS_q^* determinated at step 8. If there exists one SS_q' equal to SS_r, then λ_r is a structurally fixed mode of type (i) ; i.e., $\lambda_r \in \Lambda_{S1}$.

10-STOP. Λ : set of fixed modes.

Λ_{S1} : set of structurally fixed modes of type (i).

Λ_{S2} : set of structurally fixed modes of type (ii).

Λ_{NS} : set of non structurally fixed modes.

Example 3.12. Consider the following system :

$$A = \begin{bmatrix} 0 & 0 & 0 & 0 & 1 \\ 0 & 3 & 0 & 0 & 0 \\ 0 & 0 & 1 & 1 & 0 \\ 0 & 0 & 0 & 2 & 0 \\ 1 & 0 & 0 & 0 & 2 \end{bmatrix} \quad B_1 = \begin{bmatrix} 1 \\ 0 \\ 0 \\ 0 \\ 0 \end{bmatrix} \quad B_2 = \begin{bmatrix} 0 \\ 0 \\ 0 \\ 1 \\ 0 \end{bmatrix} \quad B_3 = \begin{bmatrix} 0 \\ 1 \\ 0 \\ 0 \\ 0 \end{bmatrix}$$

$$C_1 = (0 \quad 0 \quad 0 \quad 1 \quad 0)$$
$$C_2 = (1 \quad 0 \quad 0 \quad 0 \quad 0)$$
$$C_3 = (0 \quad 1 \quad 1 \quad 0 \quad 0)$$

For K = diag.(1,2,1), we have :

$$A+BKC = \begin{bmatrix} 0 & 0 & 0 & 1 & 1 \\ 0 & 4 & 1 & 0 & 0 \\ 0 & 0 & 1 & 1 & 0 \\ 2 & 0 & 0 & 2 & 0 \\ 1 & 0 & 0 & 0 & 2 \end{bmatrix}$$

The following results are obtained :

eigenvalues of (A+BKC)	SK	SS
4	$\begin{bmatrix} 0 & & \\ & 0 & \\ & & 1 \end{bmatrix}$	
1	$\begin{bmatrix} 0 & & \\ & 0 & \\ & & 0 \end{bmatrix}$	$\begin{bmatrix} 0 & 0 & 0 & 0 & 0 \\ 0 & 0 & 0 & 0 & 0 \\ 0 & 0 & 1 & 0 & 0 \\ 0 & 0 & 0 & 0 & 0 \\ 0 & 0 & 0 & 0 & 0 \end{bmatrix}$

-1	$\begin{bmatrix} -0.48 & & \\ & -0.24 & \\ & & 0 \end{bmatrix}$	
3	$\begin{bmatrix} 0.49 & & \\ & 0.24 & \\ & & 0 \end{bmatrix}$	
2	$\begin{bmatrix} 0.52 & 10^{-15} & \\ & -0.39 & 10^{-16} \\ & & 0 \end{bmatrix}$	$\begin{bmatrix} 0 & 0 & 0 & 0 & 0 \\ 0 & 0 & 0 & 0 & 0 \\ 0 & 0 & 0 & 0 & 0 \\ 0 & 0 & 0 & 1 & 0 \\ 0 & 0 & 0 & 0 & 1 \end{bmatrix}$

The fixed modes of the system are therefore : $\Lambda = \{1,2\}$. A structurally equivalent matrix to (A+BKC) is chosen :

$$D^* = \begin{bmatrix} 0 & 0 & 0 & 0.2 & 0.25 \\ 0 & 1.4 & 0.4 & 0 & 0 \\ 0 & 0 & 0.65 & 0.7 & 0 \\ 1.6 & 0. & 0 & 1.9 & 0 \\ 1.05 & 0 & 0 & 0 & 2.5 \end{bmatrix}$$

The eigenvalues of D^* are : $\{1.4, 0.65, -0.245, 2.62, 2.024\}$. Only the sensitivity matrix of 0.65 is found to be identically zero. Therefore, the set of fixed modes of the new system is $\Lambda^* = \{0.65\}$.

The structural sensitivity matrix of 0.65 is :

$$\begin{bmatrix} 0 & 0 & 0 & 0 & 0 \\ 0 & 0 & 0 & 0 & 0 \\ 0 & 0 & 1 & 0 & 0 \\ 0 & 0 & 0 & 0 & 0 \\ 0 & 0 & 0 & 0 & 0 \end{bmatrix}$$

which is the same as which of 1.
We obtain the following results :

$$\Lambda = \{1, 2\}$$
$$\Lambda_{NS} = \{2\}$$
$$\Lambda_{S1} = \{1\}$$

3.5.5. - Comments

The above paragraph was concerned with the concept of structurally fixed modes. Structurally fixed modes are related to the property of controllability under decentralized information structure for which several definitions have been provided. The differences between them have been clarified and related to the existence of two types of structurally fixed modes.

Structurally fixed modes of type (i) arise from the lack of connectivity of the system. The characterization given by Sezer and Siljak (SEZ-81a) (see § 3.5.3a) is the generalization, in a structural framework, of the results of Anderson and Clement (AND-81a) (see § 3.3.1) : there exists a partition of the system in two aggregated stations such that structurally fixed modes of type (i) are structurally uncontrollable by one of the stations and structurally unobservable by the other one. It has been pointed out that these fixed modes are not controllable under decentralized information pattern even if the time invariance of the controller is removed.

Structurally fixed modes of type (ii) arise from their special location at the origin. They can be made controllable under decentralized information pattern if a time-varying controller is allowed (similarly to non structurally fixed modes). The special situation of a fixed mode at the origin has been clarified and can be summarized as follows.

Given a system with a simple fixed mode at the origin, two cases must be considered :

<u>Case 1</u> :

The transfer function matrix can be put in a block-triangular form :

$$W(p) = \begin{bmatrix} W_{\alpha\alpha} & (p) & 0 \\ W_{\alpha\beta} & (p) & W_{\beta\beta} & (p) \end{bmatrix}$$

where the characteristic polynomial of $W_{\alpha\beta}$ (p) has a pole at the origine.

1- The zero block is due to the special structure of the system : $\lambda_0 = 0$ is a structurally fixed mode of type (i).

2- The zero block arises from a cancellation due to special values of the system parameters : $\lambda_0 = 0$ is a non structurally fixed mode.

Case 2 :

The transfer function matrix has no particular form (the graph of the system is strongly connected) :

$$W(p) = \begin{bmatrix} W_{\alpha\alpha}\ (p) & W_{\beta\alpha}(p) \\ W_{\alpha\beta}\ (p) & W_{\beta\beta}\ (p) \end{bmatrix}$$

the characteristic polynomial of $W_{\beta\alpha}$ (p) has a pole at p=0 and every entry of $W_{\alpha\beta}(p)$ has a zero at $\lambda_0 = 0$.

1- One or more entries of $W_{\alpha\beta}$ (p) have a zero at p=0 because of parameter cancellations (ap+b-c with b=c for example) : $\lambda_0 = 0$ is a non structurally fixed mode.

2- The entries of $W_{\alpha\beta}$ (p) have a zero at $\lambda_0 = 0$ before any cancellation : $\lambda_0 = 0$ is a structurally fixed mode of type (ii). This fixed mode is independent of the parameter values but arises, as non structurally fixed modes, from a pole-zero cancellation of the interconnection blocks $W_{\alpha\beta}$ (p) and $W_{\beta\alpha}$ (p).

Several characterizations of structurally fixed modes have been presented, specially in a graphical framework since graphs are very adequate for the study of structural properties. The graph-theoretic approach is more convenient to take into account arbitrary structural constraints on the control, but only in the case for which the graph is associated to the closed-loop system.

From the computational point of view, the algebraic characterization is not very convenient since it requires, for structurally fixed modes of type (i), to find a permutation matrix such that the system is put in the form (3.5.4) and, for structurally fixed modes of type (ii), to test all the complementary subsystems. However, the existence of structurally fixed modes of type (i) can efficiently be tested using the graphical condition given in Theorem 3.26 : first, find the strong components of the graph Γ_K, then check whether there is a strong component composed only by state vertices. This can be done in time $O(n^2)$ (PAP-84). For structurally fixed modes of type (ii), the characterization using the generic rank of several real matrices seems to be the most convenient since the test can be carried out in time $O(n^{5/2})$ (PAP-84).

Finally, a characterization in terms of structural sensitivity has been provided and the corresponding algorithm for the determination of fixed modes with distinction of their types has been derived. It is particularly interesting for structurally fixed modes of type (i) since it does not only detect their existence but it determines their value.

3.6. - GRAPH-THEORETIC CHARACTERIZATION OF FIXED MODES

3.6.1. - Preliminaries

The graph-theoretic characterizations refering to structurally fixed modes have already been presented in the last paragraph. Graph-theoretic methods are very efficient for the analysis of structural properties like that structural controllability, structural observability and structurally fixed modes. The advantages of a graph-theoretic approach, specially when we have to deal with large scale systems, are numerous. Indeed, it allows the application of all the concepts of graph-theory and leads to characterization procedures which can be easily implemented by using the existing set of algorithms. Morever, it is specially convenient when dealing with arbitrary control structures. It is now interesting to note that graph-theoretic methods can also be used in the case for which the properties to analyse are not structural but parameter value dependent. The analysis can be carried out by associating a weight to every edge of the graph ; i.e., a weighted directed graph is associated to the system.

Two general graph-theoretic characterizations of fixed modes exist, which allow the distinction between structurally and non structurally fixed modes. The first one is carried out in the frequency-domain ; i.e., a weighted directed graph is associated to the transfer function matrix of the system. The second one results from a study in the time-domain ; in this case, the weighted directed graph is associated with the state space representation of the system.

3.6.2. - Frequency-domain graph-theoretic characterization

Consider a linear time-invariant system with simple modes represented by its transfer function matrix as in (3.4.1) :

$$U(p) = W(p) \ Y(p)$$
$$U \in R^m \qquad Y \in R^r$$

with :
$$y_j(p) = \sum_{i=1}^{m} w_{j,i}(p) \ u_i(p) \qquad (j=1,\ldots,r)$$

$$w_{j,i}(p) = c_j(pI-A)^{-1}b_i$$

The control structure is defined by the following set of pairs :

$$(j,i) \in S \quad \text{if} \quad k_{i,j} \neq 0 \text{ in } K \qquad\qquad\qquad (3.6.1)$$

where K is the output feedback matrix.

The graph associated to (3.4.1) is $D_S = (V_S, L_S)$ and it is defined as follows :

V_S is a set of vertices such that : $V_S = V_{1S} \cup V_{2S}$
$$V_{1S} = \{i \ / \ (j,i) \in S \text{ for some } j\}$$
$$V_{2S} = \{j \ / \ (j-m,i) \in S \text{ for some } i\}$$
L_S is a set of edges such that : $L_S = L_{1S} \cup L_{2S}$
$$L_{1S} = \{ (i,j) \ / \ (i,j) \in V_{1S} \times V_{2S}, \ w_{j-m,i}(p) \neq 0\}$$
$$L_{2S} = \{ (i,j) \ / \ (i-m,j) \in S\}$$

Every vertex represents either an input or an output of the system which is involved at least in a feedback loop and every edge represents either a nonzero transfer function or a permitted feedback link.

A transmittance $t_{i,j}(p)$ is associated to every edge $(i,j) \in L_S$:

$$t_{i,j}(p) = w_{j-m,i}(p) \quad \text{for } (i,j) \in L_{1S}$$
$$t_{i,j}(p) = 1 \qquad\qquad \text{for } (i,j) \in L_{2S}$$

The transmittance $T_p(p)$ of any path P in D_S is defined as the product of the transmittances associated with the edges composing P.

Using the above definitions, Locatelli et al. (LOC-77) give the following characterization of fixed modes :

<u>Theorem 3.30.</u> A pole λ_0 of the system (3.4.1) is a fixed mode with respect to the feedback structure defined by S in (3.6.1) if and only if λ_0 is not a pole of any transmittance associated with a cycle of D_S.

This characterization is the graph-theoretic transcription of the algebraic characterisation of fixed modes in term of transmission zeros presented in Paragraph 3.2. Indeed, a cycle of D_S defines a subsystem W^i_j (p) (see § 3.2) for which the corresponding submatrix in K' is non singular. The set of rows selected in W(p) is given by the output vertices involved in the cycle and the set of columns by the set of input vertices involved in the cycle. The transmittance associated with the cycle is thus equal to the minor $W^i_j(p)$. It is clear that if the minor $W^i_j(p)$ has no pole at λ_0, then $\phi(p).W^i_j(p)$ has a zero at λ_0 and λ_0 is a transmission zero of the considered subsystem.

Nevertheless, we have to point out a fundamental difference which makes this graph-theoretic characterization less powerfull than the algebraic one. There is no cycle associated with a subsystem for which $W^i_j(p)$ has a triangular form even if it corresponds to a non singular submatrix of K'. This is why this characterization is only valid for the case of systems with simple poles. Indeed, consider the subsystem :

$$W^i_j\ (p) = \begin{bmatrix} W_{i,i}(p) & 0 \\ W_{i,j}(p) & W_{j,j}(p) \end{bmatrix}$$

and assume that it corresponds to a non singular submatrix in K'. Then, it is clear that $W_{i,i}(p)$ and $W_{j,j}(p)$ correspond also to non singular submatrices in K'. For a system with simple poles, two situations can arise :

* λ_0 is neither a pole of $W_{i,i}(p)$ nor of $W_{j,j}(p)$: consequently, λ_0 is not a pole of $W^i_j(p)$.
* λ_0 is a pole of $W_{i,i}(p)$ or of $W_{j,j}(p)$ but not of both : consequently, λ_0 is a pole of $W^i_j(p)$.

Therefore, all the information about $W^i_j(p)$ is contained in $W_{i,i}(p)$ and $W_{j,j}(p)$ and the subsystem $W^i_j(p)$ does not need to be taken into account.

Consider now that λ_0 is not a simple pole but a pole of order q and assume that it is a pole of order q' for $W_{i,i}(p)$ and of order q-q', for $W_{j,j}(p)$. Then :

$$\phi(\lambda_0).\ W_{i,i}(\phi_0) = 0 \quad \text{and} \quad \phi(\lambda_0).\ W_{j,j}(\lambda_0) = 0$$

i.e., λ_0 is a transmission zero of the subsystems $W_{i,i}(p)$ and $W_{j,j}(p)$. But :

$$\phi(\lambda_0) \; W_j^i(p) \neq 0$$

and λ_0 is not a transmission zero of $W_j^i(p)$. Consequently, λ_0 is not a fixed mode. In this case, we need to consider the system $W_j^i(p)$ to derive the conclusion.

Nevertheless, this characterization remains interesting. In particular, it presents the advantage to treat the cases of an arbitrary control structure very easily. Moreover, it allows some distinction between the different types of fixed modes. λ_0 is a structurally fixed mode of type (i) if there is no edge whose transmittance has λ_0 as a pole involved in the composition of a cycle. λ_0 is a non structurally fixed mode or a structurally fixed mode of type (ii) if at least one edge whose transmittance has λ_0 as a pole composes one cycle and if λ_0 is not a pole of the transmittance of this cycle due to a pole-zero cancellation. Note that the distinction between non structurally fixed modes and structurally fixed modes of type (ii) is not possible since the cancellations within the transfer functions themselves do not appear, but the question only arises for the case of fixed modes at the origin.

Example 3.13. Consider the following system with simple poles :

$$W(p) = \begin{bmatrix} \dfrac{3}{p-2} & 0 & \dfrac{p+1}{p(p-2)} \\[3mm] \dfrac{1}{p-2} & \dfrac{1}{p+2} & \dfrac{1}{p(p-2)} \\[3mm] \dfrac{1}{p+1} & \dfrac{1}{p+2} & 0 \end{bmatrix}$$

The poles of the system are $\{0,-1,-2,2\}$. The control structure is given by :

$$K = \begin{bmatrix} k_{11} & & \\ & k_{22} & \\ & & k_{33} \end{bmatrix} \qquad \text{and } S = \{(1,1) \; ; \; (2,2) \; ; \; (3,3) \}$$

The graph $D_S = (V_S, L_S)$ associated with this system is shown in Figure 3.7 :

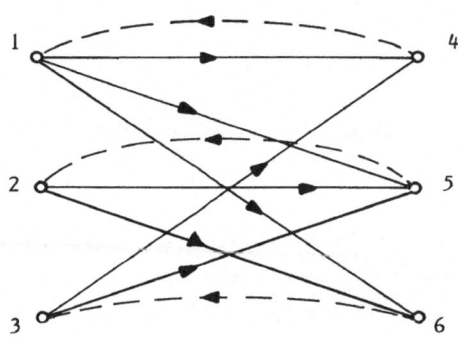

Figure 3.7

$V_{1S} = \{1,2,3\}$
$V_{2S} = \{4,5,6\}$

$L_{1S} = \{(1,4);(1,5);(1,6);(2,5);(2,6);(3,4);(3,5)\}$
$L_{2S} = \{(4,1);(5,2);(6,3)\}$

There are 5 cycles in D_S with the following associated transmittances :

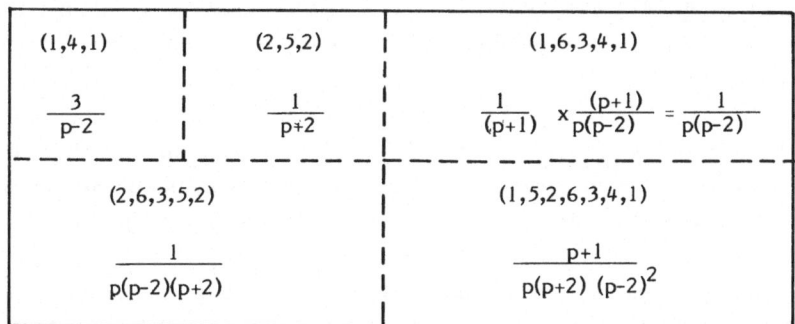

Among the poles of the system, $\lambda_0 = -1$ does not appear as a pole of any trans-mittance. This situation arises from the pole-zero cancellation for the cycle $(1,6,3,4,1)$. Therefore, $\lambda_0 = -1$ is a non structurally fixed mode (it cannot be a structurally fixed mode of type (ii) since it is not located at the origin).

3.6.3. – Time-domain graph-theoretic characterization

The system is now represented by its state-space model (3.2.1) :

$$\dot{x}(t) = A\ x(t) + B\ u(t)$$
$$y(t) = C\ x(t)$$

and we consider an arbitrarily constrained control specified by the output feedback matrix K.

This characterization is issued from the work of Mason (MAS-56) relating the structure of the graph associated with a square matrix to its determinant. We consider an nxn matrix M and we associate a weighted digraph consisting of n vertices and a set of edges. There is an edge from the vertex i to the vertex j if $m_{ji} \neq 0$ and the weight of this edge is m_{ji}. Then, we have the following :

$$\det M = \sum_{\substack{\text{even} \\ \text{permutations}}} m_{1q_1}\ m_{2q_2}\ \cdots\ m_{nq_n} - \sum_{\substack{\text{odd} \\ \text{permutations}}} m_{1q_1}\ m_{2q_2}\ \cdots\ m_{nq_n} \qquad (3.6.2)$$

where $(q_1,\ldots,q_n)$ is some permutation of $(1,\ldots,n)$.

Definition 3.10 (REI-83) (REI-84a,b). A cycle family is a set of disjoint cycles. Its weight is given by the product of the weights of all edges composing the cycle family.

Each term $(m_{1q}\ m_{2q}\ \cdots\ m_{nq})$ in (3.6.2) corresponds thus to a cycle family which touches all n vertices in the graph. If this cycle family consists of f disjoint cycles, then the sign factor of its corresponding summand in (3.6.2) is $(-1)^{n-f}$.

The approach of Reinschke (REI-84a) consists in using (3.6.2) in order to obtain a graph-theoretic characterization of the coefficients of the characteristic polynomial of the closed-loop system. Note that this approach was already used by Evans and Kruser (EVA-84) to determine the conditions of potential pole assignment using decentralized static control (see § 3.5.2c) and remember that they needed to go through the calculation of the closed-loop dynamic matrix (A+BKC). This disadvantage is avoided by Reinschke by using the following (n+m+r)x(n+m+r) matrix :

$$M = \begin{bmatrix} 0 & C & 0 \\ 0 & A & B \\ K & 0 & 0 \end{bmatrix}$$

The graph associated with M is $D_K = (U \cup X \cup Y,\ E \cup E_K)$ where $X = \{x_1,\ldots,x_n\}$ is the set of state vertices, $Y = \{y_1,\ldots,y_r\}$ is the set of output vertices and $U = \{u_1,\ldots,u_m\}$ is the set of input vertices. E is a set of edges such that $(u_i, x_j) \in E$ if $b_{ji} \neq 0$ and b_{ji} is the weight of this edge, $(x_i, x_j) \in E$ if $a_{ji} \neq 0$ and a_{ji} is the weight of this edge, $(x_i, y_j) \in E$ if $c_{ji} \neq 0$ and c_{ji} is the weight of this edge. E_K is a set of feedback edges such that $(y_i, u_j) \in E_K$ if $k_{ji} \neq 0$ and k_{ji} is the weight of this edge.

Note that this graph is the same as the one already used in Paragraph 3.5.3.b-1.

Definition 3.11 (REI-84a). The width of a cycle family in D_K is defined as the number of state vertices involved in it.

Using (3.6.2) and Definitions 3.10 and 3.11, Reinschke states the following result :

Theorem 3.31 (REI-84a). The coefficients α_i, (i=1,...,n), of the closed-loop characteristic polynomial :

$$\det (pI-A-BKC)=p^n + \alpha_1 p^{n-1} + \ldots + \alpha_{n-2} p^2 + \alpha_{n-1} p + \alpha_n$$

are given by the sum of the weights of all the cycle families of width i in D_K multiplied by a sign factor $(-1)^d$ where d is the number of disjoint cycles in the cycle family under consideration.

The characterization of fixed modes is based on the following arguments. In Theorem 3.31, the characteristic polynomial is given as a series expansion around p=0 but a series representation of the same polynomial is possible around any value λ_0 ; i.e. :

$$\det (pI-A-BKC)=\det ((p-\lambda_0)I-(A-\lambda_0 I)-BKC)$$
$$=(p-\lambda_0)^n + (p-\lambda_0)^{n-1} \alpha_1(\lambda_0) + \ldots + (p-\lambda_0) \alpha_{n-1}(\lambda_0) + \alpha_n(\lambda_0) \qquad (3.6.3)$$

and the coefficients $\alpha_i(\lambda_0)$ can be obtained using Theorem 3.31 where A is replaced by $(A-\lambda_0 I)$ when defining the graph D_K, which will be denoted by $D_K(\lambda_0)$.

Every coefficient $\alpha_i(\lambda_0)$ can be represented as :

$$\alpha_i(\lambda_0) = \alpha_i^o(\lambda_0) + g_i(\lambda_0, K) \qquad (3.6.4)$$

where $\alpha_i^o(\lambda_0)$, (i=1,...,n), are the coefficients of the open-loop characteristic polynomial in the form of a series expansion around λ_0.

$p = \lambda_0$ is a fixed mode if and only if it is a zero of the open-loop and also of the closed-loop characteristic polynomials, i.e. :

1- $\qquad \alpha_n^o(\lambda_0) = 0$

2- $\qquad \alpha_n(\lambda_0) = 0$

but from (3.6.4), (2) implies (1) and (2) is therefore sufficient. If moreover, $\alpha_i(\lambda_0)=0$, $(i=n,\ldots,n-h+1)$, but $\alpha_{n-h}(\lambda_0) \neq 0$, the multiplicity of the fixed mode at $p=\lambda_0$ is of course h.

The graph-theoretic interpretation of the above discussion leads to the following result :

Theorem 3.32 (REI-84a). The system (3.2.1) has a fixed mode of multiplicity h at $p=\lambda_0$ with respect to the control specified by K if and only if, for $j=n,\ldots,n-h+1$ (but not for $j=n-h$), one of the following two conditions holds :

1 - there are no cycle families of width j in $D_K(\lambda_0)$.

2 - there exists two or more cycle families of width j in $D_K(\lambda_0)$ which cancel each other numerically for all admissible values of the nonzero entries of K.

It is clear that condition (2) corresponds to non structurally fixed modes. First, note that there is a least one different edge belonging to E in two different cycle families. The cancellation of condition (2) arises from the particular values of the weights associated to this edges. It is clear that if the weight of one of the edges involved in one cycle family and not in the other is changed, the cancellation will no longer occur. Therefore, this fixed mode depends on the values of the parameters of the system and is non structural. In the case for which condition (1) is satisfied, the conclusion requires a deeper analysis. Consider the graph $D_K(p)$ associated with a system of dimension n for a non specified value of p and assume that it contains N cycle families of width n: N is at least equal to 1 since there is a self-loop at every state vertex with weight $(a_{ii}-p)$. For condition (1) to be satisfied, the affectation of a numerical value λ_0 to p must eliminate some self-loops. This will occur when λ_0 is set to some a_{ii}. We obtain thus a first interesting result :

Corollary 3.10. A necessary condition for λ_0 to be a fixed mode arising from condition (1) in Theorem 3.32 is that λ_0 is equal to some a_{ii}, $i \in \{1,\ldots,n\}$.

Since condition (2) of Theorem 3.32 only characterizes non structurally fixed modes (but not all of them), the following result directly follows :

Corollary 3.11. A necessary condition for a pole λ_0 to be a structurally fixed mode is that its value appears in the diagonal elements of the dynamic matrix A.

Now, condition (1) of Theorem 3.32 can be satisfied if there is at least one self-loop with weight $a_{ii}-p$ and $a_{ii}=\lambda_0$ in every cycle family of width n of $D_K(p)$. Two cases can occur :

Case 1 : There is at least 1 self-loop with weight λ_0-p in every cycle family of width n of $D_K(p)$ which corresponds to the same state vertex. Then, when we set p to λ_0, all the cycle families of width n become of width (n-1), independently of the parameter values of the system (weights of the other edges).

Case 2 : There is at least 1 self-loop with weight λ_0-p in every cycle family of width n of $D_K(p)$ and none of them corresponds to the same state vertex. Then, if any parameter in the diagonal of A corresponding to these state vertices is changed, some cycle family of width n will remain. In this case, condition (1) of Theorem 3.32 depends of the parameter values.

This discussion can be summarized in the following corollary :

Corollary 3.12. λ_0 is a structurally fixed mode if and only if there is at least 1 self-loop with weight λ_0-p in every cycle family of width n of $D_K(p)$ which corresponds to the same state vertex.

The particular case of fixed modes at the origin does not require a different analysis : λ_0 is set to 0 and the graph to be considered is thus D_K. The differentiation between structurally fixed modes of type (i) and (ii) can be carried out by using condition (i) of Theorem 3.26 : λ_0 is a structurally fixed mode of type (i) if it is not involved in any cycle family of any width including an edge of E_K.

This characterization is of a great interest since if is the only complete graph-theoretic characterization of fixed modes (the one presented in the preceding paragraph is only valid for systems with simple poles).

From the computational point of view, the well-known algorithms (KRO-67) (LIA-69) (RAO-69) for finding paths and cycles can be adapted for the computer aided determination of cycle families of prescribed width.

Example 3.14. Consider the following system :

$$A = \begin{bmatrix} 0 & 0 & 0 & 0 & 1 \\ 0 & 3 & 0 & 0 & 0 \\ 0 & 0 & 4 & 1 & 0 \\ 0 & 0 & 0 & 2 & 0 \\ 1 & 0 & 0 & 0 & 1 \end{bmatrix} \qquad B_1 = \begin{bmatrix} 1 \\ 0 \\ 0 \\ 0 \\ 0 \end{bmatrix} \quad B_2 = \begin{bmatrix} 0 \\ 0 \\ 0 \\ 1 \\ 0 \end{bmatrix} \quad B_3 = \begin{bmatrix} 0 \\ 1 \\ 0 \\ 0 \\ 0 \end{bmatrix}$$

$$C_1 = (0 \quad 0 \quad 0 \quad 1 \quad 0)$$
$$C_2 = (1 \quad 0 \quad 0 \quad 0 \quad 0)$$
$$C_3 = (0 \quad 1 \quad 1 \quad 0 \quad 0)$$

and a decentralized control specified by : $K = \text{diag.}(k_1, k_2, k_3)$. The open-loop characteristic polynomial is :

$$\det (pI-A) = (p-2)(p-4)(p-3)(p^2-p-1)$$

The graph $D_K(p)$ is shown in Figure 3.8. There are 6 cycle families of width 5 in $D_K(p)$ which are given in Figure 3.9.

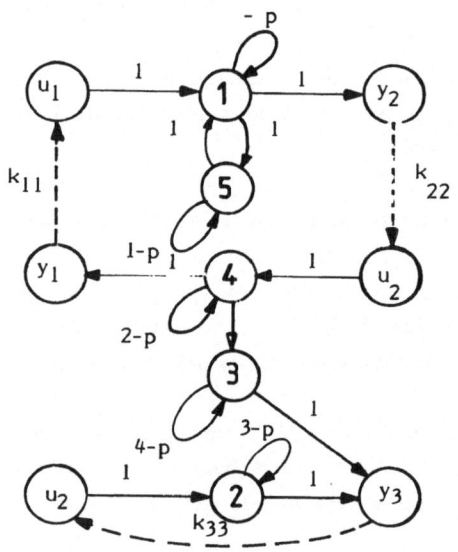

Fig. 3.8

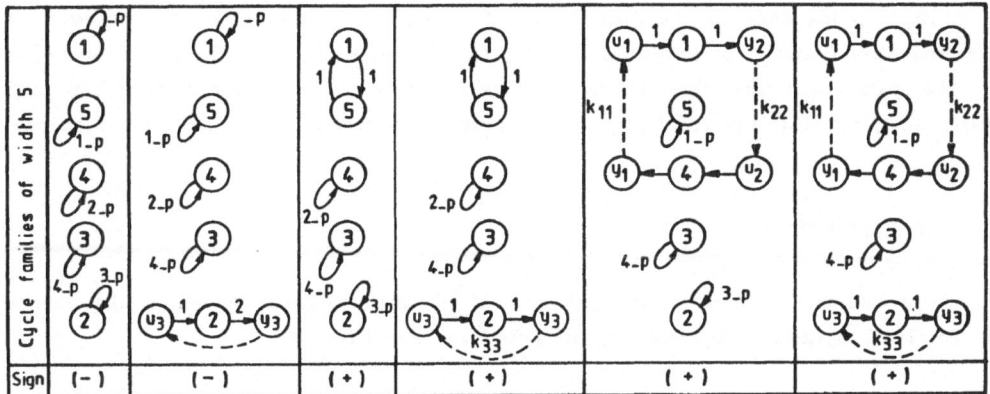

Fig. 3.9

Case 1 : $\lambda_0 = 4$

In this case, there is no more cycle families of width 5 ; moreover, the self-loop with weight 4-p appears in every cycle family of width 5 in $D_K(p)$ and it corresponds to the same state vertex x_3. Therefore $\lambda_0 = 4$ is a structurally fixed mode (of type (i), of course).

Case 2 : $\lambda_0 = 2$

In this case, the cycle families 1, 2, 3 and 4 are of width 4 and have not to be considered anymore.

The coefficient $\alpha_5(\lambda_0 = 2)$ is calculated from the cycle families 5 and 6 :

$$\alpha_5(\lambda_0 = 2) = -2k_1 k_2(1 + k_3) \neq 0$$

Therefore $\lambda_0 = 2$ is not a fixed mode.

3.6.4. - Comments

The two graphical characterizations given in this paragraph present the advantage of being complete in the sense that both structurally and non structurally fixed modes are characterized. This result is performed by using weighted directed graphs associated to the system, whose weights reflect the specific values of the system parameters. The first one, which is stated in a frequency-domain framework is, unfortunately, only valid for systems with simple poles. Contrarily, the second one, stated in a time-domain framework, can be applied to the general case of systems with multiple poles and the multiplicity of the fixed modes is easily determined. Both characterizations allow to conclude on the type of the fixed modes in an easy way. From the computational point of view, they require to determine the cycles in the digraph and, for the case of large scale systems, efficient algorithms are necessary.

3.7. - CONCLUSION

This chapter presents an overview of all the different existing characterizations of fixed modes. They can be classified in four groups :

- <u>Characterizations in terms of transmission zeros</u> of some subsystems of the system : the same result can be obtained in a frequency or in a time-domain framework since the concept of transmission zeros is as well defined in both domains.

- <u>Characterizations in the time-domain</u> : the most important one is stated in the form of a matrix rank test which points out the importance of the concept of complementary subsystems resulting from the partitioning of the system in two aggregated stations. This characterization reveals that fixed modes are simultaneously uncontrollable by one aggregated station and unobservable by the other one.

- <u>Characterizations in the frequency-domain</u> : these characterizations give a deep insight into the reasons of occurence of fixed modes. We find again the partitioning of the system in two aggregated stations and the transfer function matrix representation is specially adequate to point out the importance of the interconnection terms between the aggregated stations. This approach shows clearly that fixed modes may arise from different origins related to the interconnection terms pattern. They can be classified in two different types :

- structurally fixed modes whose existence is independent of the parameter values of the system and arise from the structure of the system.

- non structurally fixed modes whose existence depends on the parameter values and arise from some perfect parametric cancellations. These fixed modes can be removed by perturbing the parameters of the system.

A differenciation between structurally fixed modes themselves is pointed out.

It is clear that structurally fixed modes seem to have a more physical reality than non structurally fixed modes. Indeed, the perfect parametric cancellations leading to non structurally fixed modes have a very low probability in practice. Moreover, they can be considered as effective only in the case for which the mathematical parameters involved in the cancellations correspond to the same physical parameters. Otherwise, it is clear that the lack of accuracy in the modeling process makes improbable these cancellations. Nevertheless, it remains important to detect these cancellations in a mathematical framework. They reveal critical situations in which close values of the parameters may have an adverse effect on the degree of controllability under constrained information structure of a mode.

The introduction of the concept of controllability under decentralized information structure makes clearer the analogy with the case of centralized control. Fixed modes are the extension of the concepts of uncontrollable and unobservable modes and structurally fixed modes establish the relation with structurally uncontrollable and unobservable modes. A graph-theoretic approach is specially adequate for the study of all structural properties, so is for structurally fixed modes for which some graph-theoretic characterizations are presented.

- Graph-theoretic characterizations : graphs can be used for the study of structural properties like structurally fixed modes but they can also deal with quantitative data if they are weighted. Using this tool, two graph-theoretic characterizations of fixed modes (structural or not structural) are provided.

In the whole chapter, a particular effort has been made to point out the equivalences between the different characterizations. It is shown that some of them are perfectly similar, differing only by the notations and by the definitions which are used to carry out the final result. This is the case for the characterizations in term of transmission zeros presented in Paragraph 3.2. In Paragraph 3.3, it appears that all the time-domain characterizations are derived from the so-called matrix rank test characterization. Finally, those presented in Paragraph 3.4 in the frequency domain are all "included" in the so-called general transfer function characterization.

It has to be noted that graph-theoretic characterizations present the advantage to deal with arbitrary feedback structures as easily as with decentralized structures since the feedback pattern is represented in the graph.

Chapter II pointed out the problems arising from the existence of fixed modes in reference to the problem of stabilization and pole assignment. The two next chapters are concerned now with the different methods which can be used to avoid or to eliminate fixed modes preserving the objective of a "reduced" information flow. These methods are based on the characterizations presented in this chapter.

CHAPTER 4

DECENTRALIZED STABILIZATION
IN PRESENCE OF NON STRUCTURALLY FIXED MODES

4.1. - INTRODUCTION

In the previous chapter, we have shown that fixed modes can be classified in structurally controllable and uncontrollable modes under constrained information structure. The controllable modes which are a consequence of a perfect matching of system parameters are what we called non structurally fixed modes and structurally fixed modes of type (ii). The uncontrollable modes which arise from a lack of information transfer among the stations are what we called structurally fixed modes of type (i).

In this chapter, only systems with controllable fixed modes under constrained information structures are considered. It is shown that they can be eliminated by changing the control nature even when the structural constraints are maintained. This is a fundamental difference between centralized and decentralized fixed modes. Indeed, if a system has centralized fixed modes (uncontrollable or unobservable modes), then no matter what control law is used (linear or non linear, dynamic or non dynamic, distributed or finite-dimentional), the fixed modes remain in the sense that if λ_0 is such a mode, the closed-loop response for a suitable initial condition contains terms proportional to $\exp(\lambda_0 t)$. Wang and Davison (WAN-73) showed that decentralized fixed modes remain when an arbitrary linear, invariant and finite-dimensional control law is used ; and in (AND-81a) the finite-dimensionality constraint is removed without changing the result. Nevertheless, Kobayashi et al. (KOB-78) and Anderson and Moore (AND-81b) pointed out that by using a general control law of the form :

$$u_i(t) = K_i (I_i(t)) \tag{4.1.1}$$
$$\text{with } I_i(t) = \{ y_i (\tau), u_i(\nu) ; \quad \tau \in [0, t] \quad \nu \in [0, t] \}$$

the controllable fixed modes under constrained information structure may be eliminated in the sense that such controller can bring any initial state to the zero state
in a finite time.

Consequently, the decentralized stabilization in presence of unstable decentralized fixed modes can be achieved by a decentralized control law, provided that
one of the properties of linearity or time-invariance have been sacrified.

An easy interpretation of this result is given by the frequency-domain characterization of Anderson (AND-82). For simplicity, let us consider a 2-input 2-output
system with simple modes. If the system has a decentralized fixed mode at λ_0, then
its transfer matrix (or its transpose) has the following form (AND-82) :

$$W(p) = \begin{bmatrix} W_{11}(p) & \vdots & W_{12}(p) \\ \hdashline W_{21}(p) & \vdots & W_{22}(p) \end{bmatrix}$$

$$W(p) = \begin{bmatrix} \text{entry with no pole at } \lambda_0 & \vdots & \text{entry with pole at } \lambda_0 \\ \hdashline \text{entry with zero at } \lambda_0 & \vdots & \text{entry with no pole at } \lambda_0 \end{bmatrix}$$

Now, if the feedback control $u_2 = k_2 y_2$ is applied at station 2, we have :

$$\frac{y_1(p)}{u_1(p)} = W_{11}(p) + W_{12}(p) \cdot \frac{k_2}{1-W_{22}(p)k_2} \cdot W_{21}(p) \qquad (4.1.2)$$

If $W_{12}(p)$ has a pole at λ_0 (and $W_{21}(p)$ has a zero at λ_0, a pole-zero cancellation
occurs resulting in a fixed mode. But if k_2 is a time-varying or non linear element,
the cancellation will no longer occur because one cannot commute the time-varying
block with an adjacent time-invariant block, and thereby juxtapose a cancelling
pole-zero pair.

Finally, we note that if $W_{12}(p)$ or $W_{21}(p)$ is identically zero, the fixed mode
originates from the structure of the system and it is uncontrollable under decentralized constraints. The nature of the decentralized feedback is thus without effect on
the existence of such fixed mode. In this case, the only possible approach is to
relax the structural constraints on the control.

The system considered here has S local stations and is described by :

$$\begin{cases} \dot{x}(t) = A\ x(t) + \sum_{i=1}^{S} B_i\ u_i \\ \\ y_i(t) = C_i x(t) \qquad (i=1,\dots,S) \end{cases}$$

(4.1.3)

where $x \in R^n$, $u_i \in R^{m_i}$ and $y_i \in R^{r_i}$ are the state vector and the local input and output vectors of the ith station. The global description is given by :

$$\begin{cases} \dot{x} = A\ x + B\ U \\ \\ y = C\ x \end{cases} \quad \text{with} \quad \begin{aligned} B &= \begin{bmatrix} B_1 & \dots & B_S \end{bmatrix} \\ \\ C &= \begin{bmatrix} C_1' & \dots & C_S' \end{bmatrix}' \end{aligned}$$

(4.1.4)

4.2. - SAMPLE AND HOLD

Let us consider the system (4.1.3) and put a general sampler-hold, with exponential decay α and sampling period T, in serie with each input of the system. Then, the resulting discrete-time system is described by :

$$x\ ((k+1)T) = F\ x(kT) + \sum_{i=1}^{S} G_i\ u_i\ (kT)$$

$$y_i(kT) = C_i\ x(kT) \qquad (i = 1, \dots, S)$$

(4.2.1)

where

$$F = \exp(AT)$$
$$G_i = \exp(-\alpha T) \int_0^T [\exp(A+\alpha I)]\ dt\ B_i$$

From (WAN-82), the set of fixed modes of the discrete-time system (4.2.1) is defined as follows :

$$\Lambda_d = \bigcap_{K_i \in R^{r_i \times m_i}} \sigma\ (F + \sum_{i=1}^{S} G_i K_i C_i)$$

(4.2.2)

A sufficient condition for the existence of a stabilizing set of local discrete time dynamic controllers is given by the following theorem :

Theorem 4.1 (WAN-82). If there exists a sampling period T and a real number α such that the fixed modes of (4.2.1) are contained in the unit disc, then the system (4.2.1), and hence (4.1.3), can be stabilized by a set of local discrete-time dynamic controllers.

Wang (WAN-82) illustrated this approach with a simple example. However, no constructive procedure to determine α and T was proposed.

Consider the system :

$$\dot{x} = \begin{bmatrix} 0 & 1 & 0 \\ -2 & -3 & 0 \\ 0 & 0 & 1 \end{bmatrix} x + \begin{bmatrix} 0 \\ 1 \\ 0 \end{bmatrix} u_1 + \begin{bmatrix} 0 \\ 0 \\ 1 \end{bmatrix} u_2$$

$$y = \begin{bmatrix} 0 & 0 & 1 \\ -1 & 1 & 0 \end{bmatrix} x$$

which can also be represented as follows :

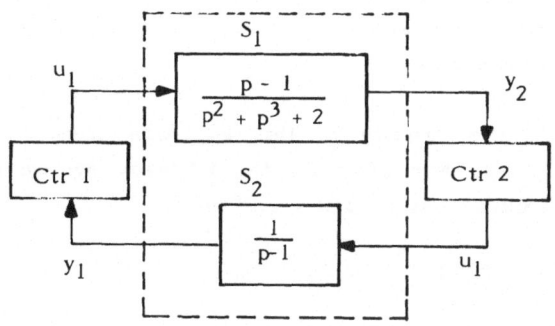

This system has a fixed mode at 1, which can be viewed as a result of pole-zero cancellation when local feedback is applied. Put a sample and hold in serie with each input and apply a unity feedback at station 2, then the overall system has the following configuration :

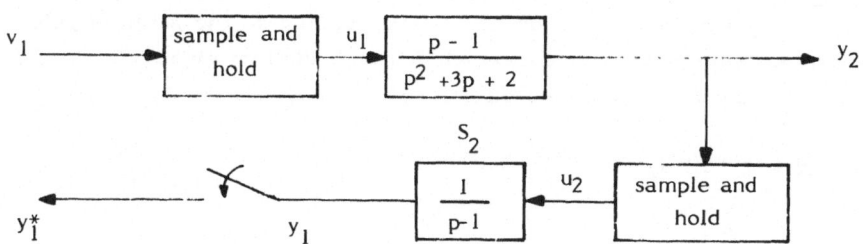

The pulse transfer functions of S_1 and S_2 in serie with a sample and hold of sampling period T are given by :

$$h_1(z) = \frac{z(-0.5 + 2 e^{-T} - 1.5 e^{-2T}) + (-1.5 e^{-T} + 2 e^{-2T} - 0.5 e^{-3T})}{(z - e^{-T})(z - e^{-2T})}$$

$$h_2(z) = \frac{e^{T} - 1}{z - e^{T}}$$

and the pulse transfer function between v_1 and y_1^* is equal to $h_1(z) \times h_2(z)$, which has no pole-zero cancellation. Hence, a discrete-time dynamic compensator at station 1 can easily be designed so that the overal system is stable. Although a zero-order hold ($\alpha = 0$) is sufficient in the above example, a sample and hold with decay ($\alpha \neq 0$) may be necessary in some cases.

On another hand, the validity of this approach seems to be restricted to systems with distinct poles. For example, the following system provided by Wang (WAN-82) cannot be stabilized using sample and hold :

$$\dot{x} = \begin{bmatrix} 1 & 0 & 0 \\ 0 & 1 & 0 \\ 0 & 0 & -6 \end{bmatrix} x + \begin{bmatrix} 0 \\ 0 \\ 1 \end{bmatrix} u_1 + \begin{bmatrix} 1 & 0 \\ 0 & 1 \\ 0 & 0 \end{bmatrix} u_2$$

$$y_1 = \begin{bmatrix} 1 & 0 & 0 \\ 0 & 1 & 0 \end{bmatrix} x$$

$$y_2 = \begin{bmatrix} 0 & 0 & -12 \end{bmatrix} x$$

This system has a simple fixed mode at 1 corresponding to the multiple pole.
For the class of systems (4.1.3) for which A has n distinct eigenvalues λ_i, i = 1, 2, ..., n and that can be transformed as :

$$\begin{cases} \dot{\bar{x}} = \lambda \bar{x} + \sum_{i=1}^{S} \bar{B}i \, \mu_i \\ y_i = \bar{C}_i \bar{x} \quad (i = 1, \ldots, S) \end{cases} \tag{4.2.3}$$

where $\lambda = \text{diag}\{\lambda_1, \lambda_2, \ldots, \lambda_n\}$

an interesting result was provided latter by Ozguner and Davison (OZG-85). Consider that the system (4.2.3) is controlled by a digital controller with a constant sampling period $T > 0$ and a zero-order hold. Then, the resultant sampled system can be described by :

$$\dot{\bar{x}} (t+T) = e^{\lambda T} \bar{x}(t) + \Gamma \sum_{i=1}^{S} \bar{B}i \, \mu_i$$

$$(4.2.4)$$

$$y_i (t) = \bar{C}_i \, \bar{x}(t) \qquad i = 1, \ldots, S$$

where $\Gamma = \text{diag.} [\Gamma_1, \Gamma_2, \ldots, \Gamma_n]$

and $\Gamma_i = T$ if $\lambda_i = 0$

$$\Gamma_i = \frac{e^{T \lambda i} - 1}{\lambda_i} \quad \text{if } \lambda i \neq 0$$

Then, we have the following result :

Theorem 4.2 (OZG-85). Assume that the system (4.1.3) (4.2.3) has p fixed modes among which the fixed modes λ_j ($j = 1, \ldots, p_s$) are structurally fixed modes of type (i). Then, the sampled system (4.2.4) has only p_s fixed modes $e^{\lambda j T}$, $j = 1, \ldots, p_s$, for almost all $T > 0$.

The interpretation of the results presented in this section becomes clearer when fixed modes (except structurally fixed modes of type (i)) are viewed as pole-zero cancellations of a specific kind in the decentralized system. The effects that sampling has on poles and zeros make that the cancellations do not occur in the model of the sampled system.

4.3. - USE OF TIME-VARYING CONTROLLERS

In this section, we present the results of Anderson and Moore (AND-81b), Purviance and Tylee (PUR-82) that use a decentralized time-varying feedback to stabilize linear invariant systems with decentralized fixed modes. First we present the particular case of systems with two control stations, then we discuss briefly the extension of Anderson and Moore to systems with S stations.

Consider a controllable and observable two-station system described by :

$$\begin{cases} \dot{x} (t) = A \, x(t) + B_1 \, u_1(t) + B_2 \, u_2(t) \\ \\ y_i(t) = C_i x_i(t) \qquad (i = 1, 2) \end{cases}$$

$$(4.3.1)$$

where $x \in R^n$, $u_i \in R^{m_i}$ and $y_i \in R^{r_i}$ are the state vector and the local input and output of station i respectively. A, B_i and C_i are constant matrices of appropriate dimensions. Suppose also that we apply a periodic time-varying control law, with period T, at the second station :

$$u_2(t) = K_2(t) \; y_2(t) \tag{4.3.2}$$

then the resulting time-varying closed-loop system is :

$$\dot{x}(t) = \left[A + B_2 K_2(t) \; C_2\right] + B_1 u_1(t)$$
$$\tag{4.3.3}$$
$$y_1(t) = C_1 \; x(t)$$

For this system, uniform controllability and observability (KAI-80) means that we can design an observer and a linear state feedback which will stabilize the system. If we denote by $\Phi_{K_2}(t, \tau)$ the transition matrix of system (4.3.3), then the observability grammian matrix is :

$$OG(\tau, \tau + T) = \int_{\tau}^{\tau + T} \Phi_{K_2}^{T}(t, \tau) \; C_1' \; C_1 \; \Phi_{K_2}(t, \tau) \; dt \tag{4.3.4}$$

and the controllability grammian matrix is :

$$CG(\tau, \tau+T) = \int_{\tau}^{\tau+T} \Phi_{K_2}(\tau, t) \; B_1 B_1' \; \Phi_{K_2}^{T}(\tau, t) \; dt \tag{4.3.5}$$

The condition of uniform controllability and observability is satisfied if the matrices CG $(\tau, \tau+T)$ and OG $(\tau, \tau+T)$ are strictly positive-definite.

4.3.1. - Piecewise constant feedback laws

Anderson and Moore (AND-81b) proposed to use a periodic piecewise constant feedback at the second station. Given the two following assumptions :

- Centralized controllability and observability, i.e. $\left[(B_1 B_2), A, (C_1' \; C_2')'\right]$ is controllable and observable.
- Connectivity assumptions, i.e. the transfer matrices between stations are not identicaly zero :

$$W_{12}(p) = C_1(pI-A)^{-1} B_2 \neq 0 \tag{4.3.6}$$
$$W_{21}(p) = C_2(pI-A)^{-1} B_1 \neq 0 \tag{4.3.7}$$

their results are expressed in the following theorem :

Theorem 4.2 (AND-81b). Consider the controllable and observable (in a centralized sense) system given by (4.3.1). Applying a periodic feedback $u_2(t) = K_2(t) \, y_2(t)$ at the second station, the system (4.3.2) is uniformly controllable and observable if the connectivity assumptions (4.3.6) and (4.3.7) hold and if $K_2(t)$ is piecewise constant taking at least $1+\max(m_2,r_2)$ distinct values over one period.

Remark 4.1 : The assumptions required by the above theorem are equivalent to the assumption of controllability and observability under decentralized information structure.

This result can be analysed as follows : if the system has a fixed mode due to a lack of observability of station 1, then by the assumption of centralized observability, station 2 observes this mode and transmits related information to station 1 through the transmission channel W_{12}. A dual analysis can be made if the fixed mode is caused by a lack of controllability of station 1.

The above theorem can be similarly derived for discrete systems (JAM-83). Moreover, the case of systems with S control stations was considered by Anderson and Moore (AND-81b) ; they showed that if the connectivity assumptions of complementary subsystems hold (system structurally controllable and observable under decentralized constraints), then the system can be made uniformly controllable and observable from station 1 by applying successively a periodic piecewise constant feedback control law $u_i(t) = K_i(t) \, y_i(t)$, $i = 2, \ldots, S$ at the other stations. For each station i, $K_i(t)$ must take at least $\prod_{j=2}^{i} 1+\max(m_j, r_j)$ distinct values over one period. It is clear that the number of different values increases dangeroulsy with the number of stations making difficult the practical implementation of this approach.

Example 4.1 (AND-81b). Consider a controllable and observable system with two stations given by :

$$\dot{x} = \begin{bmatrix} 1 & 0 & 0 \\ 0 & 1 & 0 \\ 0 & 0 & 2 \end{bmatrix} x + \begin{bmatrix} 0 & 1 \\ 1 & 0 \\ 0 & 0 \end{bmatrix} u_1 + \begin{bmatrix} 0 \\ 0 \\ 1 \end{bmatrix} u_2$$

$$y_1 = \begin{bmatrix} 0 & 0 & 1 \end{bmatrix}$$

$$y_2 = \begin{bmatrix} 0 & 1 & 0 \\ 1 & 0 & 0 \end{bmatrix}$$

The system has a decentralized fixed mode at $\lambda_0 = 1$ since

$$\text{rank} \begin{bmatrix} \lambda I - A & B_2 \\ \hline C_1 & 0 \end{bmatrix} = 2 < 3 \qquad \text{for } \lambda = \lambda_0 = 1$$

Applying the following time-varying control at the second station :

$$u_2(t) = K_2(t) \ y_2(t)$$

with

$$K_2(t) = \begin{cases} \begin{bmatrix} 0 & 1 \end{bmatrix} & \text{for } 2k < t < 2k+1 \\ \begin{bmatrix} 1 & 0 \end{bmatrix} & \text{for } 2k+1 < t < 2k+2 \\ & k = 0, 1, 2, \ldots \end{cases}$$

the observability and controllability grammian matrices (calculated analyticaly) over the range (2k, 2k+1) for all k= 0, 1, 2, ... are positive definite with a condition number[1] approximately equal to 100 and 6, respectively. So, the use of a periodic (piecewise constant) gain $K_2(t)$ provides reasonable controllability and observability. These properties can be improved by choosing another kind of time-varying feedback laws as it is shown in the following section.

4.3.2. - Sinusoïdal feedback laws

Purviance and Tylee (PUR-82) consider the particular case of a 2-input 2-output controllable and observable system with a decentralized fixed mode. They interprete the observability problem resulting in the fixed mode as which of communicating to station 1 the value of the fixed mode (which is observable by station 2) through the transfer function W_{12}, via the time-varying feedback law with gain $K_2(t)$. This problem is a standard one in communication system analysis and a good

---

[1]The condition number (CN) of a rectangular matrix with full rank is given by the ratio between the maximal and the minimal singular values of the matrix (MOO-81). Hence, it is a good measure of the effective rank of the matrix, and of course, it is in our interest to have a condition number as close to 1 as possible (if CN = 1, then the effective rank equals the actual rank).

solution is to use sinusoïdal modulation (feedback) at a frequency matched to the frequency response of the communication channel W_{12} (VAN-68).

Using a simple example, they show that if the system verifies the connectivity assumptions, then by using a sinusoïdal feeback law, the resulting degree of controllability and observability is higher (decrease of the corresponding grammian condition number) than by using the binary feedback law proposed by Anderson and Moore (AND-81b).

Example 4.2 (PUR-82). Consider a 2-input 2-output controllable and observable system described by :

$$x = \begin{bmatrix} 0 & 0 & 0 \\ 0 & -1 & 0 \\ 0 & 0 & -2 \end{bmatrix} x + \begin{bmatrix} 1 \\ 0 \\ 0 \end{bmatrix} u_1 + \begin{bmatrix} 0 \\ -1 \\ 2 \end{bmatrix} u_2$$

$$y_1 = \begin{bmatrix} 0 & 1 & 1 \end{bmatrix} x$$

$$y_2 = \begin{bmatrix} 1 & 0 & 0 \end{bmatrix} x$$

The transfer matrix of this system is :

$$W(p) = \begin{bmatrix} 0 & \dfrac{p}{(p+1)(p+2)} \\ \dfrac{1}{p} & 0 \end{bmatrix}$$

and it is clear that the system has a decentralized structurally fixed mode of type (ii) ($\lambda_0 = 0$).

Applying a sinusoïdal feedback control at the second station $K_2(t) = k \sin \omega t$, the condition number of the observability grammian matrix of the closed-loop system is shown by Figure 4.1-a and Figure 4.1-b, for different values of ω and two values of k.

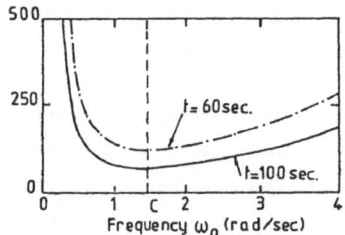

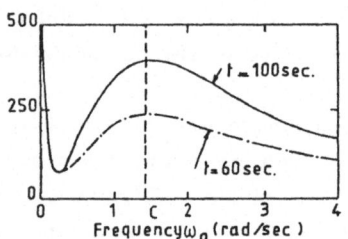

Figure 4.1-a : Condition number of   Figure 4.1-b : Condition number of
OG(t,0) for k = 0.05 (from (PUR-82))   OG(t,0) for k = 1 (from (PUR-82))

For $\omega = \omega_c = \sqrt{2}$ (ω_c is the central frequency of the bandpass of the transfer function W_{12} and we could expect to achieve optimal communication between station 2 and 1 when $\omega = \omega_c$), the condition number of the observability grammian is minimum for k = 0.05 and maximum for k = 1. This can be explained by the fact that with "large" feedback energy the fixed mode dominates the system's characteristics and destroys the balance between system modes. This balance is required for the observability grammian to have a small condition number and to achieve a good observability. This result points out the importance of energy consideration in control law design. However, with k = 0.05 and for $\omega = \omega_c$, the condition number of OG(100, 0) (see Figure 4.1-a) is 72.7.

To make a comparison with the case of piecewise constant feedback control law proposed by Anderson and Moore (AND-81b), let us apply at the second station the following binary feedback with period 2 :

$$
K_2(t) = \begin{cases} 0 & 0 \leqslant t < 1 \\ 1 & 1 \leqslant t < 2 \end{cases}
$$

The condition number of OG(100, 0) is 132.9. This example shows that the condition number has been decreased by about 54 %. Thus, it seems that sinusoïdal feedback yields better observability using less control energy. A similar conclusion can be obtained for controllability.

4.3.3. - Concluding remarks

In this section, we have shown that time-varying feedback laws can stabilize systems with fixed modes that are structurally controllable and observable under decentralized constraints. Although this approach presents the advantage of preserving the decentralized feedback structure, some difficulties are encountered when considering the practical implementation of such controllers. From the practical point of view, we think that sinusoïdal feedback laws are easier to implement than binary feedback laws, but the corresponding result has not been extended to systems with more than 2 stations. We will see in the following section that a possible extension of this approach can be made using vibrational control.

4.4. - VIBRATIONAL CONTROL

Vibrational control was introduced in 1973 by Meerkov (MEE-73) as a stabilization approach in cases where conventional control principles such that feedback or feedforward cannot be used because of a lack of available measurements on the system.

The principle of vibrational control consists in the introduction of such vibrations (with zero mean value) on the dynamic system parameters which modify the properties of the system in a desired manner.

Indeed, vibrational control does not require any measurement or additive control signals, so that it differs from feedback and feedforward principles. The control signals (vibrations) do not depend upon the states (not feedback) or upon disturbances (not feedforward), but are just functions of the independent variable t.

Since the control structure constraints can be interpreted as a lack of measurements at each local station, vibrational control seems to be a suitable control approach to stabilize systems with unstable fixed modes for which structurally constrained feedback fails. Because vibrational control does not require any measurements, its application is independent of the constraints imposed on the information pattern (TRA-85).

4.4.1. - Vibrational control principle

Vibrational control is a method of stabilization consisting in the introduction of zero mean vibrations (oscillations) on the system parameters which may modify the properties of the system, in particular its stability. Note that the introduction of vibrations into systems that are not mechanical but of an arbitrary nature can be achieved by oscillation of the technological parameters, by varying the amplification factors, etc... Unlike conventional control techniques, vibrational control does not require measurement of deviations and disturbances, i.e. it does not require neither control matrix nor observation matrix to appear in the system model. For this reason, the design method does not differ for autonomous or non autonomous systems.

Consider the following time-invariant, finite-dimensional linear system :

$$\dot{x}(t) = A\,x(t) \qquad x \in R^n$$

$$A = (a_{ij})_{\ i,j=1,\ldots,n}$$

(4.4.1)

and introduce periodic zero mean vibrations on the elements of matrix A according to the law :

$$V(t) = (v_{ij}(t))_{\ i,j=1,\ldots,n}$$

(4.4.2)

where

$$v_{ij}(t) = \alpha_{ij} \sin \omega_{ij} t$$
$$\omega_{ij} \neq \omega_{sk} \quad \text{for } ij \neq sk$$

The resulting time-varying system is :

$$\dot{x}(t) = \left[A + V(t) \right] x(t)$$

(4.4.3)

Before describing the design method, let us come up with two questions :

1. Which systems are vibrationally stabilizable (i.e. by means of vibrational control)?
2. How to determine the vibration parameters (gains and frequencies) ?

The answer to these questions is given by the following :

<u>Definition 4.1</u> (MEE-80). System (4.4.1) is vibrationally stabilizable if there exists a periodic zero mean matrix V(t) such that system (4.4.3) is asymptotically stable.

<u>Theorem 4.3</u> (MEE-80). Assume that there exists a vector-row c such that the pair (c, A) is observable, then a necessary and sufficient condition for system (4.4.1) to be vibrationally stabilizable is to have the trace of matrix A negative (Tr A $<$ 0).

<u>Remark 4.2</u>. The extension to the case Tr A $>$ 0 will be considered in a subsequent paragraph.

The analysis of the consequences resulting from the introduction of the vibrations V(t) can be achieved applying the Volosov's averaging scheme (VOL-62) (MEE-73) to system (4.4.3). For this purpose, certain conditions are required :

- the matrix V(t) has a quasi-triangular form
- α_{ij} and ω_{ij} (see formula (4.4.2)) are sufficiently large (see Theorem 4.4).

Volosov's averaging scheme consists in determining a time-invariant system such that its motions describe in "average" the motions of the time-varying system (4.4.3). This system appears in the following form :

$$\dot{z}\,(t) = (A + \overline{V})\,z(t) \qquad (4.4.4)$$

where $\overline{V}$ is a constant matrix depending on vibration amplitudes and frequencies. So, the stability conditions for the system (4.4.3) can be established by using the stability criteria for time-invariant systems on system (4.4.4). Indeed, the entries of $\overline{V}$ depend on α_{ij}'s and ω_{ij}'s : if they are chosen to verify the stability conditions for the time-invariant system (4.4.4), they will also ensure stability to system (4.4.3) in vertue of the property of Volosov's averaging scheme.

Now, assume that the vibration matrix has the following lower (or upper) quasi-triangular form[2] :

$$V(t) = \begin{bmatrix} 0 & & & & & & \underline{0} \\ (\alpha_{21}\ \sin\ \omega_{21}\ t) & 0 & & & & \\ (\alpha_{31}\ \sin\ \omega_{31}\ t)(\alpha_{32}\ \sin\ \omega_{32}\ t) & & 0 & & \\ \vdots & \vdots & & & 0 & \\ (\alpha_{n1}\ \sin\ \omega_{n1}\ t)(\alpha_{n2}\ \sin\omega_{n2}\ t) - - - - -(\alpha_{n_1,n-1}\ \sin\omega_{n_1,n-1}\ t) & & 0 \end{bmatrix} \qquad (4.4.5)$$

With this assumption, the determination of $\overline{V}$ and so of the time-invariant system (4.4.4) can be performed by the following simple procedure. Consider the following system of differential equations :

$$\dot{x}\,(t) = V(t)\,x(t) \qquad (4.4.6)$$

where $V(t)$ is the quasi-triangular matrix given by (4.4.5). Suppose the initial conditions of (4.4.6) to be :

$$x_i(t_0) = x_i^0 \qquad (i=1,\ldots,n)$$

Determine the solution of the first two equations of (4.4.6) :

---

[2] $V(t)$ is chosen in quasi-triangular form because the system cannot be analysed with any other form (MEE-80).

$$x_1(t) = x_1^0$$

$$x_2(t) = x_2^0 + \left[F_{21}(t) - F_{21}(t_0)\right] x_1^0$$

where $F_{21}(t)$ is a zero mean periodic function of t. Determine the solution of the third equation assuming that $F_{21}(t_0) = 0$, i.e. :

$$x_1(t) = x_1^0$$

$$x_2(t) = x_2^0 + F_{21}(t) x_1^0$$

Apply an analogous procedure to each equation of system (4.4.6), then at the $(i-1)^{th}$ step of the procedure we find :

$$x_k = x_k^0 + \sum_{j=1}^{k-1} \left[F_{kj}(t) - F_{kj}(t_0)\right] x_j^0 \qquad (k=1,\ldots,i-1)$$

The solution of the ith equation is sought by substituting :

$$x_k = x_k^0 + \sum_{j=1}^{k-1} F_{kj}(t) x_j^0 \qquad (k=1,\ldots,i-1)$$

where $F_{ij}(t)$ are almost periodic functions of t with a zero mean. Define the quasi-triangular matrix E such that :

$$E = (e_{ij})_{i,j=1,\ldots,n}$$

with

$$e_{ij} = \lim_{T \to \infty} \int_0^T F_{ij}^2(t)\, dt \qquad\qquad (4.4.7)$$

Finally, the matrix $\overline{V}$ of (4.4.4) is given by :

$$\overline{V} = \|\overline{v}_{ij}\|_{i,j=1,\ldots,n} = -(A' \odot E) \qquad\qquad (4.4.8)$$

where $\odot$ denotes element-by-element multiplication of the matrices, i.e. $\overline{v}_{ij} = -a_{ji} \cdot e_{ij}$.

Theorem 4.4 (MEE-80). If there exist sufficiently large positive constants α_0 and ω_0 such that $\alpha_{ij} > \alpha_0$ and $\omega_{ij} > \omega_0$ for all ij, then solutions x(t) of (4.4.3) (4.4.5) and

z(t) of (4.4.4) (4.4.8) (defined with identical initial conditions $x(t_0) = z(t_0)$) are
related by the expressions :

$$\bar{x}(t) = \left[I + F(t) \right] \quad z(t) \qquad t \in [0, \infty[$$

$$\|x(t) - \bar{x}(t)\| \leqslant 1/\alpha_0$$

(4.4.9)

where F(t) is the matrix computed through the above procedure. If the time-invariant system (4.4.4) is asymptotically stable, relation (4.4.9) holds for all $t \in [0, \infty[$,
in the opposite case, it holds only for $t \in [0, \alpha_0[$.

Theorem 4.4 shows that if the amplitudes and frequencies of the vibrations are
sufficiently large, then equations (4.4.4) (4.4.8) describe in average the motions of
the time-varying system (4.4.3) (4.4.5). In particular, if (4.4.4) is asymptotically
stable, then system (4.4.3) has also the same property. Therefore, applying any
stability criterion for time-invariant systems on system (4.4.4), one can obtain the
stability conditions for the time-varying system and determine the vibrations.

It is obvious that vibrational control remains without effect if $\bar{V} = 0$; in this
case, the dynamics of the initial and averaged equations do not differ. This remark
motivates the introduction of the following concept (see Definition 3 and Theorem 3
of (MEE-80)) : vibrationally controllable are those elements a_{ij} of the matrix A for
which there exists an almost periodic matrix V(t) (with a zero mean value) such that
$\bar{v}_{ij} \neq 0$. In view of (4.4.8), vibrationally controllable are therefore those elements a_{ij}
of the matrix A for which $i > j$ and $a_{ji} \neq 0$.

From a practical point of view, the introduction of vibrations is implemented by
making the technological parameters oscillate ; hence, the introduction of vibrations
on zero entries of matrix A is not physically possible. Consequently, vibrationally
controllable are those elements $a_{ij} \neq 0$ of A for which $i > j$ and $a_{ji} \neq 0$.

Remark 4.3.
1. To establish the analogy between vibrational control and time-varying output
feedback control considered in the previous section, we mention that for the system :

$$\begin{cases} \dot{x} = Ax + Bu \\ y = Cx \end{cases}$$

the introduction of vibrations V(t) into the matrix A has a similar effect as a time-
varying output feedback characterized by :

$$B \ K(t) \ C = V(t)$$

or

$$K(t) = B^{\dagger} \ V(t) \ C^{\dagger}$$

where $B^{\dagger}$ and $C^{\dagger}$ are the pseudo-inverses of B and C.

2. In certain cases, there is no need to introduce vibrations with large amplitudes ; the desired effect can be achieved by using only small oscillations of the parameters. See (MEE-73).

3. The design method in the case of autonomous and non-autonomous systems does not differ, but in the case of non-autonomous systems, a supplementary condition has to be considered. Let the system be given by :

$$\dot{x} = A \ x + B \ u \qquad x \in R^n \quad \text{and} \ u \in R^m$$

Suppose that the control u_i, i = 1, ..., m, are slow functions of time in the sense that $\dfrac{du_i}{dt}$ is bounded ; then the supplementary condition on the vibrations frequencies is :

$$\omega_{ij} >> \max_{\ell} \ (\sup \frac{du_{\ell}(t)}{dt}) \qquad \begin{array}{l} (i, j = 1, ... m) \\ (\ell = 1, ..., m) \end{array}$$

4. The determination of matrix $\overline{V}$ needs a heavy integration of trigonometric functions. But in the particular case for which matrix A has only one vibrationally controllable element per row, the calculation of $\overline{V}$ becomes relatively simple (TRA-84b) (TAR-85) ; indeed we have :

$$v_{ij}(t) = \alpha_{ij} \ \sin \omega_{ij} \ t$$

$$F_{ij}(t) = - \frac{\alpha_{ij}}{\omega_{ij}} \ \cos \omega_{ij} \ t$$

then

$$e_{ij} = \frac{\alpha_{ij}^2}{2 \ \omega_{ij}^2}$$

and finally :

$$\overline{v}_{ij} = - a_{ij} \ e_{ij} = - a_{ij} \frac{\alpha_{ij}^2}{2\omega_{ij}^2}$$

4.4.2. - Stabilization by vibrational control (TRA-85)

In this section, we consider a class of systems with unstable fixed modes with respect to a structurally constrained feedback control. The dynamic matrix is supposed to verify the condition of Theorem 4.3 so that the systems are vibrationally stabilizable.

To illustrate the application of vibrational control, let us consider the following example :

Example 4.3 (TAR-85). Given the two-station centrally controllable and observable system :

$$
x = \begin{bmatrix} -2 & 1 & 0 \\ 1 & 1 & 1 \\ 1 & -1 & -1 \end{bmatrix} x + \begin{bmatrix} 0 \\ -1 \\ 2 \end{bmatrix} u_1 + \begin{bmatrix} 1 \\ 0 \\ 0 \end{bmatrix} u_2
$$

$$
y_1 = \begin{bmatrix} 0 & 0 & 1 \end{bmatrix} x
$$

$$
y_2 = \begin{bmatrix} 3 & -1 & 0 \end{bmatrix} x
$$

(4.4.10)

its transfer matrix is :

$$
W(p) = \begin{bmatrix} \dfrac{2p+5}{(p+1)(p+2)} & \dfrac{p-2}{(p-1)(p+1)(p+2)} \\[4mm] \dfrac{p-1}{(p+1)(p+2)} & \dfrac{3p+2}{(p+1)(p+2)} \end{bmatrix}
$$

It is clear that the system has an unstable decentralized fixed mode at $\lambda_0 = 1$, then any decentralized output feedback fails to stabilize the system.

Consider the autonomous system associated with system (4.4.10) :

$$
\dot{x} = A x = \begin{bmatrix} -2 & 1 & 0 \\ 1 & 1 & 1 \\ 1 & -1 & -1 \end{bmatrix} x
$$

This system is vibrationally stabilizable, since for $c = (1\ 0\ 0)'$, the pair (c, A) is observable and $\mathrm{Tr}\ A = -2 < 0$. The vibrationally controllable elements of A are a_{21}

and a_{32} (or a_{12} and a_{23}) for a lower (upper) quasi-triangular vibration matrix. These considerations lead to the following matrix :

$$V(t) = \begin{bmatrix} 0 & 0 & 0 \\ \alpha_{21} \sin \omega_{21} t & 0 & 0 \\ 0 & \alpha_{32} \sin \omega_{32} t & 0 \end{bmatrix}$$

resulting in the time-varying system :

$$\dot{x} = \begin{bmatrix} A + V(t) \end{bmatrix} x \tag{4.4.11}$$

The determination of $\overline{V}$ can be performed by applying the averaging scheme described in the previous section. Matrix A has only one vibrationally controllable element in each row, then in accordance with Remark 4.3 of the previous section, the "averaged" system is :

$$z = (A+\overline{V})z = \begin{bmatrix} -2 & 1 & 0 \\ 1+\overline{v}_{21} & 1 & 1 \\ 1 & -1+\overline{v}_{32} & -1 \end{bmatrix} z \tag{4.4.12a}$$

where

$$\overline{v}_{21} = -a_{12} \frac{\alpha_{21}^2}{2\omega_{21}^2} = -\frac{\alpha_{21}^2}{2\omega_{21}^2} < 0$$

$$\overline{v}_{32} = -a_{23} \frac{\alpha_{32}^2}{2\omega_{32}^2} = -\frac{\alpha_{32}^2}{2\omega_{32}^2} < 0 \tag{4.4.12b}$$

The characteristic equation of the system (4.4.12) is :

$$\det (pI-A-\overline{V}) = p^3 + 2p^2 - (1+\overline{v}_{21} + \overline{v}_{32}) p - 2 - \overline{v}_{21} - 2\overline{v}_{32}$$

and the stability conditions for this system, established by using Routh's criterion, are :

$$\overline{v}_{21} < 0$$
$$\overline{v}_{32} < 0$$
$$\overline{v}_{21} < -2(1+\overline{v}_{32})$$

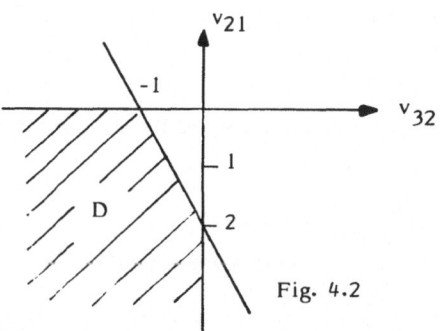

Fig. 4.2

The domain D of admissible solutions for these inequalities is given by Figure 4.2. By choosing $\bar{v}_{21} \in D$ and $\bar{v}_{32} \in D$, the time-invariant system (4.4.12) is made asymptotically stable. Then, if the amplitudes and frequencies of vibrations are chosen according to (4.4.12b) and with sufficiently large values (cf. Theroem 4.4), the time-varying system (4.4.11) is made asymptotically stable.

4.4.3. - Vibrational feedback control laws

The stabilization method used in the last section is based on introducing vibrations on the parameters of the system model. From a practical point of view, this can be performed by acting on the technological parameters of the system. This implies, first, that one knows the explicit expressions of the model parameters as functions of the technological parameters and second, that some physical entities of the system can be designed so that they have a vibratory value. The first requirement is not generally satisfied since the system model is often obtained by identification methods. Neither is the second requirement : although it may be considered to give a vibratory value to an electrical entity, this is impossible for mechanical elements. The domain of application of the above method is therefore restricted to specific cases.

The approach presented in this section uses at the same time vibrational and feedback control theory. The above difficulties are thus encompassed by introducing the vibrational control action through the feedback (RUN-85).

Consider the system (4.1.3) and, for simplicity, the case S=2 (the general case of S>2 stations will be considered subsequently). In addition to the centralized controllability and observability assumption, make the following connectivity assumption :

$$C_1 B_2 \neq 0 \ , \quad C_2 B_1 \neq 0 \tag{4.4.13}$$

which guarantees that the system has no structurally fixed modes of type (i) ((SEZ-81a) (COR-76a,b), see § 2.2.3)

Apply the following dynamic controller to the system (4.1.3) :

$$u_1(t) = \frac{\alpha}{\varepsilon} K_1 \left(\frac{t}{\varepsilon}\right) y_1(t) + v_1(t)$$

$$u_2(t) = K_2 \left(\frac{t}{\varepsilon}\right) y_2(t)$$

(4.4.14)

$$\dot{\xi}(t) = F \left(\frac{t}{\varepsilon}\right) \xi(t) + G \left(\frac{t}{\varepsilon}\right) y_1(t)$$

$$v_1(t) + H \left(\frac{t}{\varepsilon}\right) \xi(t) + L \left(\frac{t}{\varepsilon}\right) y_1(t)$$

where all the matrices have almost periodic entries and ξ is the state of the controller of dimension ν. Then the controlled system is :

$$\dot{x} = Ax + \frac{\alpha}{\varepsilon} B_1 K_1 \left(\frac{t}{\varepsilon}\right) C_1 x + B_2 K_2 \left(\frac{t}{\varepsilon}\right) C_2 x + B_1 v_1$$

(4.4.15)

$$y_i = C_i x \quad i=1,2$$

With $\psi\left(\frac{t}{\varepsilon}\right)$ being a fundamental matrix for $\frac{\alpha}{\varepsilon} B_1 K_1\left(\frac{t}{\varepsilon}\right) C_1$, define the transformation

$$x(t) = \psi \left(\frac{t}{\varepsilon}\right) z(t)$$

(4.4.16)

Consider the almost periodic, differentiable $m_1 \times r_1$ matrix $Q\left(\frac{t}{\varepsilon}\right)$ such that $(I + C_1 B_1 Q(\frac{t}{\varepsilon}))$ is non singular for all t and :

$$\frac{\alpha}{\varepsilon} K_1 \left(\frac{t}{\varepsilon}\right) = Q \left(\frac{t}{\varepsilon}\right) \left(I + C_1 B_1 Q \left(\frac{t}{\varepsilon}\right)\right)^{-1}$$

Then we have

$$\dot{z} = \psi^{-1} A\psi z + [B_1 \ B_2] \begin{bmatrix} P \, C_1 \, B_2 \, K_2 \, C_2 \, B_1 \, Q & PC_1 \, B_2 \, K_2 \\ K_2 \, C_2 \, B_1 \, Q & K_2 \end{bmatrix} \begin{bmatrix} C_1 \\ C_2 \end{bmatrix} z + \psi^{-1} B_1 v_1$$

(4.4.17)

where

$$P\left(\frac{t}{\epsilon}\right) = - Q\left(\frac{t}{\epsilon}\right)\left(I + C_1 B_1 Q\left(\frac{t}{\epsilon}\right)\right)^{-1}$$

The controller in (4.4.14) is chosen as

$$\dot{\xi} = F\,\xi + G\left(I + C_1 B_1 Q\left(\frac{t}{\epsilon}\right)\right)^{-1} y_1 \tag{4.4.18}$$

$$v_1 = \left(I + P\left(\frac{t}{\epsilon}\right) C_1 B_1\right)^{-1} H\,\xi + \left(I + P\left(\frac{t}{\epsilon}\right) C_1 B_1\right)^{-1} L\left(I + C_1 B_1 Q\left(\frac{t}{\epsilon}\right)\right)^{-1} y_1$$

where F, G, H and L are constant matrices of appropriate dimensions.

After some manipulations, (4.4.17) and (4.4.18) can be written as

$$z = \psi^{-1} A \psi z + [B_1 B_2] \begin{bmatrix} PC_1 B_2 K_2 C_2 B_1 Q & PC_1 B_2 K_2 \\ \\ K_2 C_2 B_1 Q & K_2 \end{bmatrix} \begin{bmatrix} C_1 \\ \\ C_2 \end{bmatrix} z + B_1 L C_1 z + B_1 H\,\xi \tag{4.4.19}$$

$$\dot{\xi} = F\,\xi + G\,C_1\,z$$

Taking averages[3] in (4.4.19) gives the following time-invariant system :

$$\dot{\bar{z}} = A\bar{z} + B_1 L C_1 \bar{z} + B_1 H\,\bar{\xi} \tag{4.4.20}$$

$$\dot{\bar{\xi}} = F\,\bar{\xi} + G\,C_1\bar{z}$$

where $\hat{A} = \overline{A} + B_1 B_2 \begin{bmatrix} \Gamma_{11} & \Gamma_{12} \\ \Gamma_{21} & \Gamma_{22} \end{bmatrix} \begin{bmatrix} C_1 \\ C_2 \end{bmatrix}$

---

[3] The average of an almost periodic matrix M (t) is given by :

$$M = \lim_{T \to \infty} \frac{1}{T} \int_0^T M(\tau)\,d\tau$$

$$\overline{A} = \overline{\psi^{-1} A \psi} = A + B_1\overline{P}C_1 A + AB_1\overline{Q}C_1 + B_1\overline{PC_1AB_1Q}C_1$$

$$\Gamma_{11} = \overline{PC_1B_2K_2C_2B_1Q} \qquad \Gamma_{12} = \overline{PC_1B_2K_2}$$
$$\Gamma_{21} = \overline{K_2C_2B_1Q} \qquad\qquad \Gamma_{22} = \overline{K_2}$$

The average Γ can be assigned by choosing K_1, and hence Q, and K_2. It is shown in (RUN-85) that an appropriate choice of Γ can make the triple $(\hat{A}, B_1, C_1)$ controllable and observable. Therefore, using the results of (BRA-70), the matrices L, H, G and F (where ξ is of dimension $\ell = \min (\nu_o-1, \nu_c-1)$, $\nu_o(\nu_c)$ being the observability (controllability) index of (4.4.19)) can be chosen such that the system (4.4.20) is made asymptotically stable.

Then, using the results of (MEE-73) and (BOG-61), the main result shown in (RUN-85) is that ε_0 can be selected so that the system (4.4.19) also becomes stable for all $0 < \varepsilon \leqslant \varepsilon_0$ and by (4.4.16) so does the original system (4.4.15).

The following theorem summarizes the above discussion :

Theorem 4.5 (RUN-85). Assume that (4.4.13) holds. Then there exists an $\varepsilon_0 > 0$ and a controller of the form (4.4.14) that ensures the asymptotic stability of the system (4.1.3) for all $0 < \varepsilon \leqslant \varepsilon_0$.

The subsequent algorithm (RUN-85) is then proposed for the design of the controller (4.4.14).

Step 1 : Choose feedback matrices $K_1 (\frac{t}{\varepsilon})$ and $K_2 (\frac{t}{\varepsilon})$ that achieve the desired controllability and observability properties (almost any Γ and, consequently, any $K_1(\frac{t}{\varepsilon})$ any $K_2(\frac{t}{\varepsilon})$, works(POT-79)).

Step 2 : Design a linear, time invariant, dynamic feedback controller for the averaged system (4.4.20).

Step 3 : Substitute the parameters of the controller from step 2 into the controller for the original system using (4.4.18). With ε sufficiently small, this controller ensures the asymptotic stability of (4.4.15).

Example 4.4 (RUN-85). Consider the following system :

$$A = \begin{bmatrix} 0 & 1 & 0 \\ 1 & 0 & 0 \\ 0 & 0 & 0 \end{bmatrix} \qquad B_1 = C_2' = \begin{bmatrix} 1 \\ 0 \\ 0 \end{bmatrix} \qquad B_2 = C_1' = \begin{bmatrix} 0 \\ 0 \\ 1 \end{bmatrix}$$

which has a structurally fixed mode of type (ii) at $\lambda_0 = 0$. The system is centrally controllable and observable and $C_1 B_2 = C_2 B_1 = 1$ (condition (4.4.13) is satisfied).

In this case, the system (4.4.19) is given by :

$$\dot{z} = \psi^{-1} A \psi z + [B_1 B_2] \begin{bmatrix} -q^2 K_2 & -q K_2 \\ q K_2 & K_2 \end{bmatrix} \begin{bmatrix} C_1 \\ C_2 \end{bmatrix} z + B_1 L C_1 z + B_1 H \xi \qquad (4.4.21)$$

$$\dot{\xi} = F \xi + G C_1 z$$

Choose $q = \sqrt{2} \sin \frac{t}{\varepsilon}$, $K_2 = \gamma + \sqrt{2} \beta \sin \frac{t}{\varepsilon}$ and therefore

$$K_1 = q(1 + C_1 B_1 q)^{-1} = \frac{\sqrt{2}\alpha}{\varepsilon} \cos \frac{t}{\varepsilon}$$

Since

$$\overline{A} = \psi^{-1} A \psi = \begin{bmatrix} 0 & 1 & 0 \\ 1 & 0 & \overline{q} \\ 0 & 0 & 0 \end{bmatrix} = \begin{bmatrix} 0 & 1 & 0 \\ 1 & 0 & 0 \\ 0 & 0 & 0 \end{bmatrix} = A,$$

and taking averages in (4.4.21), the system (4.4.20) is given by :

$$\dot{\overline{z}} = \begin{bmatrix} -\alpha\beta & 1 & -\gamma\alpha^2 \\ 1 & 0 & 0 \\ \gamma & 0 & \alpha\beta \end{bmatrix} \overline{z} + B_1 L C_1 \overline{z} + B_1 H \overline{\xi} \qquad (4.4.22)$$

$$\dot{\overline{\xi}} = F \overline{\xi} + G C_1 \overline{z}$$

Since the triple $(\hat{A}, B_1, C_1)$ is controllable and observable for all α, β and γ, choose for example $\alpha = \beta = \gamma = 1$. The dynamic compensator for (4.4.22) must be of order 2 ($\nu_0 = \nu_c = 3$). Set the matrices :

$$F = \begin{bmatrix} f_1 & 0 \\ 0 & f_2 \end{bmatrix} \qquad H = [h_1 \ h_2], \qquad G = \begin{bmatrix} g_1 \\ g_2 \end{bmatrix}, \ L = 1$$

then (4.4.22) can be rewritten as :

$$\begin{bmatrix} \dot{z} \\ \dot{\xi} \end{bmatrix} \begin{bmatrix} -1 & 1 & 1\text{-}1 & h_1 & h_2 \\ 1 & 0 & 0 & 0 & 0 \\ 1 & 0 & 1 & 0 & 0 \\ 0 & 0 & g_1 & f_1 & 0 \\ 0 & 0 & g_2 & 0 & F_2 \end{bmatrix} \begin{bmatrix} z \\ \xi \end{bmatrix}$$

The compensator coefficients are chosen to have the closed-loop eigenvalues at $-1 \pm j$, -1, -1, -2 :

$$f_1 = -3 + \sqrt{5} \approx -0,764 \qquad f_2 = -3 - \sqrt{5} \approx -5,236$$
$$1 = -12$$

$$h_1 = g_1 = \left[\frac{90\sqrt{5} - 105}{2\sqrt{5}} \right]^{1/2} \approx 0,146$$

$$h_2 = g_2 = \left[\frac{105 + 47\sqrt{5}}{2\sqrt{5}} \right]^{1/2} \approx 6,854$$

The resulting time-varying system (4.4.15) is

$$\begin{bmatrix} \dot{x} \\ \dot{\xi} \end{bmatrix} = \begin{bmatrix} 0 & 1 & -1+\dfrac{\sqrt{2}}{\varepsilon^2}\cos\dfrac{t}{\varepsilon} & 0,146 & 6,854 \\ 1 & 0 & 0 & 0 & 0 \\ 1+\sqrt{2}\sin\dfrac{t}{\varepsilon} & 0 & 0 & 0 & 0 \\ 0 & 0 & 0,146 & -0,764 & 0 \\ 0 & 0 & 6,854 & 0 & -5,236 \end{bmatrix} \begin{bmatrix} x \\ \xi \end{bmatrix} \qquad (4.4.23)$$

Simulations show that (4.4.23) is asymptotically stable if $\varepsilon < 0,1$ (RUN-85).

Therefore, in spite of the existence of a decentralized unstable fixed mode, the original system has been stabilized by <u>decentralized</u> vibrational feedback control.

Remark 4.4.

1. Although the above stabilization approach combines vibrational and feedback control, its feature is that the design of the correcting vibrational control action ($K_1(\frac{t}{\epsilon})$ and $K_2(\frac{t}{\epsilon})$) and the feedback control law (L, H, G, F) can be conducted independently. The feedback control design is done for a much simpler linear time-invariant system. However, in regard to the original system, the vibrational control action takes place through the feedback.

2. In the general case of S > 2 stations, (4.4.13) must be replaced by the following condition :

$$\exists\, 1 \leqslant i \leqslant S \text{ such that } C_i B_j \neq 0 \text{ and } C_j B_i \neq 0, \quad 1 \leqslant j \leqslant S, \quad j \neq i \qquad (4.4.13')$$

and the controller (4.4.14) becomes :

$$\mu_i = \frac{\alpha}{\epsilon} K_i \left(\tfrac{t}{\epsilon}\right) y_i + v_i \qquad 1 \leqslant i \leqslant S$$

$$\mu_j = K_j \left(\tfrac{t}{\epsilon}\right) y_i \qquad 1 \leqslant j \leqslant S \qquad j \neq i$$

$$\xi = F \left(\tfrac{t}{\epsilon}\right) \xi + G \left(\tfrac{t}{\epsilon}\right) y_i \qquad\qquad (4.4.14')$$

$$v_i = H \left(\tfrac{t}{\epsilon}\right) \xi + L \left(\tfrac{t}{\epsilon}\right) y_i$$

Then Theorem 4.5 remains true with (4.4.13') and (4.4.14') in place of (4.4.13) and (4.4.14), respectively.

4.5. - CONCLUSION

In this chapter, it is shown that systems with unstable non structurally fixed modes can be stabilized by using time-varying or non-linear feedback control laws which preserve the feedback structure constraints. The use of sample and hold (WAN-82) (OZG-85), the application of piecewise constant or sinusoïdal feedback laws (AND-81b) (PUR-82), and the introduction of almost periodic vibrations on the system parameters (TRA-85) or through the feedback (RUN-85) are different stabilizing approaches which are presented in this chapter.

Obviously, the sacrifice of linerarity or time invariance of the control brings difficulties in the analysis of such controlled systems, even of small dimensionality. Time-varying systems analysis methods or averaging schemes must be used.

On another hand the implementation of such control laws may be practically unconvenient. Consequently it is the authors believe that relaxing the structure constraints (if possible) is a more realistic approach. This approach is investigated in the following chapter and several algorithms are provided.

CHAPTER 5

CHOICE OF FEEDBACK CONTROL STRUCTURE TO AVOID
FIXED MODES

5.1. - INTRODUCTION

Chapter III pointed out that fixed modes can be classified according to their origines. Fixed modes arising from special configurations in the parameter values such as non structurally fixed modes or structurally fixed modes of type (ii) remain controllable under structural constraints. The existence of such fixed modes does not prevent the possibility of finding a control law satisfying the structural constraints which gives a satisfactory behaviour to the system. However, either the property of time-invariance or of linearity must be sacrified. Different approaches and methods for the design of such control laws have been presented in the previous chapter. Unfortunately, they involve analysis difficulties even for small dimension systems since they require using appropriate analytical tools for time-varying or nonlinear systems. Moreover, from the practical point of view, some physical problems can be encountered when trying implement such control laws. Consequently, the total cost resulting from this option, which allows to preserve the structural specifications (decentralization for example), may be higher than the cost associated to the design and implementation of a time-invariant, linear control law for which structure constraints have been relaxed. Note also that this is the only way to cope with the existence of structurally fixed modes of type (i) which arise from a lack of information transfer between stations and are uncontrollable under structural constraints.

Therefore, relaxing the structural constraints on the control seems to be the most convenient way to solve the stabilization or pole placement problem in presence of fixed modes. The purpose of this chapter is to present the different available methods for the design of an appropriate feedback control structure. These methods are based on the different characterizations of fixed modes which were presented in chapter 3 : one can be more appropriate than another depending on the type of situation we are dealing with. Roughly speaking, two types of situations can occur :

- the system is physically partitioned in several stations, due for example to different geographical locations of the inputs and the outputs. In this case, it is clear that the decentralized structure would be the most appropriate. Our purpose will thus be to determine the minimal information exchanges between stations required by the existence of fixed modes. The optimality criterion can be chosen as the number of feedback links between two different stations or as the cost associated with the implementation of these feedback links.

- either the system does not reflect a prespecified partitioning or the stations present the particularity that the cost of a local feedback is not neglectable with respect to the cost of a feedback between two different stations. In this case, we want to determine the minimal control structure(s) for which the system has no fixed modes. They will generally appear without all the local feedbacks.

Note that the problem resulting from the second situation is more general. Indeed, the first problem can be formulated in the same way by setting to zero the costs associated to the local feedbacks. Therefore, all the methods which will be presented in the general framework can also be used for the particular case.

5.2. - RELAXING PRESPECIFIED FEEDBACK CONSTRAINTS

5.2.1. - Preliminaries

Since the existence of fixed modes arises from the constraints imposed on the control structure, the most natural way to eliminate them is to partially relax these constraints ; i.e. to introduce additional information exchanges between stations. Consider the following system (C,A,B) :

$$\begin{cases} \dot{x}(t) = A\,x(t) + \sum_{i=1}^{S} B_i\,u_i(t) \\ y_i(t) = C_i\,x(t) \end{cases} \qquad (i = 1, \ldots, S) \qquad (5.2.1)$$

where :

$$B = [B_1, \ldots, B_S]$$
$$C' = [C'_1, \ldots, C'_S]$$
$$u'(t) = [u'_1(t), \ldots, u'_S(t)]$$
$$y'(t) = [y'_1(t), \ldots, y'_S(t)]$$

with $x(t)$ R^n and $u_i(t)$ R^{m_i} , $y_i(t)$ R^{r_i}, $i = 1, \ldots, S$.

$m = \sum\limits_{i=1}^{S} m_i$, $r = \sum\limits_{i=1}^{S} r_i$ and S is the number of control and observation stations.

If the control is subject to a decentralized structure, the output feedback matrix K appears with a block diagonal structure where each diagonal block is asso-ciated with a local feedback (K Ω_d). Suppose that the set of decentralized fixed modes is not empty ($\Lambda(C,A,B,\Omega_d) \neq 0$) ; the problem is thus to determine a new set Ω^* such that Ω_d C Ω^* and $\Lambda(C,A,B,\Omega^*) = 0$. The additional information transfer between some station j and some station i appears by the adjunction of an off-diago-nal block in the feedback matrix :

$$K \quad C \quad \Omega_d \qquad\qquad K \quad C\,\Omega^*$$

$$\Lambda\,(C,A,B\ \Omega_d) \neq \emptyset \qquad\qquad \Lambda\,(C,A,B,\Omega^*) = \emptyset$$

The new structure is determined with respect to an optimality criterion which can be either the number of supplementary feedback links between different stations or their associated cost (specified from physical considerations).

5.2.2. - Wang and Davison procedure (WAN-78a)

Given the system (5.2.1), Wang and Davison considered the problem of deter-mining a control structure allowing the stabilization of the system and minimizing the transmission cost. Consider the following set of matrices :

$$K(r_{ij}, \; i,j=1,\ldots,S) = \{K/K = \text{block } \{K_{ij}\} = \begin{bmatrix} K_{11} & \cdots & K_{1S} \\ \vdots & & \vdots \\ K_{S1} & \cdots & K_{SS} \end{bmatrix}$$

$$K_{ij} \subset R^{m_i x r_j}, \; \text{rank } K_{ij} = r_{ij} \}$$

(5.2.2)

and define as z_{ij} the cost of transmitting a scalar time function from station i to station j per unit of time. Each feedback law $u_i = K_{ij} y_j$ requires the transmission of a time function with r_{ij} components from station j to station i. The total transmission cost resulting from the implementation of u = K y, K   K(r_{ij}, i,j=1,...,S) is :

$$\mathcal{Z}(r_{ij}, \text{ i,j=1},\ldots,S) = \sum_{j=1}^{S} \sum_{i=1}^{S} r_{ij} z_{ji} \qquad (5.2.3)$$

The problem can thus be formulated as follows :

$$\underset{0 \leqslant r_{ij} \leqslant \min(m_i, r_j)}{\text{Min}} \mathcal{Z}(r_{ij}, \text{ i,j=1},\ldots,S)$$

under $\Lambda[C, A, B, K(r_{ij}, \text{ i,j=1},\ldots,S)] \subset \mathcal{C}^-$ $\qquad (5.2.4)$

where $\mathcal{C}^-$ is the half left hand side complex plane.

Every sub-matrix K_{ij} of rank r_{ij} can be written as the product of two subma-trices : $K_{ij} = L_{ij} M_{ij}$, where L_{ij} is of size $m_i \times r_{ij}$ and M_{ij} is of size $r_{ij} \times r_j$. If $\sigma_I(A) = \{\lambda_1,\ldots,\lambda_q\}$ is the set of unstable eigenvalues of A, then condition (5.2.4) is satisfied if and only if there exist L_{ij} and M_{ij} such that :

$$\det (\lambda_i I - A - B [\text{block } \{L_{ij} M_{ij}\}] C) \qquad (5.2.5)$$
$\forall \lambda_i \subset \sigma_I(A)$
is not identically zero.

Condition (5.2.5) means that none of the unstable eigenvalues of A is a fixed mode with respect of the feedback control u =   block $L_{ij} M_{ij}$ y.

At this step, Wang and Davison (WAN-78a) did not give any criteria to set apart iteratively the worse solutions from the others. Given the argument that there is only a finite number of possibilites in choosing r_{ij} 's, the minization problem can be solved in a finite number of steps. Their search procedure consists coarsely in testing all the possibilities given by the choice of the control structure (some K_{ij} can be 0) and the choice of the set (r_{ij}, i,j=1,...,S), begining by those involving the lowest cost.

This is illustrated by the following example :

Example 5.1. Consider the following system with two stations :

$$\dot{x} = \begin{bmatrix} 0 & 0 & 0 \\ 0 & 0 & 0 \\ 0 & 0 & -2 \end{bmatrix} x + \begin{bmatrix} 1 & 0 \\ 0 & 1 \\ 0 & 0 \end{bmatrix} u_1 + \begin{bmatrix} 0 \\ 0 \\ 1 \end{bmatrix} u_2$$

$$y_1 = \begin{bmatrix} 0 & 0 & 1 \end{bmatrix} x$$

$$y_2 = \begin{bmatrix} 1 & 0 & 0 \\ 0 & 1 & 0 \end{bmatrix} x$$

and define $z_{11} = z_{22} = 0$, $z_{12} = 1$ and $z_{21} = 2$.

1. Obviously, we consider first the decentralized feedback represented by the set :

$$K_1 = K \; (r_{11} = r_{22} = 1, \; r_{12} = r_{21} = 0) \; = \; \left\{ \begin{bmatrix} k_{11} & 0 & 0 \\ k_{21} & 0 & 0 \\ 0 & k_{32} & k_{33} \end{bmatrix}, k_{ij} \in R \right\}$$

The corresponding fixed polynomial of (C,A,B) is :

$$F(p; \; C,A,B,K_1) = \underset{K \in K_1}{\text{g.c.d}} \; \{ \det \; (pI - A - BKC) \} \; = p$$

Therefore with local feedbacks, the system has a fixed mode at the origin.

2. Now, consider the case in which we have a transfer from station 1 to station 2 :

$$K_2 = K \; (r_{11} = r_{22} = r_{21} = 1, r_{12} = 0) = \; \left\{ \begin{bmatrix} k_{11} & 0 & 0 \\ k_{21} & 0 & 0 \\ k_{31} & k_{32} & k_{33} \end{bmatrix}, k_{ij} \in R \right\}$$

and the cost is $\mathcal{Z} (r_{11} = r_{21} = r_{22} = 1, \; r_{12} = 0) = 1$. The fixed polynomial is also equal to p ; K_2 is not an admissible solution.

3. Finally, consider the case in which we transmit a single linear combination of the observation at station 2 to station 1 :

$$K_3 = K(r_{11} = r_{22} = r_{12} = 1, r_{21} = 0) = \begin{bmatrix} k_{11} & k_{12} & k_{13} \\ k_{21} & k_{22} & k_{23} \\ 0 & k_{32} & k_{33} \end{bmatrix} \quad k_{ij} \in R, \quad \text{rank} \begin{bmatrix} k_{12} & k_{13} \\ k_{22} & k_{23} \end{bmatrix} = 1$$

and the associated cost is $(r_{11} = r_{21} = r_{22} = 1, r_{21} = 0) = 2$.

The rank constraint can be expressed by :

$$\begin{bmatrix} k_{12} & k_{13} \\ k_{22} & k_{23} \end{bmatrix} = \begin{bmatrix} f_1 \\ f_2 \end{bmatrix} \begin{bmatrix} g_1 & g_2 \end{bmatrix}$$

where f_1, f_2, g_1 and g_2 are arbitrary real numbers. The fixed polynomial is now :

$$\underset{\substack{f_i, g_i, k_{ij} \\ i=1,2}}{g.c.d} \left\{ \det \begin{bmatrix} p - f_1 g_1 & -f_1 g_2 & -k_{11} \\ -f_2 g_1 & p - f_2 g_2 & -k_{21} \\ -k_{32} & -k_{33} & p+2 \end{bmatrix} \right\} = 1$$

such that the system has no fixed modes anymore. If we choose :

$$f_1 = f_2 = k_{11} = 1 \; ; \; g_1 = k_{21} = k_{33} = 0 \; ; \; g_2 = k_{32} = -1$$

all the poles of the system are located at - 1.

Though in the case of the example the solution is obtained in 3 steps, it is clear that the number of possibilities to be tested increases rapidly with the dimensions of the system. In consequence, this procedure cannot be used for large scale systems which are effectively those requiring structural constraints in the control and minimization of the information transmission cost.

5.2.3. - Armentano and Singh ' procedure (ARM-82)

Armentano and Singh procedure is limited to the particular class of systems (5.2.1) called interconnected systems and satisfying the specifications in (3.3.7). For such systems, they gave a characterization of fixed modes based on the block-diagonal dominance property (see § 3.3.3.b-2-, Chapter III).

Given the system (3.3.7), consider a feedback law of the form :

$$u = K \, y \quad \text{with} \quad K \in \Omega_F \qquad \text{and} \quad \Omega_d \in \Omega_F$$

where Ω_d is the set of block-diagonal matrices, then we have :

$$\Lambda (C,A,B,\Omega_F) \subseteq \Lambda (C,A,B,\Omega_d) \qquad \qquad \text{(see § 2.2.2., Chapter II).}$$

If λ_0^* is a fixed mode with respect to the two control structures Ω_d and Ω_F, the application of Corollary 3.3 to $(A + BKC)$, $K \in \Omega_F$, shows that there exists $i \in \{1,\ldots,S\}$ such that :

$$\left\| (\hat{A}_{ii} - \lambda_0 \, I)^{-1} \right\|^{-1} \; < \; \sum_{\substack{j=1 \\ j \neq i}}^{S} \left\| A_{ii} + B_i \, K_{ij} \, C_j \right\| \qquad (5.2.5)$$

$$\forall \; K_{ii} \in R^{m_i \times r_i}, \; \forall \; K_{ij} \in R^{m_i \times r_j} \quad \text{and} \quad i \neq j.$$

Therefore, as a particular case, (5.2.5) is also verified for $K_{ii} = 0$, $K_{ij} = 0$, $(i,j=1,\ldots,S)$. Define the sets :

$$\overline{S} = \{ i \in \{1,\ldots,S\} \; / \; (3.3.10) \text{ is satisfied)} \}$$
$$\widetilde{S} = \{ i \in \{1,\ldots,S\} \; / \; (5.2.5) \text{ is satisfied)} \}$$

then whe have $\widetilde{S} \subset \overline{S}$.

Consider that λ_0^* is a decentralized fixed mode. From the above discussion, we know that if λ_0^* remains a fixed mode with respect to another control structure, the K_{ij} 's can be set to zero for $i \in \widetilde{S}$ without affecting the situation. Moreover, (5.2.5) is satisfied for a subset of indices of $\overline{S}$. In order to eliminate the fixed mode, Armentano and Singh propose to add the off diagonal submatrices corresponding to the indices $i \notin \overline{S}$. It is clear that this strategy can be applied only when $\overline{S}$ is a proper subset of $\{1,\ldots,S\}$. Note also that if the system has more than one decentralized fixed mode, the same approach can be used replacing $\overline{S}$ by the union of the sets $\overline{S}_i$ corresponding to each fixed mode λ_i.

Example 5.2 : Consider the following system :

$$
\begin{bmatrix} \dot{x}_{11} \\ \dot{x}_{12} \\ \dot{x}_{21} \\ \dot{x}_{22} \\ \dot{x}_{31} \\ \dot{x}_{32} \end{bmatrix}
=
\begin{bmatrix}
0 & 1 & 0 & 0 & 0 & 0 \\
0 & 0 & 0 & 0 & 0 & 0 \\
1 & 0 & 0 & 1 & 0 & 1 \\
0 & 0 & 0 & 0 & 0 & 0 \\
0 & 0 & 0 & 1 & 0 & 1 \\
0 & 0 & 0 & 0 & 0 & 0
\end{bmatrix}
\begin{bmatrix} x_{11} \\ x_{12} \\ x_{21} \\ x_{22} \\ x_{31} \\ x_{32} \end{bmatrix}
+
\begin{bmatrix}
0 & 0 & 0 \\
1 & 0 & 0 \\
0 & 0 & 0 \\
0 & 1 & 0 \\
0 & 0 & 0 \\
0 & 0 & 1
\end{bmatrix}
\begin{bmatrix} u_1 \\ u_2 \\ u_3 \end{bmatrix}
$$

The decentralized control laws are of the form :

$$u_i = K_{ii} \ x_i = \begin{bmatrix} k_{i1} & k_{i2} \end{bmatrix} \begin{bmatrix} x_{i1} \\ x_{i2} \end{bmatrix} \qquad i=1,2,3$$

and, when applied to the system, the following closed-loop matrix is obtained :

$$A + BK = \begin{bmatrix} 0 & 1 & 0 & 0 & 0 & 0 \\ k_{11} & k_{12} & 0 & 0 & 0 & 0 \\ 1 & 0 & 0 & 1 & 0 & 1 \\ 0 & 0 & k_{21} & k_{22} & 0 & 0 \\ 0 & 0 & 0 & 1 & 0 & 1 \\ 0 & 0 & 0 & 0 & k_{31} & k_{32} \end{bmatrix}$$

The system has thus a decentralized fixed mode at the origine $\lambda_0 = 0$.

Using the following matricial norm :

$$\| A \| = \max_i \ \sum_j \ | a_{ij} |$$

$\lambda_0 = 0$ satisfies (3.3.10) for $S = 2,3$:

$$\| \hat{A}_{ii}^{\ -1} \|^{-1} = \cfrac{1}{\max \ (\cfrac{k_{i2}}{k_{i1}} + \cfrac{1}{k_{i1}}) \ , \ 1} \ \leq 1 \qquad i=2,3$$

It can easily be verified that $\lambda_0 = 0$ remains as a fixed mode for K* while it is eliminated for K** :

$$K* = \begin{bmatrix} K_{11} & 0 & 0 \\ K_{21} & K_{22} & K_{23} \\ K_{31} & K_{32} & K_{33} \end{bmatrix} \qquad K** = \begin{bmatrix} K_{11} & K_{12} & K_{13} \\ 0 & K_{22} & 0 \\ 0 & 0 & K_{33} \end{bmatrix}$$

Though the efficiency of the above procedure cannot be contested, the fact that it is based on a sufficient (and not necessary) condition for the absence of fixed modes makes it rough. Indeed, it only allows the determination of the block-rows (indices i) where the addition of all the off-diagonal matrices certifies the elimination of the fixed modes. It does not provide any information on the indices j or on the number of necessary off-diagonal submatrices. If we consider the preceding example, it is easily verified that if K_{13} remains 0 in K^{**}, the fixed mode is also eliminated. The solution obtained by using this procedure is thus obviously suboptimal. Moreover, its application is limited to the case for which $\overline{S}$ is a proper subset of $\{1,\ldots,S\}$ (which can only be concluded after the evaluation of (3.3.10) for each $i=1,\ldots,S$!), in which case none information can be drawn out. This situation is illustrated by the following example :

Example 5.3. Consider the following system :

$$
\begin{bmatrix} x_1 \\ \text{---} \\ x_2 \end{bmatrix} =
\begin{bmatrix} 1 & 0 & | & 0 & 1 \\ 0 & 0 & | & 0 & 0 \\ \text{---} & & + & \text{---} & \\ 1 & 0 & | & 0 & 1 \\ 0 & 0 & | & 0 & 0 \end{bmatrix}
\begin{bmatrix} x_1 \\ \text{---} \\ x_2 \end{bmatrix} +
\begin{bmatrix} 0 & | & 0 & 0 \\ 1 & | & 0 & 0 \\ \text{---} & + & \text{---} & \\ 0 & | & 1 & 0 \\ 0 & | & 0 & 1 \end{bmatrix}
\begin{bmatrix} u_1 \\ \text{---} \\ u_2 \end{bmatrix}
$$

$$
\begin{bmatrix} y_1 \\ \text{---} \\ y_2 \end{bmatrix} =
\begin{bmatrix} 1 & 0 & 0 & 0 \\ 0 & 1 & 0 & 0 \\ 0 & 0 & 1 & 0 \end{bmatrix}
\begin{bmatrix} x_1 \\ \text{---} \\ x_2 \end{bmatrix}
$$

With a decentralized feedback control law $K = \text{block}\ (K_{11},K_{22})$, the closed-loop matrix is given by :

$$
A + BKC =
\begin{bmatrix}
1 & 0 & 0 & 1 \\
k_{11} & k_{12} & 0 & 0 \\
1 & 0 & k_{21} & 1 \\
0 & 0 & k_{22} & 0
\end{bmatrix}
$$

and the system has a fixed mode at the origin. Unfortunately, since :

$$\left\| \, A_{11}^{\,-1} \right\|^{\,-1} = \frac{1}{\max\{\,1,\ (\frac{1}{k_{12}} + \frac{k_{11}}{k_{12}})\,\}} \qquad <= 1$$

$$\left\| \, A_{22}^{\,-1} \right\|^{\,-1} = \frac{1}{\max\{\frac{1}{k_{22}}\ ,\ (1 + \frac{k_{21}}{k_{22}})\}} \qquad = 1$$

$\overline{S} = \{1,2\}$ is not a proper subset of $\{1,2\}$ and the procedure cannot be applied.

5.2.4. – Approach based on the system modes sensitivity (TAR-84) (TAR-85)

The procedure proposed in this paragraph is based on the characterization of fixed modes using their sensitivity, which was presented in Paragraph 2.4.2, chapter II and applied to their evaluation.

Given a prespecified structural constraint on the control, a fixed mode is a closed-loop eigenvalue which remains insensible to any variations of the nonzero entries of the feedback matrix (see Definition 2.7). From the definition of the sensitivity matrix SK_r of a mode λ_r (see Definition 2.8), it was stated that λ_r is a fixed mode if and only if SK_r is identically zero (see Theorem 2.6). If the control is not subjected to any structural constraint (i.e., full feedback is allowed), then the nonzero entries of the sensitivity matrix SK_r indicate the elements of the feedback matrix which may affect the corresponding mode λ_r. The following theorem comes :

Theorem 5.1 (TAR-85). λ_r is not a fixed mode of the system with respect to a feedback pattern $K \in \Omega_F$ if and only if at least one element of the set K_{λ_r} appears in the feedback matrix K, where K_{λ_r} is defined by :

$$K_{\lambda_r} = \{k_{ij} \, / \, (sk_r)_{ij} \neq 0 \, \}$$

and $(sk_r)_{ij}$ are the entries of the sensitivity matrix SK_r.

This theorem provides a simple way to determine the set of supplementary feedback links which performs the elimination of the fixed modes of the system. Consider that the system has q fixed modes, then the set of supplementary feedback links K* must satisfy :

$$\text{Card } (K^* \cap K_{\lambda_k}) \geqslant 1 \quad (k=1,\ldots,q)$$

where $K_{\lambda_k} = \{k_{ij} \; / \; (SK_k)_{ij} \neq 0\}$ and SK_k is the sensitivity matrix of the mode λ_k evaluated with respect to a full feedback.

Remark 5.1. We recall that the evaluation of the sensitivity matrix of a multiple mode is difficult to perform (see § 2.4.2). This approach should thus be limited to systems with simple modes.

Note that K^* can generally be determined without taking into account all the sets K_{λ_k}. Indeed, if $K_{\lambda_i} \subset K_{\lambda_j}$, $i \neq j$ and Card $(K^* \cap K_{\lambda_i}) \geqslant 1$, it follows that Card $(K^* \cap K_{\lambda_j}) \geqslant 1$ so that K_{λ_j} can be suppressed. $\bar{q} \leqslant q$ sets need to be considered instead of q.

The problem is thus the following :

Problem 5.1. Find K^* such that :

$$\text{Card } (K^* \cap K_{\lambda_i}) \geqslant 1 \quad (i=1,\ldots,\bar{q} < q)$$

If we consider that a different cost is associated with every feedback link, it is clear that our interest is to determine K^*, solution of Problem 5.1, and minimizing the total cost resulting from the feedback links involved in K^*.

Consider the set of elements constituted by the union of all retained sets K_{λ_i} :

$$Z = \sum_{i=1}^{q} K_{\lambda_i}$$

$$\text{Card } Z = z \leqslant m \times r$$

Remark 5.2. Every element of Z represents a feedback link between an output and an input. So far, the notation was k_{ij} for a feedback link between the output j and the input i. For convenience, the elements of Z are renamed z_i, $(i=1,\ldots,z)$.

Associate a cost $c_i \geqslant 0$ with every feedback link z_i of Z and define the following boolean vector :

$$W = (w_1, \ldots, w_r)'$$

with

$$w_i = \begin{cases} 1 & \text{if } z_i \in K^* \\ \\ 0 & \text{otherwise} \end{cases}$$

Define also the following matrix :

$$L = (l_{ij}) \begin{smallmatrix} i=1,\ldots,\bar{q} \\ j=1,\ldots, z \end{smallmatrix} \quad \text{with} \quad l_{ij} = \begin{cases} 1 & \text{if} \quad z_i \quad K_{\lambda_i} \\ \\ 0 & \text{otherwise} \end{cases}$$

The problem of finding the minimum information pattern K^* such that the system has no fixed modes can thus be formulated by the following boolean linear program :

Problem 5.2. $\min \sum\limits_{j=1}^{z} c_j w_j$

under $\sum\limits_{j=1}^{z} l_{ij} w_j \geqslant 1 \qquad (i=1,\ldots,\bar{q})$

Now, it is interesting to notice that Problem 5.2 appears in the form of the well-known "covering set problem" of graph theory which can be reformulated in terms of graphs as follows :

Consider the unidirectional graph $G = [\, Z, K_s, \Delta \,]$ where :

$Z = \{z_1, \ldots, z_z\}$
$K_\lambda = \{K_{\lambda_1}, \ldots, K_{\lambda_q}\}$: set of parts of Z.

Δ : univoc application from K_λ to Z
$\Delta(z_i) = \{K_{\lambda_j} \, / \, z_i \in K_{\lambda_j}\} \quad (i=1,\ldots,z)$.

and the costs c_i associated with each vertex z_i, $(i=1,\ldots,z)$.

Problem 5.2 is thus brought back to the following covering set problem :

Problem 5.3. Find $H \subset Z \, / \, \bigcup\limits_{z_i \in H} \Delta(z_i) = K_\lambda$

minimizing $\sum\limits_{z_i \in H} c_i$

This problem can be solved by using any of the following algorithms existing in the literature :

- Method of the covering set (KAU-68) (ROY-70)
- Branch and Bound procedure (KAU-68) (ROY-70)
- Gomory's method (KAU-68)
- Thiriez's method (THI-71)

Example 5.4. Consider again the example 3.12 in which the system has two fixed modes : $\Lambda = \{\lambda_1 = 1, \lambda_2 = 2\}$. The associated sensitivity matrices with respect to full feedback are given by

$$SK_1 = \begin{bmatrix} 0 & 0 & -1/3 \\ 0 & 0 & -1/3 \\ 0 & 0 & 0 \end{bmatrix} \qquad SK_2 = \begin{bmatrix} 0 & 0 & 0 \\ 1/3 & 0 & 1/6 \\ 0 & 0 & 0 \end{bmatrix}$$

Then, we have :

$$K_{\lambda_1} = \{k_{13}, k_{23}\}$$

$$K_{\lambda_2} = \{k_{21}, k_{23}\}$$

and $Z = \{k_{13}, k_{23}, k_{21}\} = \{z_1, z_2, z_3\}$

$$L = \begin{bmatrix} 1 & 1 & 0 \\ 0 & 1 & 1 \end{bmatrix}$$

In this example, the boolean linear program to solve takes the following form :

1. All the costs are equal to 1, $c_i = 1$ (i=1,2,3) (minimization with respect to the number of feedback links) :

under
$$\begin{aligned} &\min (w_1 + w_2 + w_3) \\ &w_1 + w_2 \geqslant 1 \\ &w_2 + w_3 \geqslant 1 \end{aligned} \qquad w_i = \begin{cases} 1 \\ 0 \end{cases} \qquad (i = 1,2,3)$$

The solution is $w = (0\ 1\ 0)'$ and the addition of k_{23} is sufficient to eliminate the fixed modes.

2. The costs are given by $c_1 = 1$, $c_2 = 3$, $c_3 = 2$:

$$\min\ (w_1 + 3w_2 + 2w_3)$$

$$\text{under} \qquad \begin{array}{l} w_1 + w_2 \geqslant 1 \\ \\ w_2 + w_3 \geqslant 1 \end{array} \qquad w_i = \begin{cases} 1 \\ \\ 0 \end{cases} \qquad (i = 1,2,3)$$

In this case the program gives two solutions :

$$W = (\ 1\ 0\ 1\ )' \text{ corresponding to the elements } k_{13} \text{ and } k_{21}$$
$$W = (\ 0\ 1\ 0\ )' \text{ corresponding to the element } k_{23}$$

<u>Remark 5.3.</u> It is interesting to notice that this approach can also be used in the general case for which a prespecified structure for the control is not imposed. The minimal feedback control structure can thus be obtained by evaluating the sets K for every mode of the system and applying the same optimization procedure.

5.2.5 - Specified approach for structurally fixed modes of type (i) (TAR-85) (TRA-84b)

This paragraph concerns only the systems with structurally fixed modes of type (i) and characterizes the set of sufficient feedback links which must be added to a prespecified feedback pattern in order to eliminate them. This characterization is based on the algebraic characterization of fixed modes of type (i) given by Sezer and Siljak (SEZ-81 a) (see § 3.5.3a).

5.2.5a. - Use of Sezer and Siljak characterization

Sezer and Siljak (SEZ-81a) showed that the state space representation of a system with structurally fixed modes of type (i) can be put in the following special form :

$$\begin{bmatrix} \dot{X}_1 \\ \dot{X}_2 \\ \dot{X}_3 \end{bmatrix} = \begin{bmatrix} A_{11} & 0 & 0 \\ A_{21} & A_{22} & 0 \\ A_{31} & A_{32} & A_{33} \end{bmatrix} \begin{bmatrix} X_1 \\ X_2 \\ X_3 \end{bmatrix} + \begin{bmatrix} B_\alpha^1 & 0 \\ B_\alpha^2 & 0 \\ B_\alpha^3 & B_\beta^3 \end{bmatrix} \begin{bmatrix} U_\alpha \\ U_\beta \end{bmatrix}$$

$$\begin{bmatrix} Y_\alpha \\ Y_\beta \end{bmatrix} = \begin{bmatrix} C_\alpha^1 & 0 & 0 \\ C_\beta^1 & C_\beta^2 & C_\beta^3 \end{bmatrix} X$$

$$(5.2.6)$$

where the control and observation stations are partitioned in two aggregated stations α and β .

The fixed modes with respect to the feedback pattern :

$$K = \text{block-diag.} \; (K_\alpha, \; K_\beta) \tag{5.2.7}$$

are the eigenvalues of the submatrix A_{22}. These modes (see Chapter III) are simultaneously uncontrollable by one of the aggregated stations, here β , and inobservable by the other one, here α. Since the system is supposed to be globally controllable and observable, the fixed modes are controllable by the aggregated station α and observable by the aggregated station β. Consequently, the set of feedback links whose addition is sufficient to eliminate the fixed modes is given by :

$$K_{\alpha\beta} = \{ k_{ij} \; / \; ij \; \text{such that} \; u_i \subset U_\alpha \; \text{and} \; y_j \subset Y_\beta \} \tag{5.2.8}$$

and the feedback matrix becomes :

$$K' = \begin{bmatrix} K_\alpha & \vdots & K_{\alpha\beta} \\ ----&+&---- \\ 0 & \vdots & K_\beta \end{bmatrix} \tag{5.2.9}$$

If the structure of the control matrix at station α and of the observation matrix at station β are taken into account, it comes that the sufficient set can be reduced as follows :

Theorem 5.2. Given the system (5.2.6) with structurally fixed modes of type (i), the set of sufficient supplementary links is given by :

$$K_{suf} = \{ k_{ij} \; / \; i \in I, \; j \in J \}$$

where I (J) is the set of indices of the columns (rows) which are not identically zero in the matrices B* (C*), with :

$$B* = \begin{bmatrix} B_\alpha^1 \\ B_\alpha^2 \end{bmatrix} \qquad \text{and} \qquad C* = [C_\beta^2 \quad C_\beta^3]$$

Remark 5.4. When the control (5.2.7) is applied to the system (5.2.6), the closed-loop dynamic matrix takes the following form :

$$
D = \begin{bmatrix} \bar{A}_{11} + B_\alpha^1 \, K_\alpha \, C_\alpha^1 & \vdots & 0 & 0 \\ A_{32} + B_\alpha^2 \, K_\alpha \, C_\alpha^1 & \vdots & A_{22} & 0 \\ A_{31} + B_\alpha^3 \, K_\alpha \, C_\alpha^1 + B_\beta^3 \, K_\beta \, C_\beta^1 & \vdots & B_\beta^3 \, K_\beta \, C_\beta^2 & B_\beta^3 \, K_\beta \, C_\beta^3 \end{bmatrix}
$$

$$
= \begin{bmatrix} D_{11} & D_{12} \\ D_{21} & D_{22} \end{bmatrix} \qquad\qquad (5.2.10)
$$

which has the same block-triangular form as the open-loop dynamic matrix A and where the block A_{22} is not affected by the control : it results that the eigenvalues of A_{22} are fixed modes.

It is clear that the fixed modes can be eliminated by any control feedback which destroys the block-triangular structure of D by affecting the block D_{12}.

Now, consider the feedback matrix K' in (5.2.9). The closed-loop dynamic matrix is :

$$
D' = D + \begin{bmatrix} B_\alpha^1 \, K_{\alpha\beta} \, C_\beta^1 & \vdots & B_\alpha^1 \, K_{\alpha\beta} \, C_\beta^2 & \vdots & B_\alpha^1 \, K_{\alpha\beta} \, C_\beta^3 \\ B_\alpha^2 \, K_{\alpha\beta} \, C_\beta^1 & \vdots & B_\alpha^2 \, K_{\alpha\beta} \, C_\beta^2 & \vdots & B_\alpha^2 \, K_{\alpha\beta} \, C_\beta^3 \\ B_\alpha^3 \, K_{\alpha\beta} \, C_\beta^1 & \vdots & B_\alpha^3 \, K_{\alpha\beta} \, C_\beta^2 & \vdots & B_\alpha^3 \, K_{\alpha\beta} \, C_\beta^3 \end{bmatrix} = \begin{bmatrix} D'_{11} & \vdots & D'_{12} \\ D'_{21} & \vdots & D'_{22} \end{bmatrix}
$$

with :

$$
D'_{12} = \begin{bmatrix} B_\alpha^1 \, K_{\alpha\beta} \, C_\beta^2 & \vdots & B_\alpha^1 \, K_{\alpha\beta} \, C_\beta^3 \\ \cdots\cdots\cdots\cdots\cdots\cdots & \vdots & \cdots\cdots\cdots\cdots\cdots \\ A_{22} + B_\alpha^2 \, K_{\alpha\beta} \, C_\beta^2 & \vdots & B_\alpha^2 \, K_{\alpha\beta} \, C_\beta^3 \end{bmatrix}
$$

where it appears that the block-triangular structure has been destroyed. $K_{\alpha\beta}$ is therefore a sufficient set of feedback links to eliminate the fixed modes.

To show that $K_{\alpha\beta}$ can be reduced to K_{suf}, note that D'_{12} can be written as :

$$D'_{12} = D_{12} = \begin{bmatrix} B^1_\alpha \\ B^2_\alpha \end{bmatrix} K_{\alpha\beta} \ [C^2_\beta \ C^3_\beta] = D_{12} + B* \ K_{\alpha\beta} \ C*$$

$$(5.2.11)$$

$$D'_{12} = D_{12} + \sum_{i,j} (b^i)* \ k_{ij} \ (c_j)*$$

where $(b^i)*$ is the i-th column of $B*$ and $(c_j)*$ is the j-th row of $C*$ and $k_{ij} \in K_{\alpha\beta}$. The expression (5.2.11) shows that if $(b^i)*$ or $(c_j)*$ are identically zero, then k_{ij} does not affect D'_{12} and can be eliminated. The remaining k_{ij}'s are those specified in K_{suf}.

Remark 5.5.
1. Note that the set K_{suf} is not empty. Indeed, since the system is globally controllable and observable, the reachability conditions impose that $B^1_\alpha \neq 0$ and $C^3_\beta \neq 0$.

2. If there is no prespecified control structure, it can be shown by the same approach that the feedback structure :

$$K'' = \begin{bmatrix} 0 & \vdots & K_{\alpha\beta} \\ \hdashline K_{\beta\alpha} & \vdots & 0 \end{bmatrix}$$

allows to avoid structurally fixed modes of type (i). In this structure, $K_{\alpha\beta}$ is given by Theorem 5.2 and $K_{\beta\alpha}$ by :

$$K_{\beta\alpha} = \{k_{ij} \ / \ ij \text{ such that } u_i \in U_\beta \text{ and } y_j \in Y_\alpha\}$$

$K_{\beta\alpha}$ can be identically zero in the particular case for which U_α controls the whole space and Y_β observes the whole space.

Example 5.5. Consider the Example 3.12 where $a_{55}=2$ is changed by 4 (which avoids the existence of a non structurally fixed mode at 2). Given the following permutation matrix :

$$P = \begin{bmatrix} 1 & 0 & 0 & 0 & 0 \\ 0 & 0 & 0 & 0 & 1 \\ 0 & 0 & 0 & 1 & 0 \\ 0 & 0 & 1 & 0 & 0 \\ 0 & 1 & 0 & 0 & 0 \end{bmatrix}$$

the system takes the form :

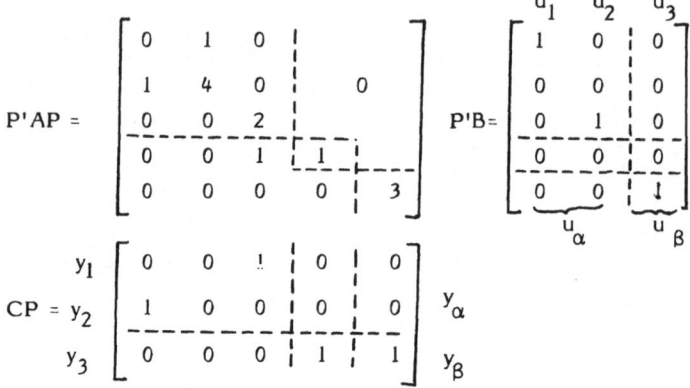

and the system has a structurally fixed mode of type (i) at 1 with respect to the decentralized control :

$$K = block\left[k_{11}, k_{22}, k_{33}\right]$$

From Theorem 5.2, we obtain the sets : $I = \{1,2\}$ and $J=\{3\}$ and, consequently $K_{suf} = \{k_{13}, k_{23}\}$. Then :

$$K'=K+K_{suf}=\begin{bmatrix} k_{11} & 0 & k_{13} \\ 0 & k_{22} & k_{23} \\ 0 & 0 & k_{33} \end{bmatrix} \qquad K''=\begin{bmatrix} 0 & 0 & k_{13} \\ 0 & 0 & k_{23} \\ k_{31} & k_{32} & 0 \end{bmatrix}$$

both guaranty the absence of fixed modes and the choice can then be made with respect to the number of feedback links for example.

5.2.5b. - Use of the sensitivity characterization to determine the aggregated stations α and β

The preceding paragraph showed that K_{suf} can be obtained in a very simple way when the state space representation of the system has the form (5.2.6) given by Sezer and Siljak' characterization. Unfortunately, in the general case, systems with structurally fixed modes of type (i) do not appear in this form and the above procedure cannot directly be applied. This difficulty can be encompassed by using the characterization of structurally fixed modes based on their sensitivity (TAR-84) (see § 3.5.4). Indeed, the structural sensitivity matrix (see Definition 3.9) allows the determination of the entries of the dynamic matrix A from which the fixed modes result ; i.e. the entries of the block A_{22} in (5.2.6). This comes from the fact that the eigenvalues of A_{22} are unsensible to any variations of the parameters except to thoses belonging to A_{22} itself. Consequently, the states X_2 corresponding to the fixed modes can be determinated. Taking into account that u_i belongs to the aggregated station U_α if it reaches X_2 ant that y_j belongs to the aggegated station Y_β if it can be reached by X_2, we can determine the "minimal" aggregated stations U_α^m and Y_β^m by using the reachability matrix R (see § 1.2) of the system which has the following form :

$$R = \begin{array}{c} \begin{array}{ccc} X & U & Y \end{array} \\ \left[\begin{array}{ccc} E & F & 0 \\ 0 & 0 & 0 \\ G & H & 0 \end{array} \right] \begin{array}{c} X \\ U \\ Y \end{array} \end{array} \qquad (5.2.12)$$

The reason why we do not obtain the aggregated stations themselves, U_α and Y_β, but rather the so-called "minimal" aggregated stations, U_α^m and Y_β^m, is that a control (observation) variable may belong to U_α (Y_β) without reaching (being reached by) X_2 because of the reachability patterns of the other state variables. Nevertheless, the set K'_{suf} obtained by applying Theorem 5.2 with U_α^m and Y_β^m is sufficient to eliminate the fixed modes since it guaranties that X_2 is involved in a loop. This approach gives thus a better solution since the number of feedback links is reduced $(K'_{suf} \subset K_{suf})$.

U_α^m and Y_β^m can be determined by the following algorithm :

Algorithm 5.1 (TAR-84).
1. Consider the set of structurally fixed modes of type (i) of the systems (C,A,B) :
$\Lambda_{S1} = \{\lambda_i, (i=1,\ldots,r).\}$

2. Determine the structural sensitivity matrix corresponding to the set of fixed modes Λ_{S1} :

$$SS = SS_1 + \ldots + SS_i + \ldots + SS_r$$

where SS_i is the structural sensitivity matrix of the mode λ_i Λ_{S1} and "+" denotes the "logic OR" operator.

3. Determine the set of state variables X_2 corresponding to the fixed modes :

$x_i \in X_2$ if there exists at least one nonzero entry in the row or column i of SS

4. Determine the reachability matrix of the system (C, A, B) (see § 1.2).

5. $U_\alpha^m = \{u_j \ / \ \text{there exists i such that } x_i \in X_2 \text{ and } f_{ij} = 1\}$

6. $Y_\beta^m = \{y_j \ / \ \text{there exists i such that } x_i \in X_2 \text{ and } g_{ij} = 1\}$

7. The set of sufficient supplementary links is given by :

$$K'_{suf} = \{k_{ij} \ / \ u_i \in U_\alpha^m \text{ and } y_j \in Y_\beta^m\}$$

Example 5.6 : Consider again the Example 3.12 with the same modification as in Example 5.5. The system has a structurally fixed mode of type (i) at $\lambda_1 = 1$.

1. $\Lambda_{S1} = \{\lambda_1 = 1\}$

2. The structural sensitivity matrix corresponding to $\lambda_1 = 1$ is :

$$SS(\lambda_1 = 1) = \begin{bmatrix} 0 & 0 & 0 & 0 & 0 \\ 0 & 0 & 0 & 0 & 0 \\ 0 & 0 & 1 & 0 & 0 \\ 0 & 0 & 0 & 0 & 0 \\ 0 & 0 & 0 & 0 & 0 \end{bmatrix}$$

3. $X_2 = \{x_3\}$

4. In the reachability matrix the row and column corresponding to x_3 in F and G are :

$$
\begin{array}{c}
\begin{array}{ccc} u_1 & u_2 & u_3 \end{array} \\
x_3 \begin{bmatrix} 0 & 1 & 0 \end{bmatrix} \leftarrow \text{3-rd row of F}
\end{array}
$$

$$
\begin{array}{c}
y_1 \\ y_2 \\ y_3
\end{array}
\begin{bmatrix} 0 \\ 0 \\ 1 \end{bmatrix}
$$

$\underset{\underline{\hspace{0.5cm}}}{\uparrow}$ 3-rd column of G

5. $U_\alpha^m = \{u_2\}$

6. $Y_\beta^m = \{y_3\}$

7. $K'_{suf} = \{k_{23}\}$. This leads to the following feedback structures ;

$$K' = \begin{bmatrix} k_{11} & 0 & 0 \\ 0 & k_{22} & k_{23} \\ 0 & 0 & k_{33} \end{bmatrix} \qquad K'' = \begin{bmatrix} 0 & 0 & 0 \\ 0 & 0 & k_{23} \\ k_{31} & k_{32} & 0 \end{bmatrix}$$

__Remark 5.6.__ Note that if one wants to determine the aggregated stations U_α and Y_β (instead of (U_α^m and Y_β^m), this can be performed by replacing the reachability matrix of the open-loop system (C,A,B) by the one of the closed-loop system (C,A+BKC,B), where K is taken as the prespecified control structure, and applying then steps 5 and 6 of the algorithm.

We want to point out that the addition of the feedback links determined in K_{suf} (K'_{suf}) provides a sufficient condition to eliminate the structurally fixed modes of type (i). However, some of them may be redundant and therefore unnecessary - the above approach does not give any information at this purpose. The interest of the procedure can be viewed in the easy way K_{suf} is determined once the system is put in form (5.2.6).

5.2.6. - Concluding remarks

This paragraph presents different approaches to eliminate fixed modes based on the idea that the structural constraints must be relaxed.

Wang and Davison approach (WAN-78a) is not convenient for large scale systems because of the high number of possibilities which must be checked before obtaining the solution. Armentano and Singh approach (ARM-82) is limited to interconnected systems. It provides only a rough solution in the sense that the solution is based on a sufficient condition to eliminate fixed modes and is therefore suboptimal. This is also the case of the procedure proposed in paragraph 5.2.5 for structurally fixed modes of type (i). The only approach including a real optimization procedure is the one based on the mode sensitivity and presented in paragraph 5.2.4 (TAR-84). But since it requires the calculation of sensitivity matrices, it can be applied only when the system has simple modes.

5.3. - CHOICE OF MINIMAL CONTROL STRUCTURES

5.3.1. - Preliminaries

The approaches presented in this section deal with systems for which a prespecified control structure is not a priori advantageous. This situation occurs either when no partitioning of the input and output arises from physical considerations (like geographical distance) or when the costs associated with local feedbacks are in the same range as those associated with feedback links between different stations.

In this cases, the problem is thus to determine the feedback pattern for which the system has no fixed modes (i.e., such that pole assignment is possible with a dynamic control law in accordance with the specified structure(s)) and minimizing a cost criterion based on the number of feedback links or the sum of their associated costs.

As was pointed out in the general introduction of this chapter, this problem is more general than the one stated with a prespecified structure. Indeed, this latter problem can be formulated in the same way by setting to zero the costs of the feedback links which are involved in the initial structure.

5.3.2. – Senning 's approach (SEN-79)

Senning 's approach is the only one which is not based on one of the characterizations of fixed modes given in Chapter 3. The problem is rather considered in the framework of optimal control theory for linear systems with a quadratic criterion.

Consider the class of systems in (5.2.1) where the input and output are partitioned in several stations and assume that no feedback pattern seems a priori advantageous. The problem is formulated in terms of the determination of a "feasibly decentralized" control.

Definition 5.1 (SEN-79). A control structure is said to be feasibly decentralized if the system is stabilizable with this control structure and the cost of information is minimal.

The problem is stated in such a way that two problems are solved simultaneously :

- the classic parametric optimization problem based on the traditional quadratic criterion for linear systems.
- the determination of an optimal control structure with respect to a criterion taking into account the partitioning of the system and the costs of the feedback links.

The solution provides a feasibly decentralized control in the form :

$$u_i = K_{ii} y_i + \sum_{\substack{j=1 \\ j \neq i}}^{S} K_{ij} y_j \qquad i=1,\ldots,S \qquad (5.3.1)$$

The extended optimization criterion is defined as follows :

$$\text{E.O.C.} : \int_0^t (x' \, Q \, x + \sum_{i=1}^{S} u_i' \, R_i \, u_i) \, dt + \sum_{i=1}^{S} m^2 \qquad (5.3.2)$$

with $Q > 0$, $R_i > 0$, $(i=1,\ldots,S)$.

The first term in the E.O.C. is the classic performance index (P.I) while the second one goes into structural considerations by a weighted measure of the non-local information. This measure is defined as the vector function norm of the non-local part of the control weighted by appropriated scalars w_{ij}, penalizing more or less the exchange of information from station j to station i.

$$m_i = \left\| \sum_{\substack{j=1 \\ j \neq i}}^{S} \gamma_{ij} \, u_{ij} \right\| = \left\| \sum_{\substack{j=1 \\ j \neq i}}^{S} \gamma_{ij} \, K_{ij} \, y_i \right\| = \left\| K_i \, \Gamma_i \quad Y \right\| \qquad (5.3.3)$$

where the norm of a matrix is defined as below :

$$\left\| M \right\|^2 = \text{tr} \int_0^t M'(t) \, M(t) \, dt$$

and with :

$$K_i = [K_{i1}, \ldots, K_{i,i-1}, \quad 0, \quad K_{i,i+1}, \ldots, K_{iS}] \qquad (5.3.4)$$

$$W_i = \begin{bmatrix} \gamma_{i1} \, I_{r_1} & & & & & \underline{0} \\ & \ddots & & & & \\ & & \gamma_{i,i-1} \, I_{r_{i-1}} & & & \\ & & & 0 & & \\ & & & & \gamma_{i,i+1} \, I_{r_{i+1}} & \\ & \underline{0} & & & & \ddots \\ & & & & & \gamma_{i,S} \, I_{r_S} \end{bmatrix}$$

$$(5.3.5)$$

I_{r_i} stands for the identity matrix of dimension $r_i \times r_i$.

Consequently, the E.O.C. becomes :

$$\text{E.O.C.} = \text{P.I} + \sum_{i=1}^{S} \left\| K_i \ \Gamma_i \ Y \right\|^2$$

and we have the following optimization task :

Find the optimal matrices K_1^* , ..., K_S^* such that :
E.O.C. $(K_1^*, ..., K_S^*)$ < E.O.C. $(K_1, ..., K_S)$
for all admissible matrices $K_1, ..., K_S$.

A necessary condition of optimality is the vanishing of the gradient of the criterion with respect to the feedback matrices K_i, (i=1,...,S) :

d E.O.C. / d K_i = 0   (i=1,...,S)

Senning gives the expression of this gradient in terms of two matrices P and X which both satisfy a Lyapunov equation and he provides the solution to the problem as follows :

<u>Theorem 5.3</u> (SEN-79). The optimal solution K_i, (i=1,...,S) satisfies the following equations :

1. $R_i \ K_i \ C \ X \ C' + K_i \ \Gamma_i \ C \ X \ C' \ \Gamma_i + B_i' \ P \ X \ C = 0$ (i=1,...,S)

2. $A_0' \ P + P \ A_0 + \Omega + C' \ \sum_{i=1}^{S} \ (K_i' \ R_i \ K_i + \Gamma_i \ K_i' \ K_i \ \Gamma_i) \ C = 0$

3. $A_0 \ X + X \ A_0' + X_0 = 0$

4. $A_0 = A + \sum_{i=1}^{S} \ B_i \ K_i \ C$

The value of the optimal extended criterion is given by :

E.O.C. $(K_1, ..., K_S)$ = tr $(P \ X_0)$
$X_0 = x_0 \ x_0'$ and $x_0 = x(t_0)$ is the initial state.

In the first step of his work, Senning shows that the control of a linear system with a dynamic compensator is equivalent to the control of an augmented system (including both the plant and the compensator dynamics) using static output feedback (for details, see (SEN-79)). The problem of determining a feasibly decentralized dynamic compensator can therefore be solved by applying Theorem 5.3 to the extended system, which makes this approach even more powerfull. This is illustraed in the following example :

<u>Example 5.7</u> (SEN-79). Consider the following system in the form (5.2.1) :

$$
A = \begin{bmatrix} -20 & 1 & -2 & 3 & -4 \\ 5 & 2 & 0 & -6 & 7 \\ -8 & 0 & 3 & 9 & -1 \\ 2 & -3 & 4 & 2 & 0 \\ -5 & 6 & -7 & 0 & 3 \end{bmatrix} \quad B_1 = \begin{bmatrix} 1 \\ 0 \\ 0 \\ 0 \\ 0 \end{bmatrix} \quad B_2 = \begin{bmatrix} 0 \\ 1 \\ 0 \\ 0 \\ 0 \end{bmatrix} \quad B_3 = \begin{bmatrix} 0 \\ 0 \\ 0 \\ 1 \\ 1 \end{bmatrix}
$$

$$
C_1 = \begin{bmatrix} 1 & 0 & 0 & 0 & 0 \end{bmatrix}
$$

$$
C_2 = \begin{bmatrix} 0 & 1 & 1 & 0 & 0 \end{bmatrix}
$$

$$
C_3 = \begin{bmatrix} 0 & 0 & 0 & 1 & 0 \end{bmatrix}
$$

The weightings for the state of the plant and those for the state of the dynamic compensators are choosen as :

$$
Q_{plant} = diag \ (100)
$$
$$
Q_{comp} = diag \ (1)
$$

and the weightings for the inputs to the plant and to the compensator as :

$$
R_i = diag \ (1), \ i=1,2,3
$$

The non-local information is weighted by a factor of 30, favorizing complete decentralized control.

The optimization yields the compensators as below :

$$
\begin{bmatrix} u_1 \\ \dot{z}_1 \\ u_2 \\ \dot{z}_2 \\ u_3 \\ \dot{z}_3 \end{bmatrix} = \begin{bmatrix} 81.6 & -56.6 & -1.76 & 0.13 & 2.06 & 0.62 \\ 146 & -107 & 0.75 & -0.12 & -1.02 & -0.31 \\ 0 & 0 & 552 & 0.19 & 0 & 0 \\ -0.32 & 0.23 & -1884 & 3.1 & 0 & 0 \\ 0.4 & -0.22 & -1.9 & -0.1 & -232 & -71.5 \\ 0.3 & -0.2 & 0.1 & 0 & 592 & 181 \end{bmatrix} \begin{bmatrix} y_1 \\ z_1 \\ y_2 \\ z_2 \\ y_3 \\ z_3 \end{bmatrix}
$$

The optimal compensator is not completely decentralized (i.e., stabilization cannot be achieved with a completely decentralized control structure) and its structure shows the following information pattern :

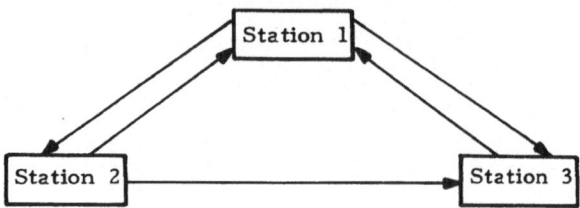

Senning 's approach is specially attractive because not only the optimal structure but also the optimal parameters are returned by the optimization. Moreover, it can apply either for the case of static output feedback either for the design of dynamic compensators by considering an augmented system. Another point of interest is that it does not require to check, in a first step, wether or not the system has unstable fixed modes for the favorized control structure (in our example, completely decentralized structure). This is possible because the optimization criterion both includes the quadratic performance index and the structural optimization criterion. It is clear that if the system has no unstable fixed modes for the favorized control structure, the optimal control will certainly have this same structure.

5.3.3. - Locatelli et al. approach : (LOC-77)

The approach of Locatelli et al. (LOC-77) is based on their frequency domain graphical characterization of fixed modes which was presented in Paragraph 3.6.2. Consider the class of linear time-invariant systems with simple modes represented by the general frequency domain model (3.4.1) where there is no partitioning in several stations. Then, Locatelli et al state the following problem : find the minimal set of feedback links $S^* \subset S$ (S as defined in (3.6.1)) such that every mode in the set Λ^* = $\{\lambda_1, \ldots, \lambda_h\}$ $\subset \sigma(A)$ can be arbitrarily assigned ; i.e., no mode in Λ^* is a fixed mode with respect to the control structure defined by S^*. The optimization is performed with respect to the following cost criterion :

$$R(S) = \sum_{(i,j) \in S} r_{i,j}$$

where $r_{i,j}$ is a cost associated with the allowed feedback link from output i to input j ($(i,j) \in S$). It is clear that this problem has a solution if and only if no mode in Λ^* is a fixed mode with respect to the control structure defined by S. The solution is then obtained by solving the following linear boolean program :

$$\min \quad \sum_{(i,j)\,\in\,L_{2S}} r_{i-m,j}\, q_{i,j}$$

under :

for g=1,...,h

(C_1^g) $\quad \sum_{(i,j)\,\in\,L_{1S}} v_{i,j}^g\, z_{i,j}^g \;\geqslant\; 1$

(C_2^g) $\quad \sum_{j/(i,j)\,\in\,L_S} z_{i,j}^g \;=\; \sum_{i/(j,i)\,\in\,L_S} z_{j,i}^g \qquad i \in V_S$

(C_3^g) $\quad z_{i,j}^g \;\leqslant\; q_{i,j} \qquad\qquad (i,j)\in L_{2S}$

with :

$$q_{i,j} = \begin{cases} 0 & \text{if } (i-m,j) \notin S^* \\[2mm] 1 & \text{if } (i-m,j) \in S^* \end{cases} \qquad (i,\,j)\in L_{2S}$$

and for g=1,...,h

$$z_{i,j}^g = \begin{cases} 1 & \text{if the edge } (i,j) \text{ is involded in a cycle whose transmittance has } \lambda_g \text{ as a} \\ & \text{pole and whose selection minimize the criterion.} \\[2mm] 0 & \text{otherwise.} \end{cases}$$

$$v_{i,j}^g = \begin{cases} 0 & \text{if for } (i,j) \in L_{1S}, \; w_{j-m,i}(\lambda_g) \neq 0, \; \infty \; (\lambda_g \text{ is neither a pole nor a} \\ & \text{zero}). \\[2mm] 1 & \text{if for } (i,j) \in L_{1S}, \; w_{j-m,i}(\lambda_g) = \infty \; (\lambda_g \text{ is a pole}). \\[2mm] -f & \text{if for } (i,j) \in L_{1S}, \; \lim\limits_{p \to \lambda_g} \dfrac{w_{j-m,i}(p)}{(p-\lambda_g)^f} = 0, \; \infty \end{cases}$$

$\qquad (\lambda_g$ is a zero of order f).

From the definition of $v_{i,j}^g$, the constraint (C_1^g) guaranties that at least one edge (i,j) whose transmittance has λ_g as a pole is selected and the constraint (C_2^g) guaranties that this edge belongs to a cycle (i.e., from Theorem 3.30, λ_g is not a fixed mode). The constraint (C_3^g) sets apart the selected edges which do not correspond to feedback links and do not affect the criterion.

The interest of this program is enforced by the fact that it can be used, with slight modifications, to provide the solution of several problems other than the one it was stated for :

1 - Determination of the fixed modes with respect to the control structure specified by S using a solvability test successively applied to $\Lambda^* = \{\lambda_i\}$, $\lambda_i \in \sigma(A)$.

2 - Minimization with respect to the number of feedback links by setting $r_{i,j} = 1$, (i,j)   S.

3 - Determination of the minimal feedback patterns avoiding fixed modes : by setting $S = \{(j,i) \ /(j=1,\ldots,r) \ ; \ (i=1,\ldots,m)\}$

$$\Lambda^* = \sigma(A)$$

4 - Determination of the minimal set of feedback links which must be added to an initial control structure specified by S_0 to eliminate the fixed modes by setting :

$$S = \{(j,i) \ / \ (j=1,\ldots,r) \ ; \ (i=1,\ldots,m)\}$$
$$r_{i,j} = 0 \quad \text{for } (j,i) \in S_0$$

<u>Example 5.8</u>. Consider the system in the example 3.13 which has a fixed mode at $\lambda_0 = -1$ for the decentralized control structure specified by $S_0 = \{(1,1);(2,2);(3,3)\}$. We want to determine the set of feedback links to add to this initial pattern in order to eliminate the fixed mode. The optimization criterion is taken as the number of supplementary links. The solution of this problem can be obtained by the previous program by setting :

$$\Lambda^* \in -1$$
$$S = \{(i,j) \ / \ i=1,2,3 \ ; \ j=1,2,3\}$$
$$r_{i,j} = 0 \quad \text{for } (i,j) \in S_0$$
$$r_{i,j} = 1 \quad \text{for } (i,j) \in S - S_0$$

The program to be solved is the following ;

$$\min q_{42} + q_{43} + q_{51} + q_{53} + q_{61} + q_{62}$$

under :

$$(C_1) \quad z_{16} \geqslant 1$$

$$(C_2) \begin{cases} z_{14} + z_{15} + z_{16} = z_{41} + z_{51} + z_{61} \\ z_{25} + z_{26} = z_{42} + z_{52} + z_{62} \\ z_{34} + z_{35} = z_{43} + z_{53} + z_{63} \\ z_{41} + z_{42} + z_{43} = z_{14} + z_{34} \\ z_{51} + z_{52} + z_{53} = z_{15} + z_{25} + z_{35} \\ z_{61} + z_{62} + z_{63} = z_{16} + z_{26} \end{cases}$$

(C_3) $z_{ij} \leqslant q_{ij}$ $i = 4, 5, 6$ $j = 1, 2, 3$

We obtain two optimal solutions :

$$S_1^* = \{(3,1)\}$$
$$S_2^* = \{(2,1)\}$$

which correspond to the following feedback structures :

$$K_1^* = \begin{bmatrix} k_{11} & 0 & k_{13} \\ 0 & k_{22} & 0 \\ 0 & 0 & k_{33} \end{bmatrix} \qquad K_2^* = \begin{bmatrix} k_{11} & k_{12} & 0 \\ 0 & k_{22} & 0 \\ 0 & 0 & k_{33} \end{bmatrix}$$

5.3.4. - Specified approaches for structurally fixed modes

The procedures presented in this paragraph are based on the graph-theoretic approaches leading to characterizations of structurally fixed modes. The problem is considered here from a structural point of view and the control structures returned by the subsequent procedures guaranty the absence of structurally fixed modes but they are not concerned by those fixed modes which arise from parameter value considerations.

We consider linear systems in the general form :

$$\begin{cases} \dot{x}(t) = A\, x(t) + B\, u(t) \\ y(t) = C\, x(t) \end{cases} \qquad\qquad (5.3.6)$$

where $x(t)$ R^n, $u(t)$ R^m, $y(t)$ R^r and A, B, C are real matrices of appropriate dimensions. We consider the general feedback pattern :

$$u(t) = K \ y \ (t) \tag{5.3.7}$$

Γ and Γ_K are the digraphs associated to the open loop and closed-loop systems, respectively, as defined in Paragraph 3.5.3b-1.

5.3.4.a. - Procedures based on the graphical characterization presented in Paragraph 3.5.3.b-1

The procedures presented below are all based on the graphical characterization of structurally fixed modes provided in (LIE-83) and (PIC-84) and formulated in Theorem 3.26.

1. - <u>Determination of the control structure to avoid structurally fixed modes of type (i)</u> : (TRA-87) : Structurally fixed modes of type (i) are characterized by condition (i) in Theorem 3.26. This condition is first expressed in terms of the concept of "loop-set" associated to a state vertex in Γ. In a second step, this formulation is used in two different ways to state the optimization problem whose solution provides the desired feedback control structure. In a first approach, the problem is brought back to the well-known "covering set problem" for which some efficient algorithms exist in the literature and which was already encountered in Paragraph 5.2.3. In a second approach, the problem is solved by using a successive "elimination" procedure.

<u>Definition 5.2</u> (TRA-87). The loop-set associated with the state vertex x_k is defined by :

$$K_{x_k} = \ k_{ij} \ / \ in \ reaches \ x_k, \ x_k \ reaches \ y_j$$

The loop-set associated to x_k is therefore the set of those feedback links which, implemented one at a time, are such that the vertex x_k is strongly connected to an input and an output vertex. Therefore, the loop-sets can easily be determined either from the graph itself either from the the reachability matrix of the system. With this definition, the following corollary is directly derived from Theorem 3.26 :

<u>Corollary 5.1</u> (TRA-87). Consider a feedback pattern in the form (5.3.7), then the following condition is sufficient for system (5.3.6) not to have structurally fixed modes of type (i) :

$$card \ (K^* \cap K_{x_k}) \geqslant 1 \quad (k=1,...,n)$$
where
$$K^* = \{ \ k_{ij} \ / \ k_{ij} \ is \ a \ nonzero \ entry \ of \ K \}$$

The condition expressed in Corollary 5.1 is the same as the one already derived in Theorem 5.1 of Paragraph 5.2.3. The only difference consists in the sets we are dealing with : in Theorem 5.1 we were concerned with the sets K_{λ_r} associated with the sensitivity matrix of the fixed modes λ_r, $(r=1,\ldots,q)$, while, in the present case, the sets are the loop-sets K_{x_k} associated to the state vertices x_k, $(k=1,\ldots,n)$. Taking into account the above remark, the problem to be solved is also Problem 5.1. The same cost criterion as in Paragraph 5.2.3 can be added to Problem 5.1 in the present case. This would lead to the boolean linear program formulated in Problem 5.2 which has been shown to be a well-known "covering set problem" of graph theory.

Example 5.9. Consider the system whose associated graph is the following :

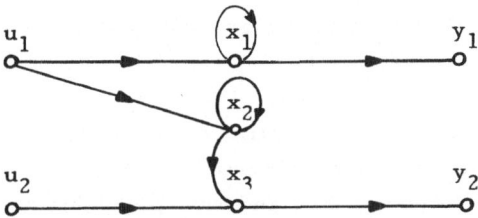

The loop-sets are given by :

$$K_{x_1} = \{k_{11}\}$$

$$K_{x_2} = \{k_{12}\}$$

$$K_{x_3} = \{k_{12}, k_{22}\}$$

Note that $K_{x_2} \subset K_{x_3}$, therefore K_{x_3} can be eliminated. If we consider an optimization criterion based on the number of feedback links, the optimization problem, which is trivial in this case, returns the solution :

$$K^* = \{k_{11}, k_{12}\}$$

Remark 5.7. The condition provided by Corollary 5.1 is only sufficient ; i.e., all the admissible solutions are not considered to find the optimal solution of Problem 5.1.

This remark is clarified by the following graphical configuration :

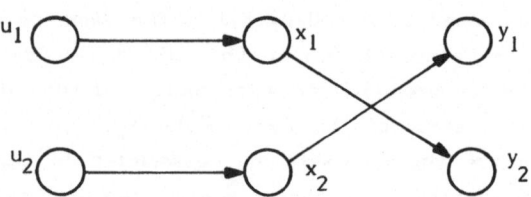

for which we have :

$$K_{x_1} = \{k_{21}\} \qquad K_{x_2} = \{k_{12}\}$$

and the unique solution for problem 5.1 is : $K^* = \{k_{21}, k_{12}\}$.

Nevertheless, if we consider the condition (i) of Theorem 3.26, the choice $\{k_{11}, k_{22}\}$ insures also that x_1 and x_2 are strongly connected to an input vertex and an output vertex in D_K.

The lack of necessity of the condition of Corollary 5.1 is due to the fact that, for some vertex x_k, condition (i) of Theorem 3.26 can be verified by the implementation of more than one feedback link and this possibility is not taken into account in our formulation.

Because of this restriction, this approach may provide a suboptimal solution. It is shown in (SEZ-83) that the number of necessary and sufficient feedback connections to satisfy Condition (i) of Theorem 3.26 is given by :

$$\mu_r = \max (u_r, y_r)$$

where u_r is the minimal number of inputs to reach every state vertex (input reachability) and y_r is the minimal number of outputs such that every state vertex reaches an output (output reachability). Therefore, μ_r can easily be determined by the inspection of the reachability matrix of the system.

In our formulation, a solution specifies α_r feedback connections, where α_r is the minimal number of input-output paths to cover the whole set of state vertices X. Therefore, a solution is optimal if and only if :

$$\alpha_r = \mu_r$$

Unfortunately, this general condition can be checked after obtaining K^* and α_r = card K^* ; i.e., after solving the optimization problem. However, the following conditions can be of help in certain cases :

- A solution is optimal if the number of loop-sets involved in the optimization is equal to μ_r. (This comes from the fact that one feedback link is sufficient to "cover" one loop-set ; i.e. $\alpha_r = \mu_r$).

- A solution is suboptimal if the number of independent loop-sets involved in the optimization is superior or equal to μ_r. (In this case, a number of feedback links superior or equal to μ_r is necessary to "cover" the independent loop-sets and some additional feedback links are required to "cover" the remaining ones ; i.e. $\alpha_r > \mu_r$).

In the case of Example 5.9, the solutions provided by the above procedure are optimal.

Though the solution may specify a suboptimal number of feedback links, this approach presents the advantage to be specially simple due to its formulation as a covering set problem. The degree of suboptimality of the solution is therefore compensated by the reduced efforts required by the procedure.

The second approach (TAR-85) presented below is restricted to the case for which the optimization criterion is the number of feedback links. The optimization problem is carried out starting from an initial non minimal set of feedback links and applying a successive elimination procedure which uses two rules.

Considering Condition (i) in Theorem 3.26, it is clear that a feedback link (y_j, u_i) such that there is no input-output path from u_i to y_j is not necessary to satisfy this condition. Consequently, we define an initial set of "usefull" feedback links as follows :

$$K_1 = \{k_{ij} \ / \ \text{a path from } u_i \text{ to } y_j \text{ exists in } \Gamma\} \qquad (5.3.8)$$

Obviously, this set is not minimal since some state vertices may belong to several input-output paths.

To every k_{ij} in K_1, associate the following boolean vector :

$$z_{ij} = \left[z_{ij}(1) \ldots z_{ij}(n) \right]'$$

$$z_{ij}(t) = \begin{cases} 1 & \text{if } k_{ij} \text{ is an element of the loop-set of } x_t, \ (t=1,\ldots,n) \\ \\ 0 & \text{otherwise} \end{cases}$$

and let us state the following definitions :

<u>Definition 5.3</u> (TAR-85). The state vertex x_t is said to be disjoint if and only if :

$z_{ij}(t) = 1$

$z_{ij}(t)$ AND $z_{qr}(t) = 0$ for all $qr \neq ij$ such that $k_{qr} \in K_1$

<u>Definition 5.4</u> (TAR-85). The vector z_{ij} is said to dominate the vector z_{qr} if :

$z_{ij}(t) = 1$ implies $z_{qr}(t) = 1$ for all $(t=1,\dots,n)$

As an extention of Definition 5.4, the set of vectors $z_{ij} \subset Z_1$ is said to dominate the set of vectors $z_{qr} \subset Z_2$ if :

$z_{ij}(t) = 1$ for some $z_{ij} \subset Z_1$ implies $z_{qr}(t) = 1$ for some $z_{qr} \subset Z_2$, for all $(t=1,\dots,n)$.

Using Definition 5.4 and Corollary 5.1, the following result comes :

<u>Corollary 5.2</u> - A set of feedback links K* is sufficient to guaranty the absence of structurally fixed modes of type (i) if :

K* = { k_{ij}/the set of vectors { z_{ij} } dominates the whole set of vectors associated to K_1 }

Therefore, this approach is concerned with determining the minimal set of vectors { z_{ij} } which dominates the whole set of vectors associated to K_1.

The two following rules are used in the elimination procedure which starts from K_1 to determine K* :

<u>Rule 1</u> : The feedback link $k_{ij} \in K_1$ can be eliminated from K_1 if its associated vector z_{ij} is dominated by at least one vector z_{qr}.

It is clear that if z_{qr} dominates z_{ij}, then every state vertex which belongs to the input-output path from u_i to y_j belongs also to the input-output path from u_q to y_r. Therefore, the presence of k_{qr} involves more state vertices in strongly connected components than the presence of k_{ij}.

<u>Rule 2</u> : The feedback link $k_{ij} \in K_1$ is necessary if for some $t \in \{1,\dots,n\}$, $z_{ij}(t) = 1$ and x_t is a disjoint state vertex.

If x_t is a disjoint vertex and $z_{ij}(t) = 1$, then k_{ij} is the only feedback link which, taken alone, makes x_t belong to a strongly connected component.

Using these rules, the following elimination procedure is proposed in (TAR-85) to determine K* :

1 - Using the digraph Γ associated to the system or its reachability matrix :

    1.1 - Determine the set K_1 defined in (5.3.8)

    1.2 - For every $K_{ij} \in K_1$ determine the associated vector z_{ij}.

2 - Set K* = { 0 }

3 - Determine the subset K_2 corresponding to the disjoint state vertices :

    $K_2 = \{k_{ij} \,/\, z_{ij}(t) = 1$ and x_t is a disjoint state vertex$\}$

    If $K_2 = 0$, go to 4, else :

    3.1 - Set K* = K* $\cup K_2$

    3.2 - If for all (t=1,...,n), there exists $z_{ij}(t) = 1$ for some $k_{ij} \in$ K*, go to 7

    3.3 - Set $z_{qr}(t) = z_{qr}(t)$ OR $z_{ij}(t)$, (t=1,...,n) for all ij $\neq$ qr such that k_{ij} K_2 and k_{qr} K_1.

    3.4 - $K_1 = \underset{K_1}{} (K_1 \cap K2)$

    3.5 - k = 1

    3.6 - $K_2 = 0$

4 - Determine $K_3 \subset K_1$ such that the set of k vectors associated to the k elements of K_3 dominates the set of remaining vectors associated to the elements in K_1.

    If $K_3 = 0$, go to 5.

    4.1 - K* = K* $\cup K_3$ (If K_3 is not unique, the solution K* is not unique). Go to 7.

5 - Determine $K_4 \subset K_1$ such that the vectors associated to the elements in K_4 are dominated by k vectors.

    If $K_4 = 0$, go to 6.

    5.1 - $K_1 = \complement_{K_1} (K_1 \cap K_4)$

6 - k = k + 1, go to 3.

7 - STOP : K* verifies Corollary 5.2.

Example 5.10 - Consider the same system as in the previous example 5.9. The set K_1 is given by : $K_1 = \{ k_{11}, k_{12}, k_{22}\}$ and the associated vectors are :

$$
\begin{array}{c}
z_{11} = \\
z_{12} = \\
z_{22} =
\end{array}
\begin{array}{ccc}
x_1 & x_2 & x_3 \\
\left[\begin{array}{ccc}
1 & 0 & 0 \\
0 & 1 & 1 \\
0 & 0 & 0
\end{array}\right]
\end{array}
$$

Since the state vertices x_1 and x_2 are disjoint, the procedure is trivial and gives the solution K* = $\{k_{11}, k_{12}\}$.

Note that the procedure does not provide all the minimal solutions included in K_1 since they are set aside in step 5. However, one could jump over step 5 if desired.

Of course, we obtain the same solution as in Example 5.9 where it was derived by using the first approach. In fact, the two approaches are based on the same optimization criterion and differ only by the way to proceed the optimization. Consequently the comments of Remark 5.3 still apply here.

2 - <u>Determination of the control structure to avoid any type of structurally fixed modes : a two step procedure</u> (TRA-87) - The first section was concerned by determining a control structure avoiding structurally fixed modes of type (i). This is sufficient in the case for which the system has no modes at the origin since it has been shown (see § 3.5, Chapter III) that structurally fixed modes of type (ii) are always located at the origin. In the other alternative, condition (ii) of Theroem 3.26 must be satisfied to guaranty the absence of structurally fixed modes of type (ii).

This approach consists in solving separately the problems corresponding to Condition (i) and Condition (ii). It is thus clear that the solutions obtained with this approach may not be optimal. One more time, this is compensated by the simplicity of the procedure which uses either of the two procedures derived in the last section.

In the following, the concept of cycle family and its width in a given a digraph it used. The reader is refered to Definitions 3.10 and 3.11.

Indeed, it is clear that Condition (ii) of Theorem 3.26 is satisfied if and only if there exists in Γ_K a cycle family of width $\geq$ n involving all the state vertices. The following procedure can thus be proposed :

1 - i=0
2 - Consider the digraph Γ.define $\Gamma_A=(X,E_A)$ and determine[1] the set $F_{n-i}=\{f_1,\ldots,f_{c_{n-i}}\}$ of cycle families of width n-i.
3 - If F_{n-i}, $\neq 0$ go to step 5.
4 - i=i+1, go to step 2.
5 - If i=0, go to step 8. Otherwise, select one $f_k \in F_{n-i}$ and consider the i vertices $\{x_{j_1},\ldots,x_{j_i}\}$ which are not involved in this cycle family.

---

[1]The determination of cycle families has been regarded as a standard problem of applied graph theory for many years. The well-known methods and algorithms for finding paths and cycles (LIA-69) (KRO-67) (RAO-69) may be comparatively easily adapted for computed-aided determination of cycle families of prescribed width.

6 - Apply either of the two procedures derived in the last section to a reduced problem, i.e. by taking only into account the state vertices determined at step 5, and adding a new constraint to guarantee disjoint cycles :

$$\text{min card } K^2$$
$$\text{card } (K^2 \cap K_i) \geq 1 \quad (i=1,\ldots,r)$$

$V\ i_1,\ i_2 \in \{1,\ldots s,\},\ w_{i_1} \cdot w_{i_2} = 0$ if s_{i_1} and s_{i_2} correspond respectively to k_{sv} and k_{1k} such that :

$$s = 1 \text{ and } v \neq k$$
or $\quad s \neq 1 \text{ and } v=k$

7 - Consider the digraph obtained from $\Gamma = (V,E)$ by adding the set of feedback edges resulting from step 6 and determine the set of cycle families of maximal width. If there is none of these cycle families involving all state vertices, go to step 5.

8 - Discard the state vertices whose loop-set contains a feedback edge belonging to K^2. Apply either of the two procedures derived in the last section to a reduced problem, i.e. by taking only into account the remaining state vertices. Let the solution be K^1.

9 - The global solution is given by : $K^* = K^1 \cup K^2$.

Example 5.11. Consider the same system as in Examples 5.9 and 5.10 which has a structural mode at the origin since the generic rank of its dynamic matrix is equal to n-1=2. The observation of its associated digraph Γ_A shows that $F_3=0$ and there is one cycle family of width 2. x_3 is the only vertex which is not involved.

Since $K_{x_3} = \{k_{12},\ k_{22}\}$ we obtain two solutions at step 6 :

$$K^2_1 = \{k_{12}\} \quad K^2_2 = \{k_{22}\}$$

which result in the two following cycle families of width 3 :

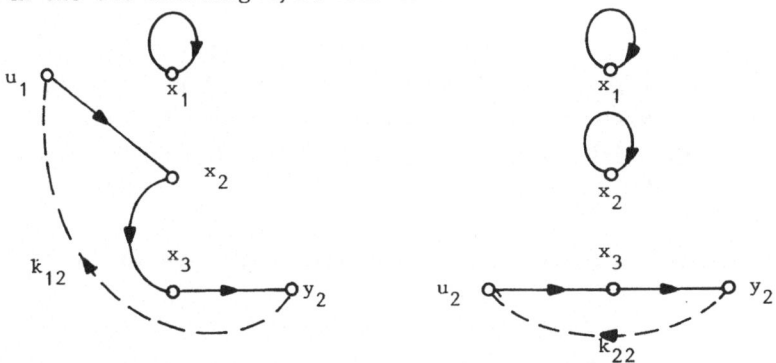

At this step, Condition (ii) of Theorem 3.26 is satisfied.

a) consider the solution $K_1^2 = \{k_{12}\}$. The reduced set of states for which the loop-sets do not contain k_{12} is $\{x_1\}$. In this case the solution is :

$$K_2^1 = \{k_{11}\}$$

and the global solution is : $K_1^2 = \{k_{11}, k_{12}\}$

b) consider the solution $K_2^2 = \{k_{22}\}$, the reduced set of states is in this case $\{x_1, x_2\}$. The solution is now $K_2^1 = \{k_{11}, k_{12}\}$ and the global solution is : $K_2^2 = \{k_{11}, k_{12}, k_{22}\}$ which is clearly worse than K_1^* since $K_1^* \subset K_2^*$.

Therefore, the solution K_1^* is retained.

Note that we obtain the same solution as in Examples 5.9 and 5.10 where only structurally fixed modes of type (i) were considered. This is due to the fact that the cycle family of width 2 contained in Γ_A is only composed of self-cycles. In one of the first procedures, when Condition (i) is satisfied for x_3, x_3 is involved in a cycle together with x_2. Because of our special configuration, we obtain a cycle family of width 3 composed by this cycle and the self-cycle at x_1. Therefore, Condition (ii) is also satisfied.

As a matter of fact, when a cycle family composed only by self-cycles exists in the set of cycle families of maximal width (Step 2), the general problem can be solved by using one of the first procedures as well. This alternative may be advantageous since the optimization task is performed in one step.

Remark 5.8. The degree of suboptimality of the solutions can be evaluated as it is shown below.

Define $U_1, \ldots, U_q$ $(Y_1, \ldots, Y_t)$ as the subsets of inputs (outputs) of minimal cardinality u_r (y_r) such that the system is input reachable (output reachable). Let B_{u_i} (i=1,...,q) $(C_{y_j}, j=1,\ldots,t)$ be the matrices composed by the columns of B (rows of C) corresponding to the inputs in U_i (outputs in Y_j) and define the integers d_i, (i=1,...,q) $(\delta_j, j=1,\ldots,t)$ as the generic rank deficiency of $(A \; B_{u_i})$ $\left(\begin{array}{c} A \\ C_{y_j} \end{array}\right)$. Define

also $d_m = \min_i d_i$ and $\delta_m = \min_j \delta_j$. Then, we have the following proposition :

Proposition 1 (TRA-87) : Given the structurally controllable, structurally observable system (5.3.6) the minimal number of feedback links such that system (5.3.6) has no structurally fixed modes is given by :

$$\mu = \max (u_r + d_m, y_r + \delta_m)$$

and the following corollary can be stated :

<u>Corollary 2</u> (TRA-87) : Given the structurally controllable and observable system (5.3.6) assume that $\mu_r = \max (u_r, y_r) = \max (m, p)$, then $\mu = \mu_r$.

It is clear that the global procedure generally requires a great amount of calculations. Moreover, the returned solution is not necessarily optimal. Therefore, this approach should better be restricted to systems without modes at the origin for which one of the procedures of the first section suffices to guaranty the absence of structurally fixed modes. For systems with modes at the origin, it will be more convenient to use one of the procedures presented in the following paragraph.

5.3.4.b. - Sezer 's procedure (SEZ-83)

Sezer 's procedure provides also the way to determine the optimal feedback pattern(s) (5.3.7) for which system (5.3.6) has no structurally fixed modes. In this approach, the optimization is proceeded with respect to the number of feedback links and it does not allow to consider different costs for the different feedback connections. It is a two step procedure which requires, first, to determine the "minimal essential" input sets and the "minimal essential" output sets as defined below :

<u>Definition 5.5</u> (SEZ-83). For any subset I (J) of the set $\{1,\ldots,m\}$ ($1,\ldots,r$), define the matrix B_I (C_J) as the submatrix of B (C) consisting of the columns (rows) with indices in I (J).

A subset of inputs (outputs) with indices I (J) is said to be <u>essential</u> if (A, B_I) $((C_J, A))$ is structurally controllable (observable), but not any $(A, B_{I'})$ $((C_{J'}, A))$ if $I' \subset I$ ($J' \subset J$).

The essential input (output) sets having a minimal number of inputs (outputs) are called minimal essential input (output) sets.

For the definitions of structural controllability and observability, we refer to Paragraph 1.3 and we remind that the system (C, A, B) is structurally controllable (observable) if and only if :

1 - (A, B) is input (output) reachable.

$$2 - gr \ (A \ B) = n \ (gr \begin{bmatrix} A \\ C \end{bmatrix} = n).$$

Given the system (5.3.6), the determination of a minimal input (output) set can be performed by using Reinschke 's procedure, which is presented below.

1 - <u>Reinschke 's procedure</u> (REI-81). Consider that a permutation of the rows and columns has been proceeded such that the matrix A of system (5.3.6) has the following block-triangular form :

$$\begin{bmatrix} A_{11} & & & \underline{0} \\ A_{21} & A_{22} & & \\ \vdots & & \ddots & \\ A_{N1} & \cdots\cdots\cdots & A_{NN} \end{bmatrix} \qquad (5.3.9)$$

where the diagonal blocks of A are irreducible matrices (corresponding to the strongly connected components in the digraph for which A is the adjacency matrix). Several algorithms exist in the literature (HAR-65) (KAU-68) (KEV-75) to put A in the above form.

Define $^z(A)$ as the submatrix consisting in the z last right columns of A and $(A)_z$ as the submatrix consisting of the z first rows of A.

The deficit of generic (structural) rank of $^z(A)$ and $(A)_z$ are denoted by ^zd and d_z, respectively, and are given by :

$$^zd = z - gr \ ^z(A)$$

$$(z=1,\ldots,n)$$

$$d_z = z - gr \ (A)_z$$

It is clear that $^nd = d_n = n - gr \ (A) = d$.

Assume that $^0d = d_0 = 0$. Then, we denote by $^1z,\ldots,^dz$ the d indices having the property $^zd = {}^{(z-1)}d + 1$, where z $\{1,\ldots,n\}$. The indices z_i, (i=1,\ldots,d), are defined in a similar way using d_z.

Example 5.12.

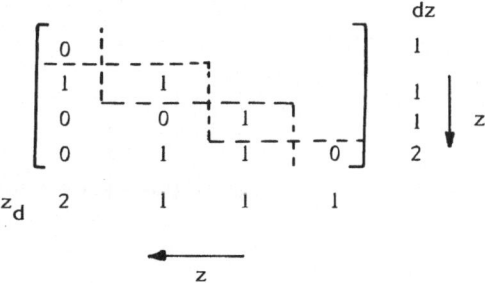

Therefore, we have d = 2 and :

$$^1z = 1, \quad ^2z = 4$$
$$z_1 = 1, \quad z_2 = 4$$

Using these definitions, Reinschke (REI-81) (REI-83) provides the following results :

Theorem 5.4. Given a matrix A in a block-triangular form where the diagonal blocks are irreducible, and with $gr(A) = n - d < n$, a minimal input matrix B which makes the pair (A,B) structurally controllable is of dimension n x d with :

1 - the d entries $b_{z_i,i}$ (i=1,...,d) are not allowed to be zero.
2 - the entries of a hyperrow of B must not be all zero if all the off-diagonal blocks of A corresponding to this hyperrow are identically zero.

If $gr(A) = n$, then the minimal input matrix B is n x 1 where the entries of the first hyperrow of B must not be all zero.

It is clear that Condition 1 implies $gr(A,B) = n$ and 2 implies that the pair (A,B) is input reachable.

Theorem 5.5. Given the same matrix A as in Theorem 5.4, a minimal output matrix C which makes the pair (C,A) structurally observable is of dimensions d x 1 with :

1 - the entries c_i, i_z (i=1,...,d) are not allowed to be zero.
2 - the entries of a hypercolumn of C must not be all zero if all the off-diagonal blocks of A corresponding to this hypercolumn are identically zero.

If $gr(A) = n$, then the minimal output matrix C is 1 x d where the entries of the last hypercolumn of C must not be all zero.

Example 5.13. For the matrix A as in Example 5.13, we obtain :

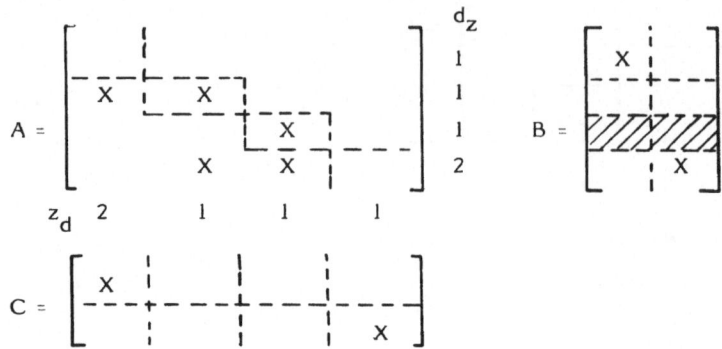

where X stands for a nonzero entry and where the shaded row of B must have at least one nonzero entry.

Note that the above procedure does not provide a unique solution.

2 - <u>Determination of the minimal control structure avoiding structurally fixed modes</u> (SEZ-83) - The following definitions are necessary to outline the procedure :

<u>Definition 5.6</u> (SEZ-83). Define the binary matrix F such that $f_{ij} = 1$ if and only if $k_{ij} = 0$ (F represents the structure of the feedback matrix).

1 - A structure F is said to be <u>favorable</u> if the system has no structurally fixed modes with respect to this feedback control structure.
2 - Given F^1 and F^2, F^1 is said to imply F^2 if $f_{ij}^1 = 1$ implies $f_{ij}^2 = 1$.
3 - A favorable structure F is said to be essential if there is no other favorable structure which implies F.
4 - Among all the essential favorable structures, the ones with minimal number of nonzero entries are said to be minimal.

We remind that a system (C,A,B) has no structurally fixed modes with respect to a control structure represented by F if and only if :

$$gr(M_F) = gr \begin{bmatrix} A & B & 0 \\ 0 & I_m & F \\ C & 0 & I_r \end{bmatrix} = n+m+r$$

and each state vertex in the digraph Γ_K belongs to a strongly connected component containing at least one edge corresponding to a feedback link.

In the following, we make the assumption that the system (5.3.6) is structurally controllable and observable, which implies that if $gr(A) = n - d$, then $d < \min(m,r)$.

Sezer stated the following result :

Theorem 5.6. Consider the sets of integers :

$$I = \{i_1, \ldots, i_k\} \qquad d < k < m$$

$$J = \{j_i, \ldots, j_q\} \qquad d < q < r$$

$$I' = \{i'_1, \ldots, i'_d\} \subset I$$

$$J' = \{j'_1, \ldots, j'_d\} \subset J$$

sucht that the system $S_{IJ} = (B_I, A, C_J)$ is structurally controllable and observable, and such that $gr(A, B_{I'}) = n$ and $gr\begin{bmatrix} A \\ C_{J'} \end{bmatrix} = n$.

If F is a structure such that $gr\ (F_{I'J'}) = d$ and such that $F_{I-I', J-J'}$ contains at least one nonzero entry in each row and column, then F is a favorable structure.

A favorable structure F satisfying Theorem 5.6 is not necessarily essential unless the sets I and J are chosen to be themselves essential. Moreover, they must be minimal if we want to obtain a minimal essential structure. Given the matrices B and C, Sezer proposes to use Reinschke's procedure in order to determine a minimal essential input set and a minimal essential output set.

Unfortunately, it will be shown in the next section that Reinschke's procedure is not perfectly adequate to this problem and that it fails in some cases. An alternative procedure will be proposed, which is specially appropriate for large scale systems since it proceeds in a sequential way.

Sezer provides thus the following procedure to determine a minimal essential structure :

1 - Determine a minimal input set I and a minimal output set J.

2 - Choose $I' \subset I$ and $J' \subset J$ such that $gr (A,B_{I'}) = n$ and

$$gr \begin{bmatrix} A \\ C_{J'} \end{bmatrix} = n$$

3 - Construct $F_{I'J'}$ such that it contains exactly d nonzero entries located in different rows and columns.

4 - Construct $F_{I-I',J-J'}$ such that it contains exactly max (k,q)- d nonzero entries located not to leave a zero row or column.

5 - Set all the other entries of F to 0.

From Theorem 5.6, F is a favorable structure. It is also essential since for some f_{ij} :

- if $i \in I-I'$ and $j \in J-J'$ then the loss of the feedback edge (y_j,u_i) leaves a strongly connected component in Γ_K without feedback edge.
- if $i \in I$ and $j \in J$, since I and J are minimal, the system without the input u_i and the output y_j is not structurally controllable neither observable and f_{ij} is necessary.

Moreover, F is minimal since I and J are themselves minimal. Note that the number of feedback links of a minimal structure is therefore equal to max(k,q), where k and q are the minimal number of inputs and outputs which guaranty structural controllability and observability to the system.

Example 5.14. Consider the following system :

$$A = \begin{bmatrix} 0 & 0 \\ 0 & 1 \end{bmatrix} \qquad B = \begin{bmatrix} 1 & 0 \\ 0 & 1 \end{bmatrix}$$

$$C = \begin{bmatrix} 1 & 0 \\ 0 & 1 \end{bmatrix}$$

We have $I = J = \{1,2\}$ and $I' = J' = \{1\}$, therefore Sezer 's procedure provides the unique solution F = diag. $(1,1)$.

It has to be noticed that :

$$F = \begin{bmatrix} 0 & 1 \\ 1 & 0 \end{bmatrix}$$

is also favorable and minimal. Consequently, this procedure does not provide all the solutions. The above example shows that a minimal, essential, favorable structure does not need to contain a set of d feedback links from y_j, j J' to u_i, i I'. This situation occurs when :

$$gr \begin{bmatrix} A & B_{I'} \\ C_{J'} & 0 \end{bmatrix} = n + \min\ (k,q)$$

in which case, $gr(M_F) = n+m+r$ for all F provided that $gr(F_{IJ}) = \min\ (k,q)$.

Example 5.15. Consider now the system described by the following structure :

$$A = \begin{bmatrix} X & & & \\ & X & & \\ & & 0 & \\ & & & 0 \end{bmatrix} \begin{bmatrix} 0 & 0 & X \\ 0 & X & 0 \\ X & 0 & 0 \\ 0 & X & 0 \end{bmatrix}$$

$$C = \begin{bmatrix} 0 & 0 & X & 0 \\ X & 0 & 0 & X \\ 0 & X & 0 & 0 \end{bmatrix}$$

We have $I = J = \{1,2,3\}$ and $I' = J' = \{1,2\}$. The above procedure provides the two following solutions :

$$F_1 = \begin{bmatrix} 1 & & \\ \hline & 1 & \\ \hline & & 1 \end{bmatrix} \qquad F_2 = \begin{bmatrix} & 1 & \\ \hline 1 & & \\ \hline & & 1 \end{bmatrix}$$

In this case, no feedback pattern F verifies $gr(M_F) = n+m+r$ unless $gr(F_{I'J'}) = d = 2$. Therefore, the procedure gives all the solutions provided all the minimal essential input and output sets have been detected.

Sezer's procedure itself does not require calculations. Nevertheless, it is based on Reinschke's procedure. Beside the fact that it fails in some cases (see the next section), it needs first, the block-triangularization of matrix A and then, the evaluation of the generic rank of 2n submatrices of A in order to determine the sets I,J which are necessary to initialise Sezer's procedure. One must also notice that this procedure performs the optimization with respect to the number of feedback links and it is not adequate in the case for which each feedback link has a different cost.

3 - <u>Sequential determination of the minimal essential input and output sets</u> (TRA-86b). The determination of the minimal input and output sets is the first step of Sezer's procedure. For this purpose, Sezer proposed to use Reinschke's procedure (presented in section 1).

However, this procedure was developed for solving the slightly different problem of determining a matrix B (C) with a minimal number of columns (rows) (minimal number of inputs (outputs)) such that the system is structurally controllable (observable). It was shown that d inputs (outputs) are sufficient (see Theorems 5.4 and 5.5). This result applies because the nonzero entries of B (C) can be arbitrarily located and can always be chosen such that the d inputs (outputs) required to fulfil the generic rank condition satisfy at the same time the connectivity condition. In the present case, matrices B and C are known and, more often than not, their entries are not located in this optimal way. Therefore, more than d inputs (outputs) are generally necessary.

These slight differences make that the procedure of (REI-81) is not perfectly suitable for solving this problem and it may fail in some cases as it is shown in the following example.

<u>Example 5.16</u>. Consider the following system (A,B) :

$$
A = \begin{bmatrix}
 & & & & & & \\
X & & & & & & \\
X & X & & & & & \\
 & & & & X & & \\
 & X & X & & X & & \\
X & & X & X & X & & X \\
 & & X & & & X &
\end{bmatrix}
\qquad
B = \begin{bmatrix}
X & \\
 & \\
 & \\
 & X \\
X & \\
 & \\
 &
\end{bmatrix}
$$

where A is already in the required form and where d_z is indicated on the right side. From Theorem 5.4, the connectivity (reachability) condition (2) is satisfied by u_1, and Condition (1) specifies that a nonzero entry is required in the first and last row of matrix B. Since the last row is empty, the wrong conclusion that the system is not structurally controllable could be stated. Nevertheless, it can be shown that this system is structurally controllable and that u_1, u_2 and u_1, u_3 are minimal essential input sets.

In this section, we present the sequential procedure of (TRA-86b) to determine the minimal essential input (output) sets of a system (5.3.6) by identifying first, the minimal input (output) sets which satisfy the connectivity condition and then the minimal input (output) sets which ensure that the generic rank condition is satisfied.

In the subsequent development, the problem is approached from the inputs point of view. Dual results can be stated for the outputs.

Define $I_{C1},\ldots,I_{Ch}$ as the sets of indices corresponding to the minimal input sets which satisfy the connectivity condition for system (5.3.6) and $I'_1,\ldots,I'_g$ as the sets of indices corresponding to the minimal input sets to have the generic rank condition satisfied.

<u>Theorem 5.7</u> (TRA-86b). U_I is a minimal essential input set for system (5.3.6) if and only if $I = I_{Ci} \cup I'_j$, i $\{1,\ldots,h\}$, j $\{1,\ldots,g\}$ and its cardinality is minimal.

The above result means that the search of the minimal essential input sets can be performed in two independent steps. For this prupose, we use the results and algorithm derived in (TRA-86a) to conclude on structural controllability (observability) of a given system (5.3.6). This algorithm can be applied, with some additional operations, to solve our problem. In an initial step, we proceed to a decomposition of the system such that the new matrix A presents the block-triangular form in (5.3.9), each diagonal block corresponding to the strong components of the graph associated with the system (this initial step is the same as in the procedure of (REI-81)).

To avoid trivialities, we make the assumption that system (5.3.6) is structurally controllable (and observable). In the opposite case however, the algorithm below would detect the uncontrollability and stop. This presents the advantage that no preliminary controllability checking is required.

With the proposed decomposition, the sets $I_{C1},\ldots,I_{Ch}$ can be determined without any calculations by using the following result :

<u>Theorem 5.8</u> (TRA-86a). The system (5.3.6) is input reachable if and only if :

$$\begin{bmatrix} B_i & A_{i1} & A_{i2} & \cdots & A_{i,i-1} \end{bmatrix} \neq 0 \qquad \forall i = 1,\ldots,N \tag{5.3.9}$$

where the matrices B_i and A_{ij} are those corresponding to (5.3.9).

In a second step, the sets $I'_1,\ldots,I'_g$ are identified by using the algorithm of (TRA-86a) and presented in Paragraph 1.3.c. When the system is structurally controllable, the algorithm returns the $f_p \times f_q$ ($f_p < f_q$) matrix F_N of full generic rank, whose columns correspond to some input and state variables. Indeed, it is shown in (TRA-86a) that the generic rank condition is satisfied if and only if F_i has full generic rank for all $(i=1,\ldots,N)$. The indices of all the inputs which are not represented in F_N constitute a subset of I'_i, $(i=1,\ldots,g)$ (TRA-86a). F_N contains all the necessary information to determine the remaining ones. Since we have $gr(F_N)=f_p$, f_q-f_p columns of F_N can be eliminated provided that at least one nonzero entry by row and by column is preserved (condition of full generic rank). If some columns of F_N correspond to input variables, it may be that some of them can be deleted. Among all possible solutions, those which delete a maximal number of columns corresponding to input variables are retained : the indices of the remaining "input columns" constitute the remaining indices of a set I'_i.

In order to determine all the sets $I'_1,\ldots,I'_g$, the problem is thus to identify all the possible sets of columns of F_N which can be deleted without affecting the full generic rank. The approach consists in determining all the square submatrices of F_N having full generic rank. This is performed in a graph-theoretic framework.

Consider a $p \times q$ $(p<q)$ structured matrix M of full generic rank and add $(q-p)$ full rows (none zero entry) at the bottom of M such that we obtain a $q \times q$ square matrix M. Associate to M a digraph $D=(V,E)$, where $V=\{v_1,\ldots,v_q\}$ is the set of vertices and E is the set of edges such that (v_i,v_j) E if and only if $m_{ji} \neq 0$.

<u>Theorem 5.9</u> (TRA-86b). Consider the matrix M and the digraph D as defined above. The number of $(p \times p)$ square submatrices having full generic rank of M is equal to the number of cycle families of width q of D. Consider any cycle family of width q of D. Then, the columns and rows of M whose entries m_{ij} correspond to edges (v_j,v_i) involved in the cycle family constitute a square $(p \times p)$ submatrix of M having full generic rank.

The above theorem provides the key for solving the problem. Identify the matrix F_N to M and build F'_N and its associated graph D_F. Then, all the square submatrices of F_N having full generic rank are determined by the search of all the cycle families of width f_q of D_F.

At this step, we have determined $I_{C1},\ldots,I_{Ch}$ and $I'_1,\ldots,I'_g$. Using Theorem 5.7, we can therefore deduce all the minimal essential input sets and know for each of them which inputs stem from I_{Ci} or from I'_j. A similar analysis must then be made for the outputs in order to obtain $J_{C1},\ldots,J_{Ck}$, $J'_1,\ldots,J'_f$ and the minimal essential output sets. This problem is easily brought back to the input problem by using the duality between observability and controllability (TRA-86a) : the algorithm of (TRA-86a) is applied on the matrix (C' A') put in the reordered form. Theorem 5.6 can thus be applied to obtain the set of all minimal essential feedback patterns for (5.3.6).

Example 5.17. Consider a system (C,A,B) where A and B are the same as in Example 5.16 and

$$
C = \begin{bmatrix} 0 & 0 & 0 & 0 & 0 & 0 & X \\ 0 & 0 & 0 & X & 0 & 0 & 0 \\ X & 0 & 0 & 0 & 0 & 0 & 0 \end{bmatrix}
$$

Using Theorem 5.8 and the dual one for output reachability (TRA-86a) (Theorem 5.9), we determine :

$$I_{C1} = \{1\} \text{ and } J_{C1} = \{1\} \quad J_{C2} = \{2\}$$

Now, we apply the algorithm of (TRA-86b) (see § 1.3.3.c) to the matrix (B A) and we obtain the matrix F_4 in a reordered form as follows :

Since u_1 is not represented in F_4, it is clear that all the sets I'_i contain the integer 1. The remaining ones are determined by using Theorem 5.9 which provides the key to extract all the square submatrices of F_4 having full generic rank.

Note that since the three last columns of F_4 are identical, they can be arbitrarily interchanged and two of them can therefore be deleted. In our case, none of them corresponds to input variables so that the deletion is arbitrary. In another case, it is clear that the ones corresponding to input variables should be deleted. We have :

The inspection of D_F shows three cycle families of width 5 involving a minimal number of edges associated with input columns (denoted by dotted lines) :

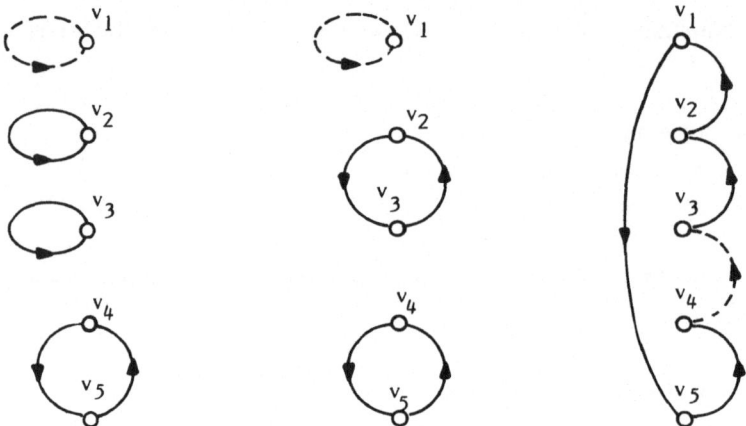

Consequently, we obtain : $I'_1 = \{1,2\}$ $I'_2 = \{1,3\}$. We have therefore two minimal essential input sets $I_1=I'_1$ and $I_2=I'_2$.

If the same problem is now approached from the outputs point of view, we get :

$$F_4 = \begin{bmatrix} X & X & & & \\ X & & X & & \\ X & X & & X & \\ & & X & & X & X \end{bmatrix} \quad y_3$$

This means that all the sets J'_i contain the indices 1 and 2. Since the two last columns of F_4 are identical, the one corresponding to y_3 is deleted and we do not need to go through the cycle families procedure to conclude that $J'_1=J_1= \{1,2\}$.

From Theorem 5.6, we obtain four minimal essential feedback patterns for this system :

$$F_1 = \begin{bmatrix} X & 0 & 0 \\ 0 & X & 0 \\ 0 & 0 & 0 \end{bmatrix} \quad F_2 = \begin{bmatrix} 0 & X & 0 \\ X & 0 & 0 \\ 0 & 0 & 0 \end{bmatrix} \quad F_3 = \begin{bmatrix} X & 0 & 0 \\ 0 & 0 & X \\ 0 & 0 & 0 \end{bmatrix} \quad F_4 = \begin{bmatrix} 0 & 0 & X \\ X & 0 & 0 \\ 0 & 0 & 0 \end{bmatrix}$$

5.3.5. - Concluding remarks

The algorithms presented in this section provide a systematic way of determining the feedback pattern(s) which guaranty the absence of fixed modes by using the fixed modes characterizations of Chapter III. An optimization criterion as the number of feedback links or the sum of their associated costs is generally adopted and, among all the admissible feedback patterns, the ones which minimize the criterion are selected.

Senning's procedure (SEN-79) is the only one which considers simultaneously the parametric and the structural optimization by using an extended optimization criterion taken as the sum of the classic quadratic criterion of linear systems and a weighted measure of the non-local information. Therefore, the solution of his algorithm provides in one step the optimal gains of the dynamic compensator and its optimal structure. All the others algorithms perform only the structural optimization and the gains must be evaluate later by using one of the parametric optimization procedures presented in the next chapter.

The different approaches are not equivalent and one must choose the one which is the most adequate to its particular problem and situation. Indeed, Locatelli et al. approach (LOC-77) is only valid for systems with simple modes and the procedures presented in Paragraph 5.3.4 do not guaranty the absence of non structurally fixed modes. However, note that these limitations are not too much constraining, specially for large scale systems, since they set apart the cases for which some unprobable perfect parametric matchings could occur.

5.4. - CONCLUSION

This chapter presents different algorithms which can be used for the design of the feedback control structure such that the system has no fixed modes and that the number of feedback links (or the sum of their associated costs) is minimized. Two situations are considered. In the first one, an initial control structure naturally arise from the partitioning of the system due to physical considerations. When the system has fixed modes with respect to this control structure, the algorithms of Section 5.2 allow to determine the supplementary information exchanges which eliminate fixed modes. In the second situation, no initial control structure is considered and the algorithms (Section 5.3) provide all the control structures for which the system has no fixed modes. Since the problem solved in the second situation is more general than the first one, the corresponding algorithms can also be used in the particular framework of the first case.

At this step, we are guaranted that the system is pole assignable by using a dynamic compensator whose structure has been determinated by one of the algorithms presented in this chapter. Except in the case for which Senning's algorithm (SEN-79) has been used, the gains of the compensator remain to determine. This problem is considered in the next chapter and several parametric optimization algorithms under structural constraints are presented.

If one wants to achieve pole assignment using static output feedback, additional conditions other than the absence of fixed modes are generally required. This problem was considered in Paragraph 3.5.2c in which the control structures of the static controller are characterized (EVA-84). However, no systematic algorithm for their design was provided and this is why this case was not integrated in this chapter.

CHAPTER 6

DESIGN TECHNIQUES - PARAMETRIC ROBUSTNESS

6.1. - INTRODUCTION

The previous chapters showed that the existence of a solution for the decentralized stabilization (or pole placement) problem depends critically on the absence of fixed modes. When a given feedback structure results in unstable fixed modes, stabilization is impossible and the solution consists generally in choosing a new feedback structure. This problem was considered in the last chapter. The remaining task is to specify numerical values for the gain matrices.

This chapter considers now the problem of synthesis of feedback gains under structural constraints and provides and overview of appropriate near-optimum design techniques. Considerations on the robustness of such controllers are also included, in the sense that uncertainties due to parameter variations or external disturbances are taken into account. The effects of structural perturbations (failure on sensors or actuator, line cuts...) will be studied in the following chapter.

6.2. - THE OPTIMIZATION PROBLEM

Consider a linear time-invariant system with S local control stations :

$$
\begin{cases}
\dot{x}(t) = A \; x(t) + \sum_{i=1}^{S} B_i u_i(t) \\
y_i(t) = C_i \; x(t) \quad (i=1,\ldots,S)
\end{cases}
\tag{6.2.1}
$$

where $x \in R^n$ is the state vector, $u_i \in R^{m_i}$ and $y_i \in R^{r_i}$ are the input and output vectors of the ith local control station. A, B_i, C_i are constant matrices of appropriate dimension. Define :

$$B = \begin{bmatrix} B_1, & B_2, & \ldots, & B_S \end{bmatrix}$$
$$C' = \begin{bmatrix} C', & C', & \ldots, & C' \end{bmatrix} \qquad\qquad (6.2.2)$$

then the equations (6.2.1) become :

$$\dot{x}(t) = A \ x(t) + B \ u(t)$$
$$y(t) = C \ x(t) \qquad\qquad (6.2.3.a)$$

where the global number of inputs and outputs are given by :

$$m = \sum_{i=1}^{S} m_i \ , \qquad r = \sum_{i=1}^{S} r_i \qquad\qquad (6.2.3.b)$$

The problem is to find S local output (or state) feedback controls such that the global system (6.2.3) is asymptotically stable and that the classical quadratic performance criterion J is minimized.

$$J = \int_{0}^{\infty} (x' \ Q \ x + u' \ R \ u) \ dt \qquad\qquad (6.2.4)$$

where Q and R are weighting matrices of appropriate dimensions, positive-semi defi-nite and positive definite, respectively.

Only static feedback control will be considered. The reason is that the syn-thesis of a dynamic controller can be brought back to the synthesis of a static controller for an augmented system (SEN-79). This is shown in the following subsec-tion.

6.2.1. - Dynamic controllers

Assume that the controller given by :

$$\dot{z}(t) = A_z \ z(t) + G \ y(t) \qquad\qquad (6.2.5.a)$$
$$u(t) = H \ z(t) + E \ y(t) \qquad\qquad (6.2.5.b)$$

where $z \in R^{n_0}$ is the state and A_z $(n_0 \times n_0)$, G $(n_0 \times r)$, H$(m \times n_0)$ and E $(m \times r)$ are constant matrices verifying the structure constraints (A_z, G, H, E are block diagonal matrices if the control is completely decentralized).

Applying the control (6.2.5) to the system (6.2.3), the closed-loop system becomes :

$$\begin{bmatrix} \dot{x} \\ \dot{z} \end{bmatrix} = \begin{bmatrix} A + BEC & BH \\ GC & A_z \end{bmatrix} \begin{bmatrix} x \\ z \end{bmatrix}$$

This equation can be rewritten as :

$$x^*(t) = (A^* + B^* K^* C^*) X^*(t)$$

where

$$A^* = \begin{bmatrix} A & 0 \\ \hline 0 & 0 \end{bmatrix} \begin{matrix} n \\ n_0 \end{matrix}, \quad B^* = \begin{bmatrix} B & 0 \\ \hline 0 & I \end{bmatrix} \begin{matrix} n \\ n_0 \end{matrix} \quad C^* = \begin{bmatrix} C & 0 \\ \hline 0 & I \end{bmatrix} \begin{matrix} l \\ n_0 \end{matrix} \quad \text{and } K^* = \begin{bmatrix} E & H \\ \hline G & A_z \end{bmatrix} \begin{matrix} m \\ n_0 \end{matrix}$$
$$\begin{matrix} n & n_0 \end{matrix} \qquad\quad \begin{matrix} m & n_0 \end{matrix} \qquad\quad \begin{matrix} n & n_0 \end{matrix} \qquad\quad\;\; \begin{matrix} l & n_0 \end{matrix}$$

Therefore, specifying the dynamic compensator (6.2.5) (i.e. the matrices E,H,G and A_z) is equivalent to specifying a static compensator $u^* = K^* y^*$ for the augmented system :

$$x^*(t) = A^* x^*(t) + B^* u^*(t)$$
$$y^*(t) = C^* x^*(t)$$

Note that the determination of the augmented system may be delicate because the problem of choosing the compensator order n_0 is not completly solved yet. However this order must be at least equal to min (ν_o, ν_c), where ν_o and ν_c are the controllability and observability indices of the system.

6.2.2. - Static controllers

Suppose now that the control $u(t)$ is generated by linear static output feed-back with time-invariant feedback gains, i.e. :

$$u(t) = K y(t) = K C x(t) \tag{6.2.6}$$

the optimization problem can be formulated as follows :

$$\min_K \quad J = \int_0^\infty (x' Q x + u' Ru) \, dt$$

subject to

$$\dot{x} = Ax + Bu \tag{6.2.7}$$
$$y = Cx$$
$$u = Ky$$

Applying the control (6.2.6) to the system (6.2.3), the closed-loop system is :

$$\dot{x}(t) = (A + BKC) x(t) = D x(t) \tag{6.2.8}$$

and x(t) is given by :

$$x(t) = [\exp (A+BKC)t] x(0) = \phi(t) \; x(0)$$

where $\phi(t)$ is the transition matrix of the system (6.2.8). Substituting x(t) and u(t) in the criterion expression, we obtain :

$$J(K) = x'(0) \{\int_0^\infty \phi'(t) \quad Q + C' K' R KC \; \phi(t) \; dt\} \; x(0) \tag{6.2.9}$$

or

$$J(K) = x'(0) P x(0) \tag{6.2.10}$$

where

$$P = \int_0^\infty \phi'(t) Q_1(K) \; \phi(t) \; dt$$
$$Q_1(K) = Q + C' K' R KC$$

If the matrix D = A + BKC has all its eigenvalues in the left half complex plan, then P satisfies the matricial Lyapunov equation (LEV-70) :

$$f = D' P + PD + Q_1 (K) = 0$$

It is well known that the solution of this problem is depending on the initial condition of the state x(0). To overcome this difficulty and to obtain a suboptimal solution valuable for a set of initial conditions we consider x(0) as a zero mean random variable (LEV-70) with given covariance $X_0 = E[x(0) \; x'(0)]$ (X = 1/n I, I identity matrix, if x(0) is supposed uniformly distributed around the origin).

Then taking the expectation of both sides of (6.2.10) and using the properties of the trace operator, i.e., tr(b'a) = tr (ab'), we get :

$$J(K) = Tr (P X_0) \tag{6.2.11}$$

where

$$X_0 = E [x(0) \; x'(0)]$$

The optimization problem can be reformulated in the following static equivalent form :

$$\min_{K} J(K) = Tr[P X_0]$$

subject to :
$$f = D'P + PD + Q_1(K) = 0$$

The constraints on the control are introduced by defining the admissible feedback matrices set :

$$K_c = \{ K/K \in R^{mxr} \text{ such that } F(K) = 0\}$$

where $F(K)$ indicates the feedback loops that are not allowed.

$$F(K) = K - \text{block diag.}(K_1,\ldots,K_S) \tag{6.2.12}$$

in the case of completly decentralized control.

Finaly, the optimization problem can be rewritten as :

$$\min_{K \in K_c} J(K) = Tr[P X_0]$$

subject to :
$$f = D'P + PD + Q_1(K) = 0 \tag{6.2.13}$$
$$Q_1(K) = C'K'RKC + Q$$
$$D = A + BKC$$
$$X_0 = E[x(0) \ x'(0)]$$

In the case of full state feedback the problem (6.2.13) becomes :

$$\min_{K \in K_c} J(K) = Tr[P X_0]$$

subject to :
$$f = D'P + PD + Q + K'RK = 0 \tag{6.2.14}$$
$$D = A + BK$$
$$X_0 = E[x(0) \ x'(0)]$$

6.2.3. - Necessary conditions for optimality :
gradient matrix calculation

The gradient matrix can be obtained analytically by applying the perturbation theory on the criterion (6.2.9) (LEV-70) (see Appendix 5) but a simpler method was provided by Geromel and Bernussou (GER-79b, BER-81, BER-82).

Let $f(K)$ be a scalar function of the matrix K which is differentiable with respect to its arguments. By definition, the matrix gradient df/dK is the matrix whose elements are $\partial f/\partial k_{ij}$.

The static case :

Theorem 6.1. Let $f(x, G(x))$ be a scalar function which is differentiable with respect to all its arguments ; x and $G(x)$ are matrices. Then :

$$\frac{df}{dX} ((x), G(x)) = \frac{\partial L}{\partial x} (x, y, z)$$

where y, z are the matrices obtained by solving the necessary conditions for the stationarity of the Lagrangian function :

$$L = f(x, y) + Tr\{z'(G(x) - y)\}$$

The dynamic case : Let the objective function be :

$$J(K) = \int_0^T f(x, K)\, dt + g(x(T))$$

where f and g are scalar functions (differentiable with respect to their arguments) ; x, K are matrices linked across the differential system :

$$\dot{x} = F(x(t), K) \quad ; \quad x(0) = x_0$$

Theorem 6.2.

$$\frac{dJ}{dK} = \int_0^T \frac{\partial H}{\partial K} (x, z, K)\, dt$$

where $H(x, z, K)$ is the Hamiltonian function :

$$H = f(x, K) + Tr[z' F(x, K)]$$

and x, z satisfy the conditions of stationarity :

$$\dot{x} = \frac{\partial H}{\partial Z} \qquad x(0) = x_0$$

$$\dot{z} = -\frac{\partial H}{\partial x} \qquad z(T) = \frac{dg}{d\,x(T)}$$

Applying Theorem (6.1) to our problem in its static form (6.2.13), with the Lagrangian function :

$$L = \text{Tr} (P \, X_0) + \text{Tr} [S' \, f]$$

we obtain the following necessary conditions :

with
$$\left\|\begin{array}{l} \dfrac{dJ}{dK} = 2 \ (RKC - B'P) \ SC' = 0 \\[2mm] f(K) = D'P + PD + Q_1 \ (K) = 0 \\[2mm] g(K,X_0) = DS + SD' + X_0 = 0 \end{array}\right.$$

(6.2.15a)

(6.2.15b)

(6.2.15c)

where
$$Q_1(K) = C'K'RKC + Q$$
$$D = A + BKC \quad \text{and} \quad X_0 = E \, [\, x(0) \ x(0)' \,]$$

Remark 6.1 - In Appendix 5, this result is obtained by using the method of variations. Hence, for a given K, the gradient matrix can be obtained by solving two linear uncoupled matricial equations (Lyapunov equations), one of them depending on the initial conditions. This is a reasonable task even for large scale system (see (HOS-77) and (DAV-68)).

If all the system states are measurable then u = Kx and the necessary conditions (6.2.15) become :

with
$$\left\|\begin{array}{l} \dfrac{dJ}{dK} = 2 \ (RK - B'P)S = 0 \\[2mm] f(K) = D'P + PD + Q + K'RK = 0 \\[2mm] g(K,X_0) = DS + SD + X_0 = 0 \end{array}\right.$$

(6.2.16a)

(6.2.16b)

(6.2.16c)

where $D = A + BK$

The optimal feedback matrix, in this case, is given by :

$$K = - R^{-1} B' P$$

(6.2.17)

where P is the symetric definite positive solution of the Riccati equation :

$$A'P + PA - PBR^{-1}B'P + Q = 0$$

(6.2.18)

Observe that the equation (6.2.16c) becomes unusefull and the control does not depend on the initial conditions.

In the general case, the otpimisation problem is thus reduced to a parametric problem which can be solved by using one of the design techniques presented in the following section.

6.3. - DECENTRALIZED CONTROL BY PARAMETRIC OPTIMIZATION

This section provides a solution to the near-optimum decentralized control problems 6.2.15 or 6.2.16. The presented algorithms are based on iterative schemes. Some use gradient matrix techniques (GER-79a,b) (JAM-83), and (CHE-84)) and others do not (GER-84).

6.3.1. - The algorithm of Geromel and Bernussou (GER-79)

Geromel and Bernussou propose a gradient method (ZOU-70) which is summarized below :

1 - Initialization by an admissible gain matrix, i.e. $K \in K_d$ (K_d is the set of block diagonal feedback matrices), such that the closed-loop system is stable.
2 - Calculation of the gradient matrix.
3 - Determination of a feasible direction G (for which the cost function decreases and the structure constraint is verified) and of the step α of progression in this direction.
4 - Convergence test. If the test is satisfied : stop ; otherwise, update K by K $-\alpha$G and go to 2.

6.3.1.a - Initialization of the procedure : Algorithm of Armentano and Singh

The problem of finding an admissible stabilizing gain was considered by many authers (AOK-73), (WAN-78b), (FES-79), (IKE-79), (SEZ-81b), (DAV-76a), (COR-76a,b), (GER-79a,b), (ARM-81)... Among those, we present the algorithm of Armentano and Singh (ARM-81) which always provides a solution (of course, if no unstable fixed modes exist).

The approach consists in minimizing the real part of the dominant eigenvalue of the closed-loop system. The gradient of an eigenvalue λ_0 of the closed-loop system with respect to the feedback gains is given by (see Chapter 3) :

$$\frac{\partial \lambda_0}{\partial k_{ij}} = \left\{ \begin{array}{ll} -W' \, b^i \, c_j \, v & i=1,\dots,m \\ & j=1,\dots,r \\ 0 & \text{if } k_{ij} = 0 \end{array} \right.$$

where v and w are the right and left eigenvectors associated with the eigenvalue λ_0, b^i is the ith column of B, and c_j is the jth row of C.

The gradient of the real part of λ_0 with respect to the feedback gain K = (k_{ij}) can be expressed as a matrix G = (g_{ij}), where :

$$g_{ij} = Re\left[\frac{\partial \lambda_0}{\partial k_{ij}}\right]$$

The algorithm is the following :

Step 1 : Choose an arbitrary $K \in K_d$.

Step 2 : Compute the dominant eigenvalue λ_d of A+BKC. If it is negative, stop. Otherwise, compute the right and left eigenvectors v_d and w' associated to λ_d.

Step 3 : Compute the gradient matrix G_d.

Step 4 : Do a steepest descent search in the direction of -G and update K = K -αG, where α is the step size. Go to step 2.

Remark 6.2.

a) If the dominant eigenvalue with positive or null real part turns out to be a local minimum, the algorithm must be started again from a different matrix.

b) The unidimensional search may be performed by using efficient algorithms like quadratic interpolation.

c) If, at any iteration, the dominant eigenvalue is multiple, perturb slightly K (WAN-73a) in such a way that it becomes simple.

6.3.1.b. - Feasible direction

Assume that $K \in K_d$ is an admissible initial condition such that the gradient matrix is not zero. The problem is to determine a matrix G in order to guarantee the existence of a step length α such that J (K - αG) < J (K) for all $0 < \alpha < \alpha$ and some $\alpha > 0$. It is easy to see that this matrix can be obtained by solving the problem (GER-79b, 82) :

$$\min \frac{1}{2} Tr [(\frac{dJ}{dK} -G)' (\frac{dJ}{dK} - G)] \tag{6.3.1}$$

subject to : F (G) = 0

If the criterion is written explicitly, we end up with the problem of the orthogonal projection of dJ/dK on the linear set defined by the constraint. It is shown in (GER-79b) that the solution is :

$$G = block\text{-}diag. (\frac{dJ}{dK}) \tag{6.3.2}$$

Therefore, the orthogonal projection onto the decentralization set is easily obtained by setting all the off-block-diagonal terms of the gradient matrix to zero.

This result can be generalized to arbitrary structure constraints and the projection of the gradient matrix is given by :

$$G = (g_{ij}) = \begin{cases} \dfrac{dJ}{dk_{ij}} & \text{if } k_{ij} \neq 0 \\ 0 & \text{if } k_{ij} = 0 \end{cases} \tag{6.3.3}$$

The existence of a feasible direction matrix is guaranted by the following lemma :

Lemma 6.1 (GER-82). The optimal solution of the problem (6.3.1) is such that :

a) if $\|G\| \neq 0$ then $\quad \bar{\alpha} > 0$ such that $J(K - \alpha G) < J(K) \qquad 0 < \alpha < \bar{\alpha}$

b) if $\|G\| = 0$ then the matrix K satisfies the Kuhn-Tucker conditions of the problem $\min_{K \in K_d} J(K)$ where K_d represents the decentralization constraint.

This lemma guarantees the existence of a step size α such that $(K - \alpha G)$ stabilizes the system, but it does not give any information about its choice. At each iteration, α may be fixed by solving the following single variable optimization problem along the search direction :

$$\begin{align} \min_{\alpha \geq 0} \quad J(K - \alpha G) \tag{6.3.4} \end{align}$$

However, in practice, it is generally advantageous to adopt a heuristic method of adaptation of the step during the iteration.

In (GER-79a), the following adaptive rule is proposed :

Let i be the iteration index, then :

$$\begin{align} \alpha^{i+1} &= \pi \, \alpha^{i} \quad \text{if } J(K^{i} - \alpha^{i} G^{i}) < J(K^{i}) \text{ and } P(K^{i} - \alpha^{i} G^{i}) > 0 \\ \alpha^{i} &= \nu \alpha^{i} \quad \text{otherwise} \tag{6.3.5} \end{align}$$

where $\pi > 1$ and $0 < \nu < 1$. The initial value of α is arbitrary.

6.3.1.c - <u>The properties of the algorithm</u>

The algorithm presented above has the two following properties :

1 - The stability of the overall system is guaranted at each iteration.

The step size α must be chosen such that :

i) $(K - \alpha G) \in K_s$ with $K_s = [\; K \; / \; A + BKC \; \text{stable}]$
ii) $J (K - \alpha G) \leqslant J (K)$ or $Tr [P(K- \alpha G) \; X_0] \leqslant Tr \; P(K) \; X_0$

It is shown in (GER-79a) that such α exists. Since the algorithm requires the calculation of the matrices P, the test (i) can be performed by means of the definite-positiveness of P, i.e. $P(K - \alpha G) > 0$.

2 - Uniforme convergence to a local optimum

The nonlinear optimization problem considered here is not generally convex with respect to K but the convergence to a local optimum is ensured (GER-79a).

The algorithm is summarized by the organigram of Figure 6.1. The corresponding Fortran routine, is provided in Appendix 6.

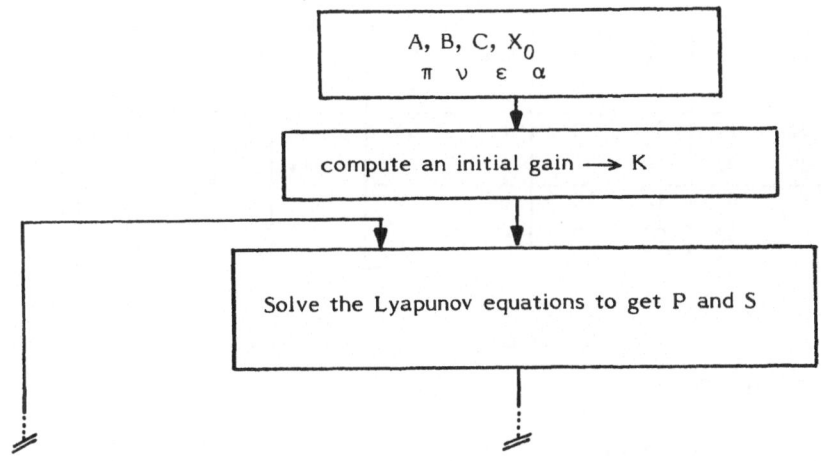

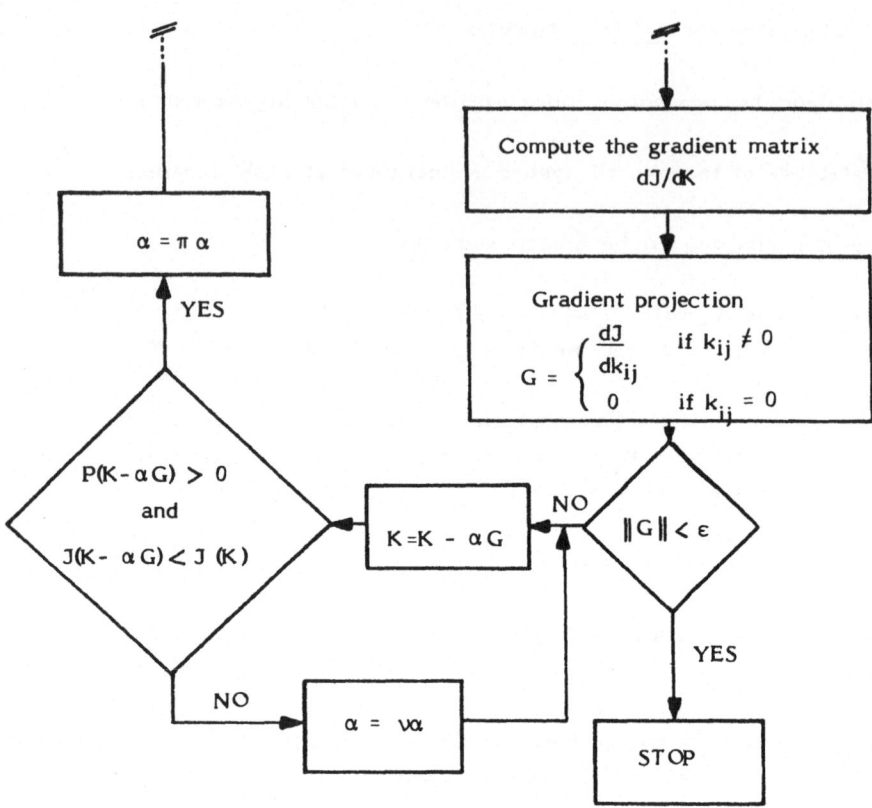

Figure 6.1

Example 6.1 (GER-79a). Consider the system (composed of three subsystems : S=3) :

$$\dot{x} = \begin{bmatrix} 0 & 1 & | & 0.5 & 1 & | & 0.6 & 0 \\ -2 & -3 & | & 1 & 0 & | & 0 & 1 \\ \hline 0.5 & 1 & | & 0 & 2 & | & 1 & 0.5 \\ 0 & 0.5 & | & 1 & 3 & | & 0 & 0 \\ \hline 1 & 0 & | & 1 & 0 & | & 0 & 1 \\ 0 & 0.5 & | & 0.5 & 0 & | & -3 & -4 \end{bmatrix} x + \begin{bmatrix} 1 & & \\ 1 & & \\ & 3 & 0 \\ & 0 & 4 \\ & & 2 \\ & & 3 \end{bmatrix} u$$

and the quadratic cost :

$$J = \int_0^\infty (x'x + u'u)\ dt$$

Consider the case of decentralized state feedback :

$$u = K x \qquad\qquad (K : \text{block diagonal matrix})$$

with $X_0 = I$ and the initial gain matrix :

$$K^1 = \begin{bmatrix} 115\,1 & -055 & & & & \\ & & 2.3 & 0.62 & & \\ & & 0.82 & 3.41 & & \\ & & & & 4.96 & 0.06 \end{bmatrix}$$

With $\alpha^1 = 0{,}01$, $\nu = 0{,}5$ and $\pi = 1{,}2$, we have :

$$K^{59} = \begin{bmatrix} 1.21 & 0.35 & & & & \\ & & 1.43 & 0.45 & & \\ & & 0.87 & 2.46 & & \\ & & & & 1.58 & 0.24 \end{bmatrix}$$

which satisfies the convergence test with $\epsilon = 0{,}001$. The cost goes from $J(K^1) = 8{,}76$ to $J(K^{59}) = 3$.

6.3.2. - The algorithm of Jamshidi (JAM-83)

Instead of using the feasible direction method, the problem (6.2.16) is solved by using the well-known Davidon-Fletcher-Powell variable metric method (FLE-63). The algorithm is the following :

Step 1 : Select an initial matrix $K = K_0$ and set $i = 0$.

Step 2 : Solve the Lyapunov equations (6.2.16b and c) using A, B, Q, R, X_0, K and the algorithm of Davison and Man (DAV-68).
Evaluate the gradient $\partial J/\partial K$ and the criterion $J(K)$.

Step 3 : Use the Fletcher-Powel method to update the gain matrix K^i:
$$K^{i\text{-new}} = K^{i\text{-old}} + \alpha D^i$$
where D^i is the search direction during the ith iteration.

Step 4 : Check whether the convergence is achieved, e.g., $\| \partial J/\partial K^i \| < \epsilon$, where ϵ is a prespecified tolerance value. If the convergence is reached, stop, otherwise go to step 2.
The following exemple illustrates the use of this algorithm :

Example 6.2 (JAM-83). Consider a fourth-order system with two subsystems :

$$\dot{x} = \left[\begin{array}{cc|cc} 0 & 0.2 & 0.25 & 1 \\ -1 & -2 & 1 & 0 \\ \hline -1 & 0,1 & 0,85 & 1 \\ 0.25 & -0,5 & 0 & -0.25 \end{array}\right] x + \left[\begin{array}{c|c} 1 & \\ 1 & \\ \hline & 0.1 \\ & 1 \end{array}\right] u$$

and the cost function :

$$J = 1/2 \int_0^\infty (x' \, Q \, x + u' \, Ru) \, dt$$

where Q = 2I and R = 2I. The problem is to find a state feedback matrix :

$$K = \left[\begin{array}{cc|cc} k_{11} & k_{12} & 0 & 0 \\ \hline 0 & 0 & k_{23} & k_{24} \end{array}\right]$$

such that J is minimized :

Take X_0 = I, and consider these two initial values for K_0:

$$K_0 = \left[\begin{array}{cc|cc} 0.1 & 0.1 & 0 & 0 \\ \hline 0 & 0 & 0.1 & 0.1 \end{array}\right] \qquad K_0 = \left[\begin{array}{cc|cc} 1 & 2 & 0 & 0 \\ \hline 0 & 0 & 3 & 4 \end{array}\right]$$

Then, the algorithm of Jamshidi converges in 7 and 13 iterations (ϵ = 0,01), respectively. The resulting averaged optimal gain K* is found to be :

$$K^* = \left[\begin{array}{cc|cc} 0,494 & 0.168 & 0 & 0 \\ \hline 0 & 0 & 0.35 & 0.58 \end{array}\right]$$

with J(K*) = 1,0985.

6.3.3. Iterative procedure of Chen et al. (CHE-84)

The algorithm of Geromel and Bernussou (see § 6.3.1) suffers of poor convergence and excessive computation per cycle due to the requirement of retaining a block-diagonal feedback matrix at each iteration.

The following iterative procedure was proposed to overcome these desadvantages. It preserves the properties of uniform convergence to a local minimum and guarantees the stability of the global system at each iteration. The procedure consits in starting the iterations with an optimal state feedback matrix (full matrix given by (6.2.17) and setting successively the off-diagonal blocks to zero. At each iteration, an optimal feedback matrix is found using a conjugate gradient method. The procedure is summarized below :

Step 1 : Determine P by solving the algebraic Riccati equation (6.2.18). Compute the full gain matrix K given by (6.2.17). Set j=1 and $K^j = K$.

Step 2 :

    a) write K^j as :

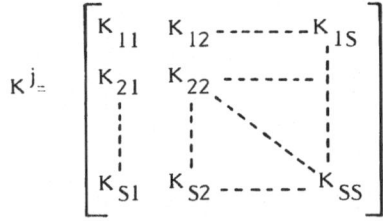

$$K^j = \begin{bmatrix} K_{11} & K_{12} & \text{------} & K_{1S} \\ K_{21} & K_{22} & \text{------} & \\ & & \diagdown & \\ K_{S1} & K_{S2} & \text{------} & K_{SS} \end{bmatrix}$$

    b) Discard one block from K^j, say K_{1S} :

$$K^i = \begin{bmatrix} K_{11} & K_{12} \text{----} K_{1,S-1} & 0 \\ K_{21} & K_{22} \text{----------} & K_{2S} \\ & \diagdown & \vdots \\ K_{S1} & K_{S2} \text{----------} K_{SS} \end{bmatrix}, \quad i = j+1$$

    c) Calculate the eigenvalues of the matrix $(A-BK^i)$. If the matrix is stable, go to step 3. Otherwise, consider K^i as the initial gain matrix for the algorithm of Armentano and Singh (see § 6.3.1a). Find a gain matrix K (with the same structure as K^i) which stabilizes (A-BK). Set $K^i = K$.

Step 3 : Using the technique of Hoskins et al. (HOS-77), solve the Lyapunov equations (6.2.16b and c) and compute the gradient (6.2.16a). Then, a conjugate gradient routine with a quadratic interpolation in the one-dimensional search is used to obtain the minimum of the performance index.

Step 4 : Discard another block from K^i, say K_{S1} :

$$K^{i+1} = \begin{bmatrix} K_{11} & K_{12} & \text{------} K_{1,s-1} & & 0 \\ K_{21} & K_{22} & \text{------------} & & K_{2s} \\ K_{s-1,1} & & & & \\ 0 & K_{s,2} & \text{-----------} & & K_{ss} \end{bmatrix}$$

Set i=i+1 and go to step 2. The procedure goes on until K^i has a block-diagonal form.

Example 6.3 (CHE-84).

Consider again the Example 6.1.

The procedure begins with :

$$K_1 = \begin{bmatrix} 0.9004 & 0.2487 & 0.1887 & 0.1823 & 0.3117 & 0.04 \\ 0.2761 & 0.29 & 1.143 & 0.4902 & 0.4006 & 0.1202 \\ 0.4608 & 0.2684 & 0.6536 & 2.219 & 0.2878 & -0.031 \\ 0.6068 & 0.1366 & 0.3873 & 0.1207 & 1.184 & 0.32 \end{bmatrix}$$

After six iterations the procedure gives :

$$K^6 = \begin{bmatrix} 1.145 & 0.3162 & 0 & 0 & 0 & 0 \\ 0 & 0 & 1.264 & 0.4981 & 0 & 0 \\ 0 & 0 & 0.7892 & 2.224 & 0 & 0 \\ 0 & 0 & 0 & 0 & 1.366 & 0.2269 \end{bmatrix}$$

The criterion values are $J(K^1) = 2.718$ and $J(K^6) = 3.007$.

Remark 6.3.

1. Since there is no constraint in eliminating off-diagonal blocks, the procedure is applicable for an arbitrary control structure without modifications.

2. The procedure can be used for output feedback. If C^{-1} exists, one can take the initial gain matrix $K = R^{-1}B'PC^{-1}$ (P being the solution of the Riccati equation (6.2.18)) and replace the equation (6.2.16) by (6.2.15). If C^{-1} does not exist, then

the initial gain matrix can be found by using the algorithm of Armentano and Singh (see § 6.3.1a).

3. This procedure was compared with the algorithm of Geromel and Bernussou (see § 6.3.1) (CHE-84). Both algorithms were implemented on a PDP-10 and the recorded CPU times for Example 6.3 were :

Geromel and Bernussou algorithm

Initial block-diagonalization by the technique

of Geromel and Bernussou (GER-79a)............................ 53.2 sec

suboptimal block-diagonalization................................ 101.26 sec

Procedure of Chen et al

Initialization (optimal full gain matrix) ;..................... 2.56 sec

Suboptimal block-diagonalization .... 35.81 sec

The numbers show a significant computational improvement.

6.3.4. Iterative procedure of Geromel and Peres (GER-84)

The problem (6.2.16) under structural constraints was considered by Geromel and Peres in (GER-84). The set of admissible feedback matrices is defined by :

$$K_c = \{ K \ / \ K \in R^{mxr} \text{ such that } F(K) = 0 \} \qquad (6.2.11)$$

where $F(K)$ is the structure constraint given by :

$$F(K) = K[\ I - C' \ (CC')^{-1} \ C \ ] \qquad (6.2.12a)$$

in the case of centralized output feedback control and

$$F(K) = K - \text{block-diag.} \ (K_1, \ K_2, \ ..., \ K_N) \qquad (6.2.12b)$$

in the case of decentralized control.

We recall that the optimal solution for the state feedback control problem is given in (6.2.17) by :

$$K^* = R^{-1} B'P$$

where P is the symetric positive solution of the Riccati equation (6.2.18).

$$A'P + PA - PBR^{-1}B'P + Q = 0$$

which can be, for simplicity, written as :

$$\pi \ (P) = Q$$

The iterative procedure of Geromel and Peres (GER-84) is based on the property that if a gain matrix K satisfies :

$$K + L = R^{-1}B'P \qquad\qquad (6.3.6)$$

where P is the positive definite solution of the Riccati equation :

$$\pi \ (P) = Q + L'RL \qquad\qquad (6.3.7)$$

then, the matrix (A-BK) is asymptotically stable for any matrix L.

Since L is arbitrary, it can be chosen such that $K \in K_c$. Geromel and Peres showed that L is given by :

$$L = F \ (R^{-1}B'P) \qquad\qquad (6.3.8)$$

The degree of suboptimality of the solution is defined by:

$$d_{so} = (J - J^*) \ / \ J^*$$

where J^* is the cost value corresponding to the optimal full gain matrix given by (6.2.17) and J the cost value at the convergence.

The following procedure was proposed :

Step 1 : Set the iteration index to zero : i=0.
Step 2 : Solve the Riccati equation :
   $$\pi \ (P_i) = Q + L_i' \ R \ L_i$$
   if i = 0, calculate also $J^* = 1/2 \ Tr \ (P_0)$
Step 3 : Calculate $L_{i+1} = F \ (R^{-1}B'P_i)$.
Step 4 : If $\|L_{i+1} - L_i\| < \epsilon$ (ϵ being a positive small real number), go to step 5, otherwise set i = i+1 and go to step 2.
Step 5 : Calculate $J = 1/2 \ Tr \ (P_i)$ and $K \in K_c$ by setting :
   $$K = R^{-1}B'P_i - L_{i+1}$$
   The degree of suboptimality is given by :
   $$d_{so} = (J - J^*)/ \ J^*$$

Remark 6.4.

1. The performance of the procedure depends mainly on the numerical method used to solve the Riccati equation at step 2. We propose the algorithm of Kleinman (KLE-68).

2. At the convergence we have :

$$\pi(P_i) = Q_i = Q + L_i' R L_i$$

Where Q_i is a symmetric semi-positive definite matrix. The method can therefore be viewed as a linear-quadratic design which determines the weighting state matrix in order to get $K \in K_c$.

Because $Q_i \geqslant Q$, the objective function increases, and the control will be near-optimal. This is the "price" for an optimal control with the desired structure.

3. The solution is independent of the initial condition $X(0)$ and is such that the closed-loop system is asymptotically stable.

4. It is easy to take into account an arbitrary control structure.

6.3.5. Comments

The design techniques presented in this section have two disadvantages :

- They converge to a local minimum of the criterion. The solution is therefore su optimal.

- They are not adequate for very large scale systems (e.g. n=100). Indeed, they are based on an optimization in the parameter space and require calculations involving the global system (resolution of matricial equations (Lyapunov or Riccati)).

However, for many problem of practical interest, they represent viable design techniques.

6.4. - DESIGN OF ROBUST DECENTRALIZED CONTROLLERS

This section is concerned with the synthesis of decentralized controllers when uncertainties due to parameter variations must be taken into account.

Three aspects of robustness are considered : control synthesis with a prescribed degree of stability (AND-71), and under small (TAR-85) or large (PEK-83) parameter variations (without disturbances). The presence of external disturbance signals will be considered in the next section dealing with the study of the "robust decentralized servo mechanism problem".

6.4.1. Controllers with a prescribed degree of stability

This section discusses the problem of pole placement in a prescribed region of the complex plane which is more general than the stabilization problem. From a practical point of view, it is not essential to fix precisely the eigenvalues of the closed-loop system but rather to ensure that these eigenvalues are within a certain region of the complex plane. Typical desired regions are shown in Figure 6.2.

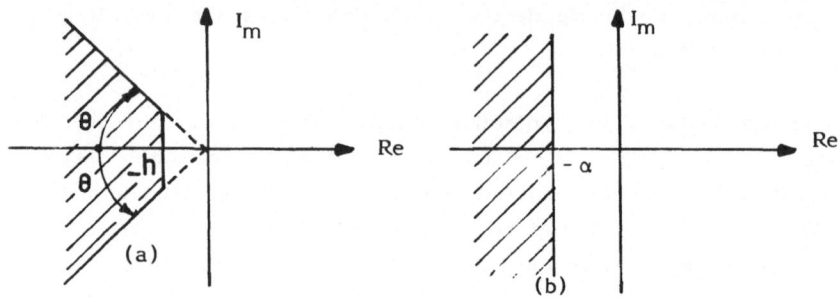

Fig. 6.2. Two cases of practical stability margins

Here, we are concerned by regions as shown in Figure 6.2b (For details concerning Figure 6.2a, see (KAW-81)).

The optimization problem is :

$$\min_{K} \quad J = \int_{0}^{\infty} (x'Qx + u'Ru) \, dt$$

$$\text{subject to } x = Ax + Bu$$
$$u = - Kx$$

and such that the eigenvalues of the closed-loop system are located in the dashed region of Figure 6.2b (i.e. eigenvalues have real parts less than $-\alpha$).

Anderson and Moore (AND-71) showed that this problem is equivalent to :

$$\min_{K} \quad J^* = \int_{0}^{\infty} e^{2\alpha t} (x' Qx + u' Ru) \, dt$$

$$\text{subject to } x = Ax + Bu$$
$$u = - Kx$$

$$\left.\right\} \qquad (6.4.1a)$$

Which can itself be brought back to a standard LQ-problem by using the following variable transformation :

$$\bar{x} = e^{\alpha t} x$$
$$\bar{u} = e^{\alpha t} u$$

Then, the problem becomes :

$$\min_K J^* = \int_0^\infty (\bar{x}' Q \bar{x} + \bar{u}' R\bar{u}) \, dt$$

subject to $\dot{\bar{x}} = A^* \bar{x} + B \bar{u}, \quad A^* = A + \alpha I$

$$\bar{u} = - K \bar{x}$$

$\left.\right\}$ (6.4.1b)

Its solution must fulfill the following necessary conditions (see § 6.2).

$$\frac{\partial J}{\partial K} = 2 (RK - B' P) S$$

$$(A^* - BK)' S + S (A^* - BK) + K' RK + Q = 0$$

$$(A^* - BK) P + P (A^* - BK) + X_0 = 0$$

The existence of a solution is guaranted if the pair (A^*, B) is controllable and the pair $(A^*, Q^{1/2})$ is observable. Anderson and Moore showed that if (A, B) is controllable and $(A, Q^{1/2})$ is observable, then (A^*, B) and $(A^*, Q^{1/2})$ are also controllable and observable.

Observe that the solutions of (6.4.1a) and (6.4.1b) are the same, i.e. $\bar{u} = -K\bar{x}$; $u = - Kx$. This solution ensures with degree α, the asymptotic stability of the original system, and the corresponding values of the criteria J and J^* are equal. This means that to garanty that the real part of the dominant eigenvalue of the closed-loop system is less than $-\alpha$, it suffices to replace A by $(A + \alpha I)$ in the original optimization problem.

The above result is still valid for output feedback or for decentralized feedback control if the system has no decentralized fixed modes out of the desired region. In this case, we can use one of the algorithms presented in the last section to solve equations (6.2.15) after changing A by $(A + \alpha I)$.

6.4.2. Optimal control with performance index sensitivity reduction (TAR-85)

The system parameters may vary or be uncertain and it is clear that the optimal control determined for a set of nominal parameter values is not generally optimal for a different set of values. This section considers the problem of taking into account the uncertainties arising from small parameter variations.

In a stochastic approach the uncertainties are modelled in general by white Gaussian noises. The control strategy provided in this section does not model the uncertainties but reduces the criterion sensitivity with respect to system parameter variations. If a parameter deviates from its nominal value, the reduced sensitivity of the plant performance index ensures that the plant behaviour will not change dramatically. The measure of this sensitivity is obtained by (TAR-85) and alternate approach is in (YAH-77).

6.4.2a. Formulation of the problem (TAR-85)

Let q be a scalar parameter of the system. If q changes to q+dq the performance index can then be approximated by :

$$J\ (q + \Delta q) = J(q) + dJ(q) = J(q) + dq\ .\ dJ/dq$$

Hence, the variation of J is proportional to the derivative of J with respect to q. The sensitivity can be measured by the norm of this derivative.

Consider the system (C,A,B) of (6.2.3) and the optimization problem :

$$\min_{K}\ J_1(A,B,C,K) = \int_0^\infty (x'Qx + u'Ru)\ dt$$

subject to $\dot{x} = (A - BKC)\ x = Dx$

where Q and R are semi-positive and positive definite matrices, respectively. It is shown (see § 6.2) that this problem can be written in the following static form :

$$\min_{K}\ J_1\ (A,B,C,K,P) = Tr\ (PX_0)$$

subject to :

$$f\ (A,B,C,K) = D'P + PD + Q_1 = 0 \tag{6.4.2}$$

with $Q_1 = Q + C'\ K'\ RKC$

$$X_0 = E\ (x(0)\ x(0)')$$

The sentivities of J_1 (A,B,C,K,P) with respect to system parameters can be calculated using Theorem 6.1 (we obtain the derivatives using the method of variations (TAR-85), this result is given in Appendix 5). The Lagrangian function of Theorem 6.1 applied to the problem (6.4.2) is written :

$$L_1(A,B,C,K,P,S) = Tr\ (PX_0) + Tr\ (S'\ f\ (A,B,C,K))$$

The sensitivities of the performance index can be obtained analytically :

$$J_A = \frac{\partial J_1}{\partial A} = \frac{\partial L_1}{\partial A} = 2\,PS$$

$$J_B = \frac{\partial J_1}{\partial B} = \frac{\partial L_1}{\partial B} = -\,2\,PSC'K'$$

$$J_C = \frac{\partial J_1}{\partial C} = \frac{\partial L_1}{\partial C} = 2\,K'(RKC - B'P)S$$

$$J_K = \frac{\partial J_1}{\partial K} = \frac{\partial L_1}{\partial K} = 2\,(RKC - B'P)SC'$$

$$\frac{\partial L_1}{\partial S} = f\,(A,B,C,K,P) = D'P + PD + Q_1(C,K) = 0$$

$$\frac{\partial L_1}{\partial P} = g\,(A,B,C,K,S) = DS + SD + X_0 = 0$$

$$(6.4.3)$$

Since we want to keep theses sensitivities near to zero and maintain the performance index J as small as possible, we define a new performance index as :

$$J_2(K) = J_1 + \frac{1}{4}\,Tr\,(J'_A \cdot L \cdot J_A) + \frac{1}{4}\,Tr\,(J'_B \cdot M \cdot J_B) + \frac{1}{4}\,Tr\,(J_C \cdot E \cdot J'_C) + \frac{1}{4}\,Tr(J_K \cdot F J'_K)$$

where L, M, E, F are symmetric semi-definite matrices of appropriate dimension.

Observing that the equations (6.4.3) can be rewritten as follows :

$$J_A = 2\,PS$$
$$J_B = -\,J_A \cdot C'\,K'$$
$$J_C = K' \cdot J_{KC} \qquad\qquad \text{with } J_{KC} = (RKC - B'P)S$$
$$J_K = J_{KC} \cdot C'$$

we note that if $J_A = 0$ (or $J_A < \epsilon_A$) and $J_{KC} = 0$ (or $J_{KC} < \epsilon_{KC}$) all the sensitivities are zero (or sufficiently small). Therefore, the problem can be simplified by considering only the sensitivities J_{AB} and J_{KC}. This is equivalent to assuming that

A and B in one hand, and K and C, in the other hand, vary simu taneously. Then, applying Theorem (6.1) we obtain :

$$J_{AB} = \frac{\partial J}{\partial A . \partial B . K . C} = 2 \ PS = J_A$$

$$J_{KC} = \frac{\partial J}{\partial (KC)} = 2 \ (RKC - B'P)S$$

where J_{AB} (J_{KC}) is the sensitivity of the criterion with respect to simultaneous variations of A and B (K and C). These results are obtained by the method of variations in Appendix 5 (TAR-85). The new performance index J_3, is given by :

$$J_3 (K) = J_1 + \frac{1}{4} \ Tr \ (J_{AB} . \ L . \ J'_{AB}) + \frac{1}{4} \ Tr \ (J_{KC} . \ F . \ J'_{KC})$$

where L and F are appropriate weighting matrices. The global optimization problem can then be formulated as follows :

$$\min_{K} J_3(K) = Tr \ (PX_0) + Tr \ (SPLPS) + Tr \ ((RKC-B'P)SFS \ (RKC-B'P)')$$

subject to $f(K) = D'P + PD + Q_1 \ (K) = 0$

$\qquad\qquad g(K,X_0) = DS + SD' + X_0 = 0$ (6.4.4)

where the matrices A, B and C assume their nominal value. The Lagrangian function is given by :

$$L_3 = J_3 + Tr \ (V' \ f) + Tr \ (W' \ g)$$

Applying Theorem (6.1) to problem (6.4.4), necessary conditions of optimality are :

$$\frac{\partial L_3}{\partial K} = \frac{\partial J_3}{\partial K} = 2 \ [(RKC - B'P)V + R \ (RKC - B'P) \ SFS - B'WS]C' \qquad (6.4.5a)$$

$$\frac{\partial L_3}{\partial V} = f(K,P) = D'P + PD + Q_1 \ (K) = 0 \qquad (6.4.5b)$$

$$\frac{\partial L_3}{\partial W} = g \ (K,S) = DS + SD' + X_0 = 0 \qquad (6.4.5c)$$

$$\frac{\partial L_3}{\partial S} = f_1 \ (K,P,S,W) = D'W + WD + Q_2 \ (K,P,S) = 0 \qquad (6.4.5d)$$

$$\frac{\partial L_3}{\partial S} = g_1 \ (K,P,S,V) = DV + VD' + Q_3 \ (K,P,S,X_0) = 0 \qquad (6.4.5e)$$

where $D = A - BKC$

$\quad Q_1(K) = Q + C' K' RKC$

$\quad Q_2(K,P,S) = SPLP + PLPS + (RKC-B'P)' (RKC-B'P)SF +$

$\qquad\qquad\qquad FS(RKC-B'P)'(RKC-B'P)$

$\quad Q_3(K,P,S,X_0) = X_0 + SSPL + LPSS - B(RKC-B'P) SFS - SFS (RKC - B'P)'B'$

Particular case

In the case of state feedback ($C = I$ and $F = 0$) the equation (6.4.5) becomes :

$$\frac{\partial J_3}{\partial K} = (RK - B'P) V - B'WS$$

where $f(K,P) = D'P + PD + Q + K'RK = 0$

$\quad g(K,S,X_0) = DS + SD' + X_0 = 0$

$\quad f_1(K,P,S,W) = D'W + WD + SPLP + PLPS = 0$

$\quad g_1(K,P,S,V,X_0) = DW + VD' + X_0 + SSPL + LPSS = 0$

The optimal feedback matrix, is given by :

$$Kopt = R^{-1}B'P + R^{-1}B'WSV^{-1}$$
$$\quad = \quad K_1 \quad + \quad K_2$$

where K_1 is the optimal solution for the nominal values of A and B, and K_2 is the part of Kopt which minimizes the sensitivites, i.e., it minimizes ($J_3 - J_1$). Observe that Kopt depends on the initial conditions through X_0.

6.4.2b. Solution of the problem (TAR-85)

The solution of the system of equations (6.4.5) can be obtained by using a gradient technique : guess an initial feedback matrix $K_0 \in K_c$ (K_c is the set of admissible feedback matrices) such that the closed-loop system is stable. We can calculate P and S from (6.4.5b) and (6.4.5c), then calculate V and W from (6.4.5d) and (6.4.5e), and the gradient from (6.4.5a). The feedback matrix can be updated by successive iterations until the gradient $\frac{\partial J_3}{\partial K} = 0$, giving a local minimum.

For this, one of the algorithms presented in § 6.3 can be used. Consider, for example, the algorithm of Geromel and Bernussou (see § 6.3.1) :

Step 1 : Find $K^1 \in K_c$ such that the closed-loop system is stable by using the algorithm of Armentano and Singh (see § VI.3.1a).

Step 2 : Solve the Lyapunov equations (6.4.5b), (6.4.5c), (6.4.5d), and (6.5.4e) :

(6.4.5b) : $f = 0 \longrightarrow P$

(6.4.5c) : $g = 0 \longrightarrow S$

(6.4.5d) : $f_1 = 0 \longrightarrow W$

(6.4.5e) : $g_1 = 0 \longrightarrow V$

and determine the matrix gradient

Step 3 : Find a feasible direction $G \in K_c$. If $\|G\| \in$ STOP, otherwise go to 4.

Step 4 : Update $K \longleftarrow K - \alpha G$ (α : step size) and go to 2.

Taking into account parameter variations introduces two supplementary Lyapunov equations at step 2.

The gradient projection on the constraint (step 3) is the same as in section (6.3.1b), and consists in setting to zero all the terms corresponding to $k_{ij} = 0$.'

The algorithm has the same properties as the algorithm of Geromel and Bernussou, i.e., uniform convergence to a local minimum and stability of the closed-loop system at each iteration. It requires more CPU time because of the two supplementary Lyapunov equations at Step 2. This is the "price" for robustness. A Fortran routine corresponding to this algorithm is provided in Appendix 6.

Example 6.4 (TAR-85). Consider again the Example 6.1, with the initial data :

$X_0 = I$, $L = I$, $F = 0$

and K^1 given by (6.3.7). After 32 iterations we obtain :

$$K^{32} = \begin{bmatrix} 1.549 & 0.437 & & & & \\ \hline & & 1.526 & 0.57 & & \\ & & 0.993 & 2.33 & & \\ \hline & & & & 1.91 & 0.275 \end{bmatrix}$$

The convergence test is satisfied with $\epsilon = 0.001$. $J(K^1) = 8.935$, $J_1(K^{32}) = 3.049$ and $J_3(K^{32}) = 3.232$.

Comparing these results with those of the Example 6.1, we note that the criterion value $J_3(K^{32})$ is 1.077 times superior to $J(K^{59})$ and that the CPU time is twice longer.

Remark 6.5. Performance index sensitivity reduction can be achieved with a prescribed degree of stability :

If the feedback control structure allows pole assignment, the results of sections (6.4.1) and (6.4.2) can be used together. The equations (6.4.5) must be solved after changing the dynamic matrix A by A + αI, where α is the desired degree of stability.

6.4.3. Robust control with respect to large perturbations in the system dynamics

This section provides conditions on the perturbation bounds such that the stability of decentralized control system is not affected.

Consider again the system (6.2.1) with S control stations which is given in compact form by :

$$(S) \quad \begin{cases} \dot{x}(t) = A\ x(t) + B\ u(t) \qquad x(t_0) = x_0 \\ \\ y(t) = C\ x(t) \end{cases} \qquad (6.4.6)$$

Let the decentralized output feedback be given by :
$$u(t) = K\ y(t) \quad (K = \text{block-diagonal matrix}) \qquad (6.4.7)$$
and suppose that the closed-loop system :
$$\dot{x}(t) = (A + BKC)\ x(t) = D\ x(t)$$
is stable.

When the decentralized control (6.4.7) is applied to the system (6.4.6) the performance index :
$$J = \int_0^\infty (x'\ QX + u'\ Ru)\ dt$$
can be written as :
$$J = x(0)'\ P\ x(0)$$
where P satisfies the matricial equation :
$$D'P + PD + Q_1 = 0 \qquad (6.4.8)$$
with $D = A + BKC$
$$Q_1 = C'\ K'\ RKC + Q \qquad (6.4.9)$$

For convenience, it is assumed that the matrix Q_1 is a nonsingular matrix, which guarantees that P is positive definite.

6.4.3.a. Characterization of the perturbations

The model of the perturbated system and control (large parameter variations, modelling errors...), is taken as :

$$(S^*) \begin{cases} \dot{x}_p = A\ x_p + B\ u_p + A_p\ x_p + B_p\ u_p \qquad x^*\ (t_0) = x_0 \\ \\ u_p = KC\ x_p \end{cases} \qquad (6.4.10a)$$

where the matrices A, B and C are those of the nominal system (S). All the perturbations are lumped into the matrices Ap and Bp.

One way to characterize the perturbations is (PEK-83, 84a,b) to combine the
information concerning their physical nature and their mathematical modelisation. One
defines the directions which are the most probable from the physical point of view.
The perturbations are decomposed in two components, one of which lies along the
given direction in the space of all perturbation matrices A_p and B_p :

$$A_p = q(t)\ A^* + dA$$
$$B_p = q(t)\ B^* + dB \qquad\qquad (6.4.10b)$$

where q(t) is a scalar function, and dA and dB represent the perturbations in
system dynamics which lie out of the directions A* and B*.

The above characterization can be motivated by the fact that a designer,
following his intuition and experience, has enough information to select the most
appropriate directions in the space of all perturbation matrices. Other interpretations
can be given and for example, for weakly coupled systems, the perturbation directions are defined by (PEK-79) :

$$A^* = \begin{bmatrix} 0 & A_{12} \cdots\cdots A_{1N} \\ A_{21} & 0 \cdots\cdots\cdots \\ \vdots & \vdots \ddots \\ A_{N1} & A_{N2} \cdots\cdots 0 \end{bmatrix} \qquad\qquad B^* = \begin{bmatrix} 0 & B_{12} \cdots\cdots B_{1N} \\ B_{21} & 0 \cdots\cdots\cdots \\ \vdots & \vdots \ddots \\ B_{N1} & B_{N2} \cdots\cdots 0 \end{bmatrix}$$

and for singularly perturbed systems by (KOK-76) :

$$C^* = \begin{bmatrix} 0 & 0 \\ A_{21} & A_{22} \end{bmatrix} \qquad\qquad B^* = \begin{bmatrix} 0 \\ B_2 \end{bmatrix}$$

Another reason for considering large perturbations is that it may be necessary
to change the parameter values during the operation of the system. In this case, the
explicit expressions of the matrices A* and B* are generally known.

6.4.3.b. - Robustness characterization

Before giving the general results, let us consider some particular cases :

1. - <u>Particular case 1</u> (PEK-84a). Consider the particular case for which $q(t) = 0$, then the perturbation matrices are given by : $A_p = dA$ and $B_p = dB$ and the closed-loop perturbed system becomes :

$$
(S*) \quad
\begin{aligned}
x_p &= (A + BKC) \, x_p + (dA + dB.KC) \, x_p \\
&= D \, x_p + dD \, x_p = (D + dD) \, x_p
\end{aligned}
$$

The bounds of the perturbations, preserving the stability of the perturbed system are given by the following theorem :

<u>Theorem 6.3.</u> If the perturbation matrices dA and dB satisfy the inequality :

$$Q_1 - dD'. P - P. \, dD > 0 \tag{6.4.11a}$$

which in turn is satisfied if :

$$dD. \, Q_1. \, dD' < 1/4 \, P^{-1} \, Q_1 \, P^{-1} \tag{6.4.11b}$$

then the perturbed system S* is asymptotically stable. The matrices P and Q_1 are given by (6.4.8) and (6.4.9) respectively.

If the matrices dA and dB are unknown, but one has some knowledge about their size, the following theorem can be used :

<u>Theorem 6.4.</u> If the perturbation matrices dA and dB satisfy one of the following inequalities :

I) $\quad \|dD\|_s < \dfrac{\lambda_{min}(Q_1)}{2 \, \lambda_{max}(P)}$

II) $\quad \|dD\|_s < [\dfrac{\lambda_{min}(P^{-1}. \, Q_1. \, P^{-1})}{4 \, \lambda_{max}(Q_1)}]^{1/2}$

for all $t \in [0, \infty[$, then the perturbed system remains asymptotically stable. P and Q_1 are given by (6.4.8) and (6.4.9) and $\|dD\|_s = \|dA\|_s + \|dB\|_{2s} - \|KC\|_s$
where $\|\cdot\|_s$, λ max (.) and λ min (.) denote spectral norm[2], maximum and minimum eigenvalue of (.), respectively.

<u>Lemma 6.4.</u> The perturbation bounds defined in Theorem 6.4 and the dominant closed-loop pole of the nominal umperturbed system are linked through :

$$\|dD\|_s < - \text{Re} \, [\lambda \text{ max (D)}]$$

---

[2] The spectral norm is $\|Y\|_s = [\text{max} \, (YY')]^{1/2}$.

2. - <u>Particular case 2</u> (PEK-84b) : Suppose that the perturbation matrices A_p and B_p lie along the directions A^* and B^*, i.e. $dA = 0$ and $dB = 0$. Now, the closed-loop perturbed system, say S_2^*, can be represented by :

$$(S_2^*) \quad \dot{x}_p = (A + BKC) \, x_p + q \, (t) \, (A^* + B^*KC) \, x_p = D \, x_p + q(t) \, D^* \, x_p$$
$$= (D + q(t) \, D^*) \, x_p$$

Then, the following theorem can be established :

<u>Theorem 6.5.</u> If the following inequalities are satisfied :

$$\lambda_{min}^{-1} \, (H) = q_{min} < q(t) < q_{max} = \lambda_{max}^{-1} \, (H)$$

for all $t \in [\, 0, \, \infty[$, then the perturbed system (S^*) remains asymptotically stable. $q(t)$ is a memoryless, time-varying non linearity, and

$$H = Q^{-1} \, (D^{*\prime}P + P \, D^*)$$

P and Q_1 are given by (6.4.8) and (6.4.9).

<u>Corollary 6.1.</u> If λ_{min} (H) is not negative, then the bound q_{min} does not exist and $q(t) \in \,] -\infty, \, q_{max}]$. If λ_{max} (H) is not positive, then the bound q_{max} does not exist and $q(t) \in [\, q_{min}, \, + \infty[$

Refering to Theorem 6.5, Petkovski (PEK-84b) proposes the following procedure to determine the largest positive number q_{max} and the smallest negative number q_{min} such that the perturbed system (S_2^*) remains stable :

<u>Step 1</u> : Using Theorem 6.5 determine $(q_{min})_0$ and $(q_{max})_0$.
<u>Step 2</u> : Consider the perturbed system (Sj) as unperturbed, i.e. :
$$A = A + (q_{min})_{j-1} \, A^*$$
$$B = B + (q_{min})_{j-1} \, B^*$$
and determine $(q_{min})_j$ using Theorem 6.5.
<u>Step 3</u> : If the closed-loop system (Sj) is stable, go to Step 2. Otherwise,

$$q_{min} = \sum_{i=0}^{j} \, (q_{min})_i.$$

<u>Step 4</u> : The smallest q_{min} is given by :

$$\overline{q}_{min} = \sum_{i=0}^{\infty} (q_{min})_i$$

Step 5 : consider the perturbed system $(\hat{S}j)$ as unperturbed i.e. :

$$A = A + (q_{max})_{j-1} A^*$$
$$B = B + (q_{max})_{j-1} B^*$$

and determine $(q_{max})_j$ using Theorem 6.5.

Step 6: If the closed loop system $(\hat{S}j)$ is stable, go to step 5. Otherwise,

$$\overline{q}_{max} = \sum_{i=0}^{j} (q_{max})_i.$$

Step 7 : The largest $\overline{q}_{max}$ is given by :

$$\overline{q}_{max} = \sum_{i=0}^{\infty} (q_{max})_i$$

The following lemma provides an alternative expression for the bounds of the scalar function $q(t)$ (observe the similarity with Theorem 6.4).

Lemma 6.5. If $q(t)$ satisfies the condition :

$$q(t) \quad . \quad \| D^* \|_s < \frac{\lambda_{min} (Q1)}{2 \lambda_{max} (P)}$$

for all $t \in [0, \infty[$, the closed-loop perturbed system is asymptotically stable. P and Q_1 are given by (6.4.8) and (6.4.9) and

$$\| D^* \|_s = \| A^* \|_s + \| B^* \|_s \cdot \| KC \|_s$$

The following lemma is the analog of Lemma 6.4.

Lemma 6.6. The perturbation bounds of $q(t)$ and the dominant closed-loop pole of the nominal unperturbed system (S) are linked through :

$$|q(t)| \cdot \| D^* \|_s \leqslant - Re [\lambda_{max} (D)]$$

3. - General case (PEK-83). In the case of general perturbations (6.4.10b), the closed-loop perturbed system S^* is given by :

$$S^* \quad \begin{aligned} \dot{x}_p &= (A+BKC) x_p + q(t) (A^* + B^* KC) x_p + (dA + dB.KC) x_p \\ &= D x_p + q(t) D^* x_p + dD . x_p \\ &= (D + q(t) D^* + dD) x_p \end{aligned}$$

Theorem 6.6. If the perturbation matrix dD satisfies the inequality :

$$\| dD \|_s \leqslant \frac{\lambda_{min}[Q_1 - q(t) (D^{*'}P + PD^*)]}{2 \lambda_{max} (P)}$$

for all $t\epsilon[\ 0,\ \infty[$, then the perturbed system S^* remains asymptotically stable. P and Q_1 are given by (6.4.8) and (6.4.9) and $\| dD \|_s = \| dA \|_s + \| dB \|_s \cdot \| KC \|_s$

4. - <u>Robustness characterization with a prescribed degree of stability</u> (PEK-84b).
When the decentralized control is determined such that the unperturbated closed-loop system is asymptoticaly stable with degree α, the above results (Theorems 6.3, 6.4, 6.5, 6.6, Lemmas 6.4, 6.5, 6.6, and Corollary 6.1) hold if the matrices P and Q_1 are given by :

$$(D+ \alpha I)' P + P (D + \alpha I) + Q + C' K' RKC = 0$$
$$Q_1 = Q + C' K' RKC + 2 \alpha P$$

We can observe that if the system is stable with degree α, then $- Re[\lambda_{max} (D)] \geqslant \alpha$ Therefore, Lemma 6.4 (or Lemma 6.6) establishes a relationship between the acceptable perturbations and the prescribed degree of stability α. Consequently, the parameter α, can be used as a design parameter.

6.5. - ROBUST DECENTRALIZED SERVOMECHANISM PROBLEM

This section gives an overview of the results obtained by Davison in reference to the so called "Decentralized servomechanism problem", considered in various forms (DAV-76a,b,c,d, 77b, 78a, 79a, 82). Our attention focuses on the results obtained within a robust control approach (DAV-76c, 77b, 78a, 79a, 82). The systems under consideration can be perturbated by large variations of the plant parameters and dynamics and by external disturbances. The problem consists in designing a decentralized controller such that the closed-loop perturbated system remains stable and that satisfactory tracking or regulation occurs.

6.5.1. - Problem formulation

Consider a linear time-invariant system, with S stations, described by :

$$\begin{aligned}
\dot{x} &= A x + \sum_{i=1}^{S} B_i u_i + E \omega \\
y_i &= C_i x + D_i u_i + F_i \omega , \quad (i=1,\ldots,S) \\
y_i^m &= C_i^m x + D_i^m u_i + F_i^m \omega , \quad (i=1,\ldots,S) \\
e_i &= y_i - y_i^d , \quad (i=1,\ldots,S)
\end{aligned}$$

(6.5.1)

261

where $x \in R^n$ is the state, $u_i \in R^{m_i}$, $y_i \in R^{r_i}$, $y_i^m \in R^{r_i^m}$ $(r_i^m \leqslant r_i)$ are the input, the output to be regulated, and the measurable output at station i. $\omega \in R^q$ is the disturbance vector which may or may not be measurable, y_i and e_i are the desired reference output and the output local error at station i. Define :

$$B = [B_1, \ldots, B_S]$$
$$D = \text{block-diag. } (D_1, \ldots, D_S)$$
$$D^m = \text{block-diag. } (D_1^m, \ldots, D_S^m) \tag{6.5.2}$$

$$C = \begin{bmatrix} C_1 \\ C_S \end{bmatrix} \quad C_m = \begin{bmatrix} C_1^m \\ C_S^m \end{bmatrix} \quad F = \begin{bmatrix} F_1 \\ F_S \end{bmatrix} \quad F^m = \begin{bmatrix} F_1^m \\ F_S^m \end{bmatrix} \quad y^d = \begin{bmatrix} y_1^d \\ y_S^d \end{bmatrix} \quad e = \begin{bmatrix} e_1 \\ e_S \end{bmatrix}$$

and assume that ω belongs to the following class of systems :

$$\dot{z}_1 = A_1 z_1$$
$$\omega = H_1 z_1 \tag{6.5.3}$$

where $z_1 \in R^{n_1}$ and $z_1(0)$ may or may not be known, and the desired reference output arises from the following class of systems :

$$\dot{z}_2 = A_2 z_2$$
$$z^d = H_2 z_2 \tag{6.5.4}$$
$$y^d = G z^d$$

where $z_2 \in R^{n_2}$ and $z_2(0)$ is known. It is also assumed without loss of generality that :

$$\text{rank} \begin{bmatrix} E \\ F \end{bmatrix} = \text{rand } H_1 = q$$
$$\text{rank } G = \text{rank } H_2 = \dim (z^d)$$

and that (H_1, A_1), (H_2, A_2) are observable. In addition, we assume that the systems (6.5.3) and (6.5.4) are unstable to avoid triviality.

The "robust decentralized servomechanism problem" is defined in (DAV-76c) as follows :

Find a decentralized linear time-invariant controller (S local controllers) for the system (6.5.1) - (6.5.4) such that :

* The closed-loop system is asymptotically stable.

* Asymptotic tracking, in presence of disturbances, occurs independently of all arbitrary perturbations in the plant model (6.5.1) (e.g. plant parameters or plant dynamic including changes in model order) which do not affect the stability of the resultant closed-loop system, i.e. $\lim_{t\to\infty} e(t) = 0$ $\forall x(0) \in R^n$, $\forall z_1(0) \in R^{n_1}$, $\forall z_2(0) \in R^{n_2}$ and for all controller initial conditions.

6.5.2. - Existence of a solution

The conditions under which a robust decentralized controller exists are provided.

6.5.2.a. - General case (DAV-76c, 77b)

Define $r = \sum_{i=1}^{S} r_i$, $m = \sum_{i=1}^{S} m_i$ and $r^m = \sum_{i=1}^{S} r_i^m$ and the matric C_m^* of dimension $(r^m + r) \times (n+r)$ as follows :

$$C_m^* = \quad C_1^{*\prime}, \ C_2^{*\prime} , \ \ldots, \ C_S^{*\prime} \quad ' \tag{6.5.5a}$$

where the C_i^*'s are given by :

$$C_1^* = \begin{bmatrix} C_1^m & 0 & 0\ldots\ldots 0 \\ 0 & I_{r_1} & 0\ldots\ldots 0 \end{bmatrix}$$

$$C_2^* = \begin{bmatrix} C_2^m & 0 & 0\ldots\ldots 0 \\ 0 & 0 & I_{r_2}\ldots\ldots 0 \end{bmatrix} \tag{6.5.5.b}$$

$$C_3^* = \begin{bmatrix} C_S^m & 0 & 0\ldots\ldots 0 \\ 0 & 0 & 0\ldots\ldots I_{r_S} \end{bmatrix}$$

The minimal polynomials of A_1 and A_2 of (6.4.3) (6.4.4) are denoted by $\varphi_1(s)$ and $\varphi_2(s)$. The least common multiple of $\varphi_1(s)$ and $\varphi_2(s)$ (multiplicity included) is given by :

$$\prod_{i=1}^{r} (s - \lambda_i) = s^g + p_g\, s^{g-1} + p_{g-1}\, s^{g-2} + \ldots + p_2\, s + p_1 \tag{6.5.6}$$

where

$\lambda_1, \lambda_2, \ldots, \lambda_g$ are its zeros.

<u>Theorem 6.7</u> (DAV-76c, 77b). A solution to the robust decentralized servomechanism problem for the system (6.5.1) - (6.5.6) exists if and only if the following conditions all hold :

(i) The system (C_m, A, B) has no unstable decentralized fixed modes.

(ii) The set of decentralized fixed modes of the g systems :

$$\left\{ C_m^* \ , \ \begin{bmatrix} A & 0 \\ C & \lambda_j I \end{bmatrix} \ \begin{bmatrix} B \\ D \end{bmatrix} \right\} \quad (j=1,\ldots,g), \text{ do not contain } \lambda_j, \ (j=1,\ldots,g),$$

respectively.

(iii) The output y_i is contained in y_i^m, (i=1,...,S), i.e., y_i is physically measurable.

In the case for which $m_i = r_i$, (i=1,...,S), we have this simpler condition :

<u>Corollary 6.2</u> (DAV-78a). Assume that $m_i = r_i$, (i=1,...,S), then there exists a solution to the decentralized robust servomechanism problem for the system (6.5.1) - (6.5.6) if and only if :

$$\text{rank} \begin{bmatrix} A - \lambda_i I & B \\ C & D \end{bmatrix} = n + r \qquad (i=1,...,g)$$

The condition of the above corollary means that no eigenvalue λ_j (j=1,...,g) of (6.5.6) coincides with a transmission zero of the system (see Appendix 1).

6.5.2.b. - Particular case of interconnected systems (DAV-76c, 79a). The plant considered here is a composite system, consisting of interconnected subsystems :

$$\dot{x}_i = A_i x_i + B_i u_i + E_i \omega + \sum_{\substack{j=1 \\ j\neq i}}^{S} A_{ij} x_j$$

$$y_i = C_i x_i + D_i u_i + F_i \omega \qquad (6.5.7)$$

$$y_i^m = C_i^m x_i + D_i^m u_i + F_i^m \omega$$

$$e_i = y_i - y_i^d \qquad (i=1,\ldots S)$$

where $x_i \in R^{n_i^*}$ is the state, and u_i, y_i, y_i^m, y_i^d and ω are defined as in the last section. By assumption the interconnection matrix is given by the general model :

$$A_{ij} \triangleq H_{ij} K_{ij} M_{ij} \quad \substack{(i,j=1,\ldots,S) \\ i\neq j} \qquad (6.5.8)$$

where K_{ij} denotes the interconnection gain connecting the subsystems i and j. The ith subsystem is obtained by setting $A_{ij} = 0$, (j=1,...,S) and i≠j, in (6.5.7).

Theorem 6.8 (DAV-76c, 79a). Assume that there exists a solution to the robust centralized servomechanism problem (DAV-75) for each subsystem of (6.5.7).

(i) Then there exists a solution to the robust decentralized servomechanism problem for the composite system (6.5.7) if the interconnection gains K_{ij} are "small enough".

(ii) Assume, in addition, that (C_i^m, A, B_i) is controllable and observable for (i=1,...,S), then there exists a solution to the robust decentralized servomechanism problem for the composite system (6.5.7) for almost all interconnection gains K_{ij}.

(iii) Assume that the interconnection matrices A_{ij} of (6.5.8) have the property that $H_{ij} = B_i$, $M_{ij} = C_j$ and $D_i = 0$ for (i=1,...,S), then there exists a solution to the robust decentralized servomechanism problem for the composite system (6.4.7) if and only if there exists a solution to the robust centralized servomechanism problem for each subsystem of (6.4.7).

6.5.3. - Robust decentralized controller design

6.5.3.a. - Controller structure

Consider the system (6.5.1) and assume that there exists a robust decentralized controller, then any decentralized controller which regulates (6.5.1) has the following structure (DAV-76c, 77b) :

$$u_i = K_i \, v_i + K_i^* \, w_i \quad (i=1,...,S) \tag{6.5.9}$$

where $v_i \in R^{r_i^s}$ is the output of a decentralized servo-compensator, and $w_i \in R$ is the output of a decentralized stabilizing compensator.

Consider the system (6.5.1) - (6.5.4), then a decentralized servo-compensator for (6.5.1) (DAV-76c) is a controller with input $e_i \in R^{r_i}$ and output $v_i \in R^{r_i^s}$ given by :

$$\dot{v} = \overline{C}_i \, v_i + \overline{B}_i \, e_i \tag{6.5.10}$$

$$\overline{C}_i = \text{block-diag.} \; (\underbrace{\overline{C}_*, \; \overline{C}_*, \ldots \; \overline{C}_*}_{\text{i matrices}})$$

$$\overline{B}_i = \text{block-diag.} \ (\underbrace{\overline{B}_*, \ \overline{B}_*, \ \ldots, \ \overline{B}_*}_{i \ \text{matrices}})$$

$\overline{C}_*$ and $\overline{B}_*$ are the (rxr) companion matrix and the (qx1) matrix, defined below :

$$\overline{C}_* = \begin{bmatrix} 0 & 1 & & & \\ 0 & 0 & 1 & & \\ \vdots & \vdots & & \ddots & \\ \vdots & \vdots & 0 & & \ddots \\ \vdots & \vdots & \vdots & & 1 \\ -P_1 & -P_2 & -P_3 & \cdots & -P_g \end{bmatrix} \qquad \overline{B}_* = \begin{bmatrix} 0 \\ \vdots \\ \vdots \\ \vdots \\ 0 \\ 1 \end{bmatrix}$$

P_i, (i=1,...,g), are given by (6.5.6).

The decentralized stabilizing compensator (DAV-76c), with inputs y_i^m, v_i, u_i and output w_i, (i=1,...,S), is given by :

$$\dot{z}_i = G_i^0 z_i + G_i^1 y_i^m + G_i^2 v_i$$

$$w_i = G_i^3 z_i + G_i^4 y_i^m + G_i^5 v_i \qquad\qquad (6.5.11)$$

where

$$\hat{y}_i^m = y_i^m - D_i^m u_i.$$

The controller structure as described above is illustrated in Figure 6.3.

The gain matrices K_i, K_i^*, G_i^0, G_i^1, G_i^2, G_i^3, G_i^4 and G_i^5 can be determinated through the decentralized stabilization scheme of Wang and Davison (WAN-73) in order to stabilize and give the desired behaviour to the following augmented system :

$$\begin{bmatrix} x \\ \dot{v}_1 \\ \vdots \\ \vdots \\ \dot{v}_S \end{bmatrix} = \begin{bmatrix} A & 0 & \cdots & \cdots & 0 \\ \overline{B}_1 C_1 & \overline{C}_1 & \cdots & \cdots & 0 \\ \vdots & & \ddots & & \\ \overline{B}_S C_S & 0 & \cdots & \cdots & \overline{C}_S \end{bmatrix} \begin{bmatrix} x \\ v_1 \\ \vdots \\ \vdots \\ v_S \end{bmatrix} + \begin{bmatrix} B \\ \text{block-diag.} \ (\overline{B}_1 D_1, \ldots \overline{B}_S D_1) \end{bmatrix} U$$

$$(6.5.12.a)$$

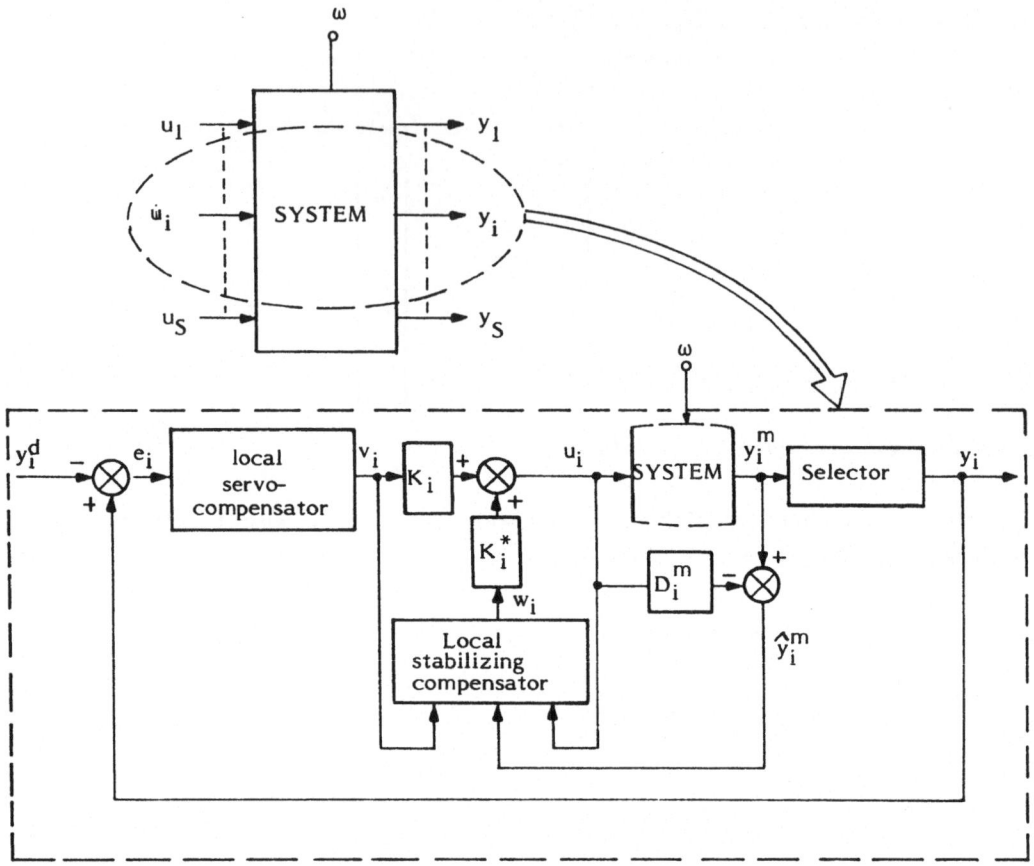

Fig. 6.3 : Controller structure

$$\begin{bmatrix} y_i^m \\ \\ v_i \end{bmatrix} = \begin{bmatrix} C_i^m \, X \\ \\ v_i \end{bmatrix} \qquad (i=1,...,S)$$

$(6.5.12b)$

The system (6.5.12) has decentralized fixed modes equal to the decentralized fixed modes of (C_m, A, B) (if any).

6.5.3.b. Controller optimization

In general, the optimization of the decentralized stabilizing compensator must guarentee :

(i) fast response

(ii) low interaction in the system, i.e., when a reference output signal changes, the other outputs should remain as close as possible of their previous values.

The parameter optimization method proposed by Davison et al. (DAV-73, 79a, 81, 82) minimizes a quadratic performance index of the form :

$$J = E \int_0^\infty (x' Qx + u' Ru) \, dt$$

where E denotes the expectation operator, subject to any imposed engineering constraints. In particular Davison and Chang (DAV-82) showed that, if the system (6.5.1) is open-loop stable and if Re $(\lambda_i) = 0$, $(i=1,...,g)$, where the λ_i' s are given by (6.5.6), (e.g. we have polynomial-sinusoidal type of disturbances and reference signals), then there always exists an initial feasible starting point for this parameter optimization problem.

6.5.3.c. Some properties of the controller (DAV-76c)

1. Using the robust controller described before, one can locate the eigenvalues of the closed-loop system in any nonempty symmetric region of the complex plane (the decentralized fixed modes of (C_m, A, B) (if any) must be in the desired region).

2. A robust decentralized controller exists generically (WAN-73) for "almost all" plants (6.5.1) provided that :

(i) $m_i \geqslant r_i$ ($i=1, ..., S$)

(ii) the output y_i is physically measurable at station i.

If either (i) or (ii) do not hold, then a solution to the robust decentralized servomechanism problem never exists.

6.5.4. - Sequentially stable robust controller design

A realistic situation is to consider that no central authority is allowed for calculating the local controllers, and that a complete knowledge of the mathematical model of the plant is not necessarily available at any control station. The problem is thus to find a solution to the robust decentralized servomechanism problem for system (6.5.1) under the two following constraints :

(i) The controller synthesis must be carried out in a sequential stable way (DAV-79b), i.e., the controllers can be connected to the system one after another resulting at any time in a stable closed-loop system.

This is motivated by physical constraints like time lags associated with the controllers connection, lack of communication hard structure... etc. If this property is achieved with a connection sequence $(1,2,\ldots,S)$, the controller is said to be se-quentially stable with respect to control station order $(1,2,\ldots,S)$.

(ii) No central authority must be used in decentralized decision making, and each control station possesses only a limited knowledge of the mathematical model of the system (typically, each station of a large scale system possesses only a local model (DAV-82)).

6.5.4.a. - Existence of a controller

Consider the system (6.5.1) with $\omega=0$ and $y^d=0$ given by :
$$\dot{x} = Ax + \sum_{i=1}^{S} B_i u_i$$

$$y_i^m = C_i^m x + D_i^m u_i \tag{6.5.13}$$

$$y_i = C_i x + D_i^m u_i \qquad (i=1,\ldots,S)$$

Apply the control :
$$u_i = \hat{K}_i \hat{y}_i^m + \hat{K}_i^* v_i^o \qquad (i=1,\ldots,S) \tag{6.5.14a}$$

where
$\hat{K}_i \in R^{r_i \times r_i^m}$ and $\hat{y}_i^m = y_i^m - D_i^m u_i$, and where the following controllers have already been applied to control stations $(1,2,\ldots,i-1)$, $i \geq 2$:
$$v_j^o = K_j v_j + K_j^* w_j \qquad (j=1,2,\ldots,i-1) \qquad i \in \{2,\ldots,S\} \tag{6.5.14b}$$
v_j is the output of a decentralized servocompensator given by (6.5.10) and w_j is the output of a decentralized stabilizing compensator given by (6.5.11).

The minimal state realization of the resultant closed-loop system obtained by applying the controller (6.5.14a,b) to the system (6.5.13) for control station i (with input v_i^o and output y_i^m) is called the ith station's local model of the system.

The problem of finding a robust decentralized servomechanism control with sequential stability, when each station possesses only a local model of the system and when the central decision making authority is not allowed is called the local model robust decentralized servomechanism problem.

It is assumed (DAV-79b, 82) that each control station knows the disturban-ces/reference signals poles, i.e. $\lambda_1,\ldots,\lambda_g$ of (6.5.6), and has the same performance criterion, here stability or pole assignability (except for the decentralized fixed modes, if any) of the closed-loop system.

Theorem 6.9 (DAV-82). Consider the system (6.5.1) in which A is assumed to be asymptotically stable. Then there exists a solution to the local model robust decentralized servomechanism problem if and only if there exists a solution to the robust decentralized servomechanism problem (see Theorem 6.7).

6.5.4.b. - Controller synthesis

Assuming that Theorem 6.9 holds, the following algorithm provides a synthesis procedure.

<u>Algortihm 6.1.</u> (Decentralized synthesis solution) (DAV-82).

<u>Step 1</u> : Apply the output feedback control :

$$u_i = \hat{K}_i \; \hat{y}_i^m + \hat{K}_i^* \; v_i^o \quad (i=1,\ldots,S)$$

where $\hat{K}_i \in R^{m_i x r_i^m}$, $\hat{K}_i \in R^{m_i x r_i}$ are arbitrary non zero matrices with rank $\hat{K}_i^* = r_i$, and where the $\hat{K}_i$'s are chosen "small enough" so as to maintain the stability of the closed-loop system.

<u>Step 2</u> : Using a centralized synthesis method (DAV-75) and the knowledge of station 1's local model of the system, apply the servocompensator (6.5.10) with i=1 to the terminals of control station 1 and apply the stabilizing compensator : $v_1^o = K_1 \, v_1 + K_1^* \, w_1$ (v_1 is given by (6.5.10) and w_1 is given by (6.5.11) so that the resulting closed-loop system is stable and has a desired dynamic response. The resulting system has thus the property of having y_1 regulated.

<u>Step 3</u> : Repeat sequentially step 2 for (i=2,3,...,S) until all the stations have regulated outputs.

If pole assignment is desired, the above algorithm can be modified as follow (DAV-82).

<u>Algorithm 6.2.</u> (Pole assignment decentralized synthesis) (DAV-82).

Assume with no loss of generality that the control synthesis is proceed in the control station order 1, 2, ..., S.

<u>Step 1</u> : i=1

<u>Step 2</u> : Using a centralized synthesis method and the knowledge of station i's local model of the system (i.e. the minimal realization of the system (6.5.13) already controlled at stations (1,2,...,i-1) with respect to the input u_i and the output $\hat{y}_i^m$), apply the stabilizing compensator ;

$$u_i = K_i \, \hat{y}_i^m + K_i^* z_i$$

$$\dot{z}_i = G_i \, z_i + G_i^* \, y_i^m \qquad\qquad (6.5.15)$$

to station i so that :

$$\begin{bmatrix} A + \sum\limits_{j=1}^{i} B_j K_j C_j & B_1 K_1^* \ldots \ldots B_j K_j^* \\[2mm] G_1^* \, C_1^m & G_1 \cdots\cdots\cdots 0 \\ \vdots & \vdots \quad \ddots \quad \vdots \\ \vdots & \vdots \quad \quad \ddots \quad \vdots \\ G_i^* \, C_i^m & 0 \cdots\cdots\cdots G_i \end{bmatrix}$$

has all its eigenvalues contained in $C\bar{g}$ (except the decentralized fixed modes of $\left\{ \begin{bmatrix} C_1^m \\ C_i^m \end{bmatrix}, A, (B_1, \ldots, B_i) \right\}$ which lie outside of $C\bar{g}$, if any). $C\bar{g}$ is a specified region of $\mathcal{C}$. This is always possible for almost all K_j, G_j, (j=1,2,...,i-1) (DAV-82).

Step 3 : If i=S, stop, otherwise, i=i+1, go to Step 2.

Remark 6.6.

1. If Theorem 6.9 holds, then for almost all gains chosen in steps 1 to 3 of Algorithm 6.1, it is always possible to carry out the synthesis (DAV-82).

2. If the sequential stability constraint is relaxed, then Algorithms 6.1 and 6.2 are still applicable for the case of unstable open-loop systems.

3. Note that the controllers obtained by Algorithms 6.1 and 6.2 are, generally, not unique with respect to the control agent sequence.

4. If $D_i^m = 0$, $D_i = 0$ (i=1,...,S), then the results of this section hold for the general case for which the information flow between control stations is arbitrarily constrained (not necessarily decentralized) (DAV-82). Indeed, as it is pointed out in (WAN-73b), a reordering of the outputs can always be performed to form an equivalent standard decentralized control problem.

6.5.5. - Robust decentralized controller for unknown systems

In this section, we consider that the system (6.5.1) that we want regulate, is not completly known. The only information on the system is the following ;

(i) The system is described by a finite dimensional linear time-invariant model.

(ii) The system is open-loop asymptotically stable.

(iii) The disturbances affecting the system and the tracking reference signals are of polynomial/sinusoidal type, i.e. Re $(\lambda_i)=0$, $(i=1,\ldots,g)$ in (6.5.6).

(iv) The system inputs can be excited, and the system outputs to regulate can be measured, i.e. $y_i = y_i^m$.

With this sole information, it is desired to find a decentralized controller which solves the robust servomechanism problem.

The question is to know whether or not there exists a finite set of experiments (taking into account noisy measurements) to perform on the plant, such that the necessary and sufficient conditions for the existence of a solution to the above problem can be expressed in terms of these experiments. If a solution exists, the following question is to know whether there exists a controller synthesis procedure (using on-line tuning methods) which satisfies the decentralized controller tuning synthesis constraints above :

(i) At any time, one controller can be implemented on one control station only.

(ii) After a controller has been implemented on a given control station, this controller is fixed and cannot be reactualized.

(iii) The resultant closed-loop system must remain stable any time of the controller synthesis.

This problem is called the robust decentralized servomechanism problem for unknown systems.

6.5.5.a. Existence of a solution

Recall that K_d is the set of block-diagonal matrices $K_d = \{ K/K = \text{block-diag.}$ $[K_1,\ldots,K_S], K_i \in R^{m_i \times r_i}, (i=1,\ldots,S).\}$

Definition 6.1 (DAV-78).

1. The steady-state tracking gain parameters $T_k(i,j)$, $(i,j=1,\ldots,S), (k=1,\ldots,g)$ of the system (6.5.1) for the case $m_i = r_i$, $(i=1,\ldots,S)$, are given by :

$$T_k(i,j) \triangleq \begin{cases} C_j (\lambda_k I - A)^{-1} B_i & \text{if } i \neq j \\ \\ C_i (\lambda_k I - A)^{-1} B_i + D_i & \text{if } i=j \end{cases} \qquad (6.5.16)$$

2. The steady-state tracking gain parameters $T_k(i,j;K_i)$, $(i,j=1,\ldots,S)$, $(k=1,\ldots,g)$, of the system (6.5.1) with respect to the input matrices $K_i \in R^{m_i \times r_i}$, $(i=1,\ldots,S)$, where rank $K_i = r_i$, $(i=1,\ldots,S)$, for the case $m_i \geqslant r_i$, $(i=1,\ldots,S)$, are given by :

$$T_k(i,j;K_i) \triangleq \begin{cases} C_j (\lambda_k I - A)^{-1} B_i K_i & \text{if } i \neq j \\ \\ C_i (\lambda_k I - A)^{-1} B_i K_i + D_i K_i & \text{if } i=j \end{cases} \qquad (6.5.17)$$

It is clear that for the case $K_i = I$ and $m_i = r_i$ ($i=1,\ldots,S$), $T_k(i,j) = T_k(i,j;K_i)$. It is worth noticing that the steady-state tracking gain parameter $T_k(i,j)$ is equal to the transfer function matrix between the input u_i and the output y_j. Davison (DAV-76d, 78) suggested algorithms, called "experiments", to evaluate the parameters $T_k(i,j)$ and $T_k(i,j;K_i)$.

Theorem 6.10 (DAV-78a). Consider the system (6.5.1) for which $m_i = r_i$, $i \{1,\ldots,S\}$, if $i \neq i_1, i_2,\ldots,i_d$, and $m_i > r_i$ if $i=i_1,i_2,\ldots,i_d$, and a set of $m_i \times r_i$ input matrices K_i, $i=i_1,\ldots,i_d$, with rank $K_i=r_i$. Then a necessary and sufficient condition for the existence of a solution to the robust decentralized servomechanism problem for unknown systems is that there exists a list of distinct integers (s_1, $s_2,\ldots s_S$) (not necessarily unique), $s_i \in \{1,\ldots,S\}$, such that the following S successive rank conditions hold :

1. $\quad \text{rank} [T_k(s_1, s_1 ; \overline{K}_{s_1})] = s_1 \qquad (k=1,\ldots,g)$

2. $\quad \text{rank} \begin{bmatrix} T_k(s_2, s_2 ; \overline{K}_{s_2}) & T_k(s_2, s_1 ; \overline{K}_{s_2}) \\ \\ T_k(s_1, s_2 ; \overline{K}_{s_1}) & T_k(s_1, s_1 ; \overline{K}_{s_1}) \end{bmatrix} = r_{s_1} = r_{s_2} \qquad (k=1,\ldots g)$

$\vdots$

N. $\quad \text{rank} \begin{bmatrix} T_k(s_S, s_S ; \overline{K}_{s_S}) & \cdots & T_k(s_S, C_1 ; \overline{K}_{s_S}) \\ \\ T_k(s_1, s_S ; \overline{K}_{s_1}) & \cdots & T_k(s_1, s_1 ; \overline{K}_{s_1}) \end{bmatrix} = \sum_{i=1}^{S} r_{s_i} \qquad (k=1,\ldots g)$

where

$$K_{s_i} = \begin{cases} I r_{s_i} & \text{if} \quad s_i \neq i_1,\ldots,i_d \\ \\ K_{s_i} & \text{if} \quad s_i = i_1,\ldots,i_d \end{cases}$$

Assuming that Theorem 6.10 holds, an algorithm is given in (DAV-78) for carrying out the decentralized controller synthesis in terms of the parameters $T_k(i,j)$, using one dimensional on-line tuning methods.

Remark 6.7.

1. If $m_i \geqslant r_i$, (i=1,...,S), Theorem 6.10 holds for almost all (C,A,B,D) systems. On the other hand, if $m_i < r_i$ for some $i \in \{1,...,S\}$, then Theorem 6.10 does not hold, and no solution exists.

2. It is interesting to note that the local controllers synthesis must be carried out in specified sequence (not necessarily unique). If this sequence is not respected then, in general, no controller synthesis can be performed. However, this is not the case if assumption (ii) is relaxed in the tuning synthesis constraints (DAV-79b).

6.6. - DECENTRALIZED CONTROL BY HIERARCHICAL CALCULATION

This section is concerned with the hierarchical calculation methods of a decentralized control for the class of large-scale linear interconnected systems. Two types of algorithms are presented : three-level calculation algorithms (HAS-78a,b, 79) and two-level calculation algorithms (XIN-82).

6.6.1. - Three-level calculation algorithms

This subsection presents the algorithm of Hassan and Singh (HAS-78b) and its extension to the case of robust decentralized control (HAS-79).

6.6.1.a. - Decentralized near-optimal controller (HAS-78b)

Consider the large-scale linear interconnected system described by :

$$\dot{x}_i = A_i x_i + B_i u_i + \sum_{i=1}^{S} A_{ij} x_j \qquad (6.6.1)$$

or, in a compact form, by :

$$\begin{cases} \dot{x} = Ax + Bu + Cz \\ z = Lx \end{cases} \qquad (6.6.2)$$

where A, B and C are appropriate block-diagonal matrices with S blocks, and L is a full matrix representing the interconnections between the systems. We want to control the system (6.6.2) by decentralized state feedback minimizing a quadratic performance index. The optimization problem can be written :

$$\min_{K} \quad J = 1/2 \int_0^T (x' \, Qx + u' \, Ru) \, dt$$

subject to :

$$\dot{x} = Ax + Bu + Cz$$

$$z = Lx \qquad\qquad\qquad (6.6.3)$$

$$u = -Kx$$

where Q and R are appropriate weighting matrices.

It is shown in (SIN-76) that the solution of the above problem has the following form :

$$u = -Gx - Tx \qquad\qquad\qquad (6.6.4)$$

where G is a block-diagonal matrix obtained by solving the decomposed Riccati equation, and T is a full matrix obtained by hierarchical calculation.

Now, substituting (6.6.4) into the criterion, we obtain :

$$J_{opt} = 1/2 \int_0^T (x' \, Qx + x' \, W^* \, x) \, dt$$

with $W^* = (G+T)' \, R \, (G+T)$

Since it is desired to obtain a decentralized control, we constrain T to be a diagonal matrix T_d, and the optimization problem becomes :

$$\min_{T_d} \quad J = 1/2 \int_0^T [(x' \, Qx + x' \, Wx)] dt$$

subject to $\dot{x} = (A-BG) \, x + Cz - B \, T_d \, x$

$$z = Lx \qquad\qquad\qquad (6.6.5)$$

$$W = (G + T_d)' \, R \, (G + T_d)$$

where B is a nxn matrix (if, in practice, B is of lower dimension than nxn, we can introduce additional fictitious controls).

Let G_d (G_0), A_d (A_0), $Q_d(Q_0)$, and B_d (B_0) be the matrices composed of the diagonal (off-diagonal) elements of the matrices G, (A-BG), Q, and B, respectively, then the matrix W can be written :

$$W = (G_d + T_d + G_0)' \, R \, (G_d + T_d + G_0)$$
$$= (F + G_0)' \, R \, (F + G_0)$$

where $F = G_d + T_d$ is a diagonal matrix.

The optimization problem can be rewritten as :

$$\min J = 1/2 \int_0^T [x' Q_d x + x'F' RFx + g (x,F,G_0)] dt$$

with $g (x,F,G_0) = x' (Q_0 + F' R G_0 + G_0 RF + G_0' R G_0) x$

subject to :

$$\dot{x} = A_d x - B_d T_d x + y (x,z,T_d)$$

with $y (x,z,T_d) = A_0 x + Cz - B_0 T_d x$

$$(6.6.6.)$$

To solve this problem, Hassan and Singh (HAS-78b) use a prediction method which consists in adding certain additional linear constraints in order to decompose the optimization problem into a number of independent subproblems for some fixed trajectories supplied by the second level. These trajectories are then improved using a fixed point type algorithm. Let us introduce the additional linear constraints :

$$x^* = x$$
$$T_d^* = T_d$$

$$(6.6.7)$$

Substituting (6.6.7) into (6.6.6), the optimization problem becomes :

$$\min J = 1/2 \int_0^T [x' Q_d x + x^{*'} F' RF x^* + g (x^*, F^*, G_0)] dt$$

with $g (x^*, F^*, G_0) = x^{*'} (Q_0 + F^{*'} R G_0 + G_0' RF^* + G_0' R G_0) x^*$

subject to :

$$\dot{x} = A_d x - B_d T_d x^* + y (x^*, z, T_d^*)$$
$$z = Lx$$
$$T_d^* = T_d$$
$$x^* = x$$

with $\quad y (x^*,z, T_d^*) = A_0 x^* + Cz - B_0 T_d^* x^*$

In order to solve this problem, let us write the Hamiltonian :

$$H = \frac{1}{2} x' Q_d x + \frac{1}{2} x^{*'} F' RF x^* + \frac{1}{2} g (x^*, F^*, G_0) +$$

$$+ \gamma' [A_d x - B_d T_d x^* + y (x^*, z, T_d^*)] + \pi' (Lx - z) +$$

$$+ \beta' (x - x^*) + \sum_{i=1}^n v_i (T_{d_i} - T_{d_i}^*)$$

where π, β, v_i are Lagrange multipliers, and γ is the costate variable.

The necessary conditions for optimality can be written as :

$$\frac{\partial H}{\partial \pi} = 0 \quad \longrightarrow \quad z = Lx \qquad\qquad (6.6.8)$$

$$\frac{\partial H}{\partial z} = 0 \quad \longrightarrow \quad \pi = C' \gamma \qquad\qquad (6.6.9)$$

$$\frac{\partial H}{\partial \beta} = 0 \quad \longrightarrow \quad x^* = x \qquad\qquad (6.6.10)$$

$$\frac{\partial H}{\partial \gamma_i} = 0 \quad \longrightarrow \quad T^*_{d_i} = T_{d_i} \qquad\qquad (6.6.11)$$

$$\frac{\partial H}{\partial T^*_d} = 0 \quad \longrightarrow \quad V = \text{diag} \left[(R \, G_o \, x^* - B'_o \, \gamma) \, x^{*'} \right] \qquad\qquad (6.6.12)$$

$$\frac{\partial H}{\partial x^*} = 0 \quad \longrightarrow \quad \beta = (F'RF + Q_o + F^{*'}RG_o + G'_o RG_o) \, x^* + (A'_o - T^{*'}_d \, B'_o - T'_d \, B'_d) \, \gamma$$

$$\qquad\qquad (6.6.13)$$

Suppose now that x^*, T_d^*, β and V have been provided by the second level, then the Hamiltonian can be decomposed such that each subproblem has only one variable T_{d_i}.

$$\frac{\partial H}{\partial T_{d_i}} = 0 \quad \longrightarrow \quad x_i^{*2} \, (G_{d_i} + T_{d_i}) \, R_i - B_{d_i} \, \gamma_i \, x_i^* + v_i = 0$$

$$\text{or} \quad T_{d_i} = - G_{d_i} - \frac{1}{R_i \, x_i^{*2}} \, (- B_{d_i} \, \gamma_i \, x_i^* + v_i)$$

$$\frac{\partial H}{\partial \gamma_i} = \dot{x}_i = A_{d_i} x_i - B_{d_i} \left[- G_{d_i} - \frac{1}{R_i \, x_i^{*2}} (v_i - B_{d_i} \gamma_i x_i^*) \right] x_i^* + y_i \, (x^*, z, T_d^*)$$

$$\qquad = A_{d_i} x_i - B_{d_i} G_{d_i} x_i^* - \frac{B_{d_i}}{R_i x_i^*} v_i - \frac{B_{d_i}^2}{R_i} \gamma_i + y_i \, (x^*, z, T_d^*) \qquad\qquad (6.6.14)$$

$$\frac{\partial H}{\partial x_i} = - \dot{\gamma}_i = Q_{d_i} x_i + A_{d_i} \gamma_i + k_i + \beta_i$$

with $\quad k_i = L'_i \pi_i$

Let $\gamma_i = P_i \, x_i + \eta_i$, then after minor manipulations we obtain :

$$\dot{P}_i = -2 A_{d_i} P_i + \frac{B_{d_i}^2}{R_i} P_i^2 - Q_i \qquad \text{with } P_i(T) = 0 \qquad (6.6.15)$$

$$\dot{\eta} = (- A_{d_i} + \frac{P_i B_{d_i}^2}{R_i} \eta_i) - P_i [B_{d_i} G_{d_i} x_i^* + \frac{B_{d_i} v_i}{R_i x_i^*} + y_i (x^*, z, T_d^*)] - k_i - \beta_i (6.6.16)$$

Hassan and singh suggest the following three-level algorithm.

<u>Algorithm 6.3</u> (HAS-78b).

<u>Step 0</u> : Guess the initial trajectories z^h and k^h at level 3 for the initial index h=1

<u>Step 1</u> : Guess the initial trajectories x^{*j}, T_d^* , β^j, v^j at level 2, and set the iteration index j=1.

<u>Step 2</u> : Using x^{*j}, T_d^{*j}, β^j, v^j obtained from step 1, calculate P_i, η_i from (6.6.15) and (6.6.16), x from (6.6.14), and γ (from $\gamma = Px + \eta$). Calculate also T_d.

<u>Step 3</u> : Substitute x and γ obtained at level 1, π and z obtained at level 3 into the right sides of (6.6.10)-(6.6.13) to obtain x^{*j+1}, T_{di}^{*j+1}, β^{*j+1}, and v^{j+1}. If the integral of the norm of the differences $(x^{*j+1} - x^{*j})$, $(T_d^{*j+1} - T_d^{*j})$, $(\beta^{j+1} - \beta^j)$, and $(v^{j+1} - v^j)$ are not sufficiently small, go to step 2. Otherwise, go to level 3 and calculate new k^{h+1} and z^{h+1} from (6.6.9) and (6.6.8). If the norm of the differences $(k^{h+1} - k^h)$ and $(z^{h+1} - z^h)$ are small, record T_d as the decentralized gain matrix, otherwise go to step 1 using k^{h+1}, z^{h+1} as the new guesses.

<u>Remark 6.8.</u>

1. The Algorithm 6.3 is a prediction type algorithm, and its convergence can be proved using a similar technique to the one used by Hassan and Singh (HAS-76) for nonlinear systems.

2. The entire calculation is done off-line, and only the decentralized gains are used on-line to compute and implement the optimal decentralized control.

3. The desadvantage of the algorithm is that T_d is initial-state dependent although it is not sensitive to small variations of the initial conditions.

6.6.1.b. - Robust decentralized near-optimal controller (HAS-79).

Hassan, Singh and Titli (HAS-79) extended the approach of the above section to provide a robust decentralized control which ensures exponential stability with a prescribed degree α (in the sense of Anderson and Moore (AND-71), see § 6.4.1) and takes into account external disturbances and structural perturbations (SIL-73, 75,76).

Consider an interconnected dynamical system composed by S subsystems described by :

$$\dot{\overline{x}}_i = \overline{A}_i \, \overline{x}_i + B_i \, \overline{u}_i + \sum_{j=1}^{S} e_{ij} \, A_{ij} \, \overline{x}_j + \overline{d}_i \quad (i=1,\ldots,S)$$

where the e_{ij}'s are the elements of the interconnection matrix E, which are introduced to incorporate any structural perturbation which may occur during the operation of the system. E is continuous in time, with $0 \leqslant e_{ij}(t) \leqslant 1$, $(i,j=1,\ldots,S)$.

We saw in section (6.4.1) that to ensure that the system is stable with degree α, it suffices to consider the performance index :

$$J = \sum_{i=1}^{S} 1/2 \int_0^T e^{2\alpha t} \left[(\overline{x}_i - \overline{x}_i^d)' \, Q_i \, (\overline{x}_i - \overline{x}_i^d) + \overline{u}_i' \, R_i \, \overline{u}_i \right] \, dt$$

where T equals at least 4 times the time constant of the system. With an appropriate variable transformation, we can put this problem into a standard "Linear Quadratic" form, i.e. :

$$\min J = \sum_{i=1}^{S} 1/2 \int_0^T \left[(x_i - x_i^d)' \, Q_i \, (x_i - x_i^d) + u_i' \, Ru \right] dt$$

subject to :

$$\dot{x}_i = A_i \, x_i + B_i \, u_i + \sum_{j=1}^{S} e_{ij} \, A_{ij} \, x_j + d_i(t)$$

with $\quad A_i = \overline{A}_i + \alpha I$

$$d_i(t) = \overline{d}_i(t) \, e^{\alpha t}$$

The optimal control for this system can be written as :

$$u_i = - G_i x - \sum_{j=1}^{S} e_{ij} \, T_{ij} \, x_j - s_i$$

where $G_i = - R_i^{-1} \, B_i' \, P_i$

P_i is the solution of the local (i.e. decomposed) Riccati equation for the ith subsystem. Let $T = [e_{ij} \, T_{ij}]$, then the optimal control, in its global form, becomes :

$$u = - Gx - Tx - S$$

Now, with the same approach and notations as in the last section, the optimization problem can be written.

$$\min J = 1/2 \int_0^T (x-x^d)' \, Q(x-x^d) + x'F' \, RFx + 2 \, x'G_d'RS +$$
$$+ 2 \, x' \, T_d' \, RS + S' \, RS + g \, (x, F, G_0) \qquad (6.6.17)$$

subject to : $\dot{x} = A_d x - B_d T_d x + y \, (x,z,T_d)$

$$z = Lx$$

where $g(x,F,G_0) = (x-x^d)' Q_0 (x-x^d) + x' F' R G_0 x + x'$
$$+ x' G_0' RFx + x' G_0' RG_0 x$$
$$y(x,z,T_d) = A_0 x + Cz - B_0 T_d x + D$$
$$D = d - BS$$

This problem is similar to the problem (6.6.6) and can be solved by using the Algorithm 6.3 after changing the optimality conditions (6.6.8)-(6.6.16) by the appropriate ones.

6.6.2. - Two-level calculation algorithm (XIN-82) -

Xinogalas,Mahmoud and Singh (XIN-82) considered the following optimization problem :

$$\min_{K_i} \quad J = \frac{1}{2} \sum_{i=1}^{S} \int_0^\infty (x_i' Q_i x_i + u_i' R_i u_i) dt$$

subject to $\quad \dot{x}_i = A_{ii} x_i + B_i u_i + \sum_{i=1}^{S} A_{ij} X_j$

$$u_i = - K_i x_i \qquad (i=1,...,S)$$

This problem can be written in a global form as :

$$\min_{K \in K_d} J = 1/2 \int_0^\infty (x' Qx + u' Ru) \, dt$$

subject to :
$$\dot{x} = Ax + Bu \qquad\qquad (6.6.19)$$
$$u = - Kx$$

where B, Q and R are diagonal matrices, and K_d is given by :

$$K_d = \{K/K = \text{block-diag.}[K_1,...,K_S] \, , \, K_i \in R^{m_i x r_i} \, , \, (i=1,...,S)\}$$

It is easy to show that (6.6.19) can be brought back to :

$$\min_{K \in K_d} J = \text{Tr} [(Q + K'RK) S)]$$

subject to :
$$g(S,x_0) = S (A-BK)' + (A-BK)S + X_0 = 0 \qquad\qquad (6.6.20)$$

with $X_0 = E[x(0) \ x(0)'] = \text{diag.} (x_i)$

Let $A_d = \text{diag.} (A_{ii})$

$A_0 = A - A_d$

An alternative formulation of the optimization problem (6.6.20) is given by :

$$\min_{K \in K_d} J = \text{Tr} [(Q + K'RK)S]$$

subject to :

$$g(S, X_0) = S(A_d - BK)' + (A_d - BK)S + X_0 + Z = 0$$
$$Z = A_0 S + SA_0'$$

The corresponding Lagrangian function can be formed as :

$$L = \text{Tr} [(Q + K'RK)S] + \text{Tr} [P \ g(S, X_0)] + \text{Tr} [T(A_0 S + SA_0' - Z)]$$

For this static optimization problem, the necessary conditions for optimality are :

$$\frac{\partial L}{\partial T} = 0 \quad \longrightarrow \quad Z = A_0 P + PA_0'$$

$$\frac{\partial L}{\partial Z} = 0 \quad \longrightarrow \quad T = P$$

$$\frac{\partial L}{\partial P} = 0 \quad \longrightarrow \quad (A_d - BK) S + S (A_d - BK)' + X_0 + Z = 0 \qquad (6.6.21)$$

$$\frac{\partial L}{\partial S} = \quad \longrightarrow \quad (A_d - BK)' T + T (A_d - BK) + Q + K'RK + A_0' P + PA_0 = 0 \qquad (6.6.22)$$

$$\frac{\partial L}{\partial K} = 0 \quad \longrightarrow \quad K = R^{-1} B' M_d S_d^{-1}$$

where $\quad M_d = \text{diag.} (TS)$

$S_d = \text{diag.} (S)$

To solve the above optimality conditions Xinogalas, Mahmoud and Singh (XIN-82) propose the following tow-level algorithm.

Algorithm 6.4 (XIN-82).

Step 1 : Guess an initial value of the decentralized gain matrix K^q.

Step 2 : Compute the eigenvalues of the matrix $(A_d - BK^q)$. If all the eigenvalues have negative real parts, go to step 3. Otherwise, use the algorithm of Armentano and Singh (ARM-81) (see § 6.3.1.a) to compute a stabilizing decentralized feedback matrix K^q, i.e. such that $(A_d - BK^q)$ is asymptoticaly stable.

Step 3 : Start the two-level hierarchical computation structure with guessed values for the matrices Z^q and T^q and send these values, together with the gain matrix K^q, to the first level. Set q=1.

Step 4 : At the first level, (6.6.21) and (6.6.22) are solved using the technique of Bartels and Stewart (BAR-72). The matrices S^q and T^q are conveyed to the second level.

Step 5 : New predictions of the matrices Z, P and K are calculated according to :

$$Z^{q+1} = A_0 S^q + S^q A_n{}'$$
$$P^{q+1} = T^q$$
$$K^{q+1} = R^{-1} B' M_d^q (S_d^q)^{-1}$$

If the conditions :
$$\sqrt{\|Z^{q+1}\| - \|Z^q\|} < \varepsilon_Z$$
$$\sqrt{\|P^{q+1}\| - \|P^q\|} < \varepsilon_P$$
$$\sqrt{\|K^{q+1}\| - \|K^q\|} < \varepsilon_K$$

are satisfied, regard K^{q+1} as the optimal solution and finish the iterative scheme. Otherwise, update the matrices Z^{q+1}, P^{q+1} and K^{q+1} using the rules :
$$Z^{q+1} = c_1 Z^q + d_1 Z^{q+1}$$
$$\quad = c_2 P^q + d_2 P^{q+1}$$
$$K^{q+1} = c_3 K^q + d_3 K^{q+1}$$

where the constants c_j and d_j satisfy $c_j + d_j = 1$ for j = 1,2,3. (The quantities ε_Z, ε_P, ε_K are small preselected tolerance values). Go to step 2.

6.7. - CALCULATION METHODS USING AN INTERCONNECTION MODEL

In this section, we present an alternate approach to the optimization problem, which uses a simple reduced model for the interactions between the subsystems (HAS-78a) (HAS-80) (CHE-81).

6.7.1. - The general interconnection model (HAS-78a)

The optimization problem considered here is :

$$\min J = 1/2 \sum_{i=1}^{S} \int_0^\infty [(x_i - x^d)' Q_i (x_i - x^d) dt + u_i' Ru_i] dt$$

subject to :
$$\dot{x}_i = A_i x_i + B_i u_i + C_i z_i + d_i \qquad (i=1,\ldots,S) \qquad (6.7.1)$$

where $Q_i \geqslant 0$, $R_i > 0$ are appropriate weighting matrices, and z_i is the interconnection vector which is assumed to be a linear combination of the states of the other subsystems, i.e.

$$z = Lx \quad (L : \text{full matrix})$$

The optimal control of each subsystem is given by :

$$u_i = - P_i^{-1} B_i' P_i x_i - R^{-1} B_i' s_i \tag{6.7.2}$$

where P_i is the solution of decomposed Riccati equation (i.e. Riccati equation for the ith subsystem) and s_i is a solution of a linear differential equation and depends on the states of the other subsystems (SIN-76) :

$$s_i = \sum_{i=1}^{S} T_{ij} x_j + \theta_i \quad (j=1,\ldots,S) \tag{6.7.3}$$

If the interactions between the subsystems are ignored, the control in (6.7.2) becomes :

$$\begin{aligned} u_i = u_{1i} &= - R_i^{-1} B_i' P_i x_i - R_i^{-1} B_i \theta_i \\ &= - K_i x_i - w_i \end{aligned} \tag{6.7.4}$$

The control (6.7.2) is decomposed into two components :

$$u_i = u_{1i} + u_{2i} \tag{6.7.5}$$

where u_{1i} is the control given by (6.7.4), and u_{2i} will be computed as shown in the following.

Now, if the interconnections are considered as unknown disturbances, the optimization problem can be written as :

$$\min J = 1/2 \sum_{i=1}^{S} \int_0^{\infty} [(x_i' - x_i^d)' Q_i (x_i' - x_i^d) + \psi_i (u_{2i}, \dot{u}_{2i})] \, dt$$

subjecto to :

$$\dot{x}' = A_i^* x_i' + C_i z_i' + D_i \tag{6.7.6}$$

$$\text{with} \quad A_i^* = A_i - B_i K_i$$
$$D_i = d_i - B_i w_i$$

where $\psi_i (u_{2i}, \dot{u}_{2i})$ is an appropriate function of the control u_{i2} and its derivative.

Let the vector γ_i be chosen to minimize the Euclidean norm of the vector $B_i u_{i2} - C_i \gamma_i$, then we have :

$$\gamma_i = (C'C_i)^{-1} \; C_i' \; B_i \; u_{i2} = \mathcal{B} \; u_{2i} \tag{6.7.7}$$

Substituting $B_i \; u_{2i}$ by $C_i \; \gamma_i$, we have :

$$\begin{aligned}
\dot{\hat{x}}_i &= A_i^* \; \hat{x}_i + C_i \; (\mathcal{B} \; u_{2i} + \hat{z}_i) + D_i \\
&= A_i^* \; \hat{x}_i + C_i \; y_i + D_i
\end{aligned} \tag{6.7.8}$$

with $\quad y_i = \mathcal{B} \; u_{2i} + \hat{z}_i \tag{6.7.9}$

where $\hat{x}_i$ is the approximate solution of the states resulting from these substitutions.

Let $\hat{z}_i$ be approximated by the following linear differential equation :

$$\dot{\hat{z}}_i = A_{zi} \; \hat{z}_i + d_{zi}$$

where A_{zi} is the part of A_G (A_G is the dynamic matrix of the overall system), which corresponds to the elements of $\hat{z}_i$. The derivative of (6.7.9) can then be written :

$$\begin{aligned}
\dot{y} &= A_{zi} + \mathcal{B} \; \dot{u}_{2i} - A_{zi} \; \mathcal{B} \; u_{2i} + d_{zi} \\
&= A_{zi} \; y_i + v_i + d_{zi}
\end{aligned} \tag{6.7.10}$$

with $\quad v_i = \mathcal{B} \; \dot{u}_{2i} - A_{zi} \; \mathcal{B} \; u_{2i} \tag{6.7.11}$

Define the function ψ_i in (6.7.6) as :

$$\psi_i \; (u_{2i}, \; \dot{u}_{2i}) = v_i' \; R_i \; v_i$$

Then the optimal decentralized control problem can be rewritten as :

$$\min J = 1/2 \sum_{i=1}^{S} \int_0^\infty (\tilde{x}_i - \tilde{x}_i^d)' \; Q_i(\tilde{x}_i - \tilde{x}_i^d) + v_i' \; R_i \; v_i \quad dt$$

subject to :

$$\dot{\tilde{x}}_i = \tilde{A}_i \; \tilde{x}_i + \tilde{B}_i \; v_i + \tilde{D}_i$$

where

$$\tilde{x}_i = \begin{bmatrix} \hat{x}_i \\ y_i \end{bmatrix} \qquad \tilde{x}_i^d = \begin{bmatrix} x_i^d \\ 0 \end{bmatrix} \qquad \tilde{A}_i = \begin{bmatrix} A_i^* & C_i \\ 0 & A_{z_i} \end{bmatrix} \qquad \tilde{B}_i = \begin{bmatrix} 0 \\ I \end{bmatrix}$$

$$\tilde{D}_i = \begin{bmatrix} D_i \\ d_{z_i} \end{bmatrix} \qquad \text{and} \qquad \tilde{Q}_i = \begin{bmatrix} Q_i & 0 \\ 0 & 0 \end{bmatrix}$$

and the optimal solution of this problem is given by :

$$v_i = - \tilde{R}_i^{-1} \tilde{B}_i' \tilde{P}_i \tilde{x}_i - \tilde{R}_i^{-1} B_i' \tilde{s}_i$$
$$= - G_{i1} \hat{x}_i - G_{i2} y_i - \tilde{R}_i^{-1} B_i' \tilde{s}_i \qquad (6.7.12)$$

where $\tilde{P}_i$ is the solution of the Riccati equation :

$$\dot{\tilde{P}}_i = - \tilde{P}_i A_i - \tilde{A}_i' \tilde{P}_i + \tilde{P}_i \tilde{B}_i \tilde{R}_i^{-1} \tilde{B}_i' \tilde{P}_i - \tilde{Q}_i$$

and $\tilde{s}_i$ is the solution of the linear-vector differential equation :

$$\dot{\tilde{s}}_i = - (\tilde{A}_i - \tilde{P}_i \tilde{B}_i \tilde{R}_i^{-1} \tilde{B}_i') \tilde{s}_i - \tilde{P}_i \tilde{D}_i + \tilde{Q}_i \tilde{x}_d$$

Substituting v_i in (6.7.10), 6.7.11), we obtain :

$$\dot{x}^* = A_i^* x_i^* + C_i y_i^* + D_i \qquad (6.7.13)$$
$$\dot{y}^* = (A_{zi} - G_{i2}) y_i^* - G_{i1} x_i^* + d_{zi}^* \qquad (6.7.14)$$
$$d_{zi}^* = - \tilde{R}_i^{-1} \tilde{B}_i \tilde{s}_i + d_{zi}$$

From the equation (6.7.11), we have :

$$\mathbb{B} \, \dot{u}_{2i} = v_i + A_{zi} \mathbb{B} \, u_{2i}$$

Again, by making a similar approximation to that made in equation (6.7.7), i.e. by calculating the two vectors α_i, β_i which minimize the Euclidean norms of the two vectors :

$$\mathbb{B} \, \alpha_i - A_{zi} \mathbb{B} \, u_{2i}$$
and $\quad \mathbb{B} \, \beta_i - v_i$

and substituting them in equation (6.7.8), we have :

$$\dot{u}_{2i}^* = A_{ui} u_{2i}^* - H_i x_i^* - F_i y_i^* - S_i \qquad (6.6.15)$$
with
$$A_{ui} = (\mathbb{B}' \mathbb{B})^{-1} \mathbb{B}_i' A_{zi} \mathbb{B}_i$$
$$H_i = (\mathbb{B}_i' \mathbb{B})^{-1} \mathbb{B}_i' G_{i1}$$
$$F_i = (\mathbb{B}_i' \mathbb{B}_i)^{-1} \mathbb{B}_i' G_{i2}$$
$$S_i = (\mathbb{B}_i' \mathbb{B}_i)^{-1} \mathbb{B}_i' R_i^{-1} \mathbb{B}_i' s_i$$

and the decentralized control is given by (6.6.13) (6.6.14) and (6.6.15). The diagram of Figure 6.5 shows the controller structure.

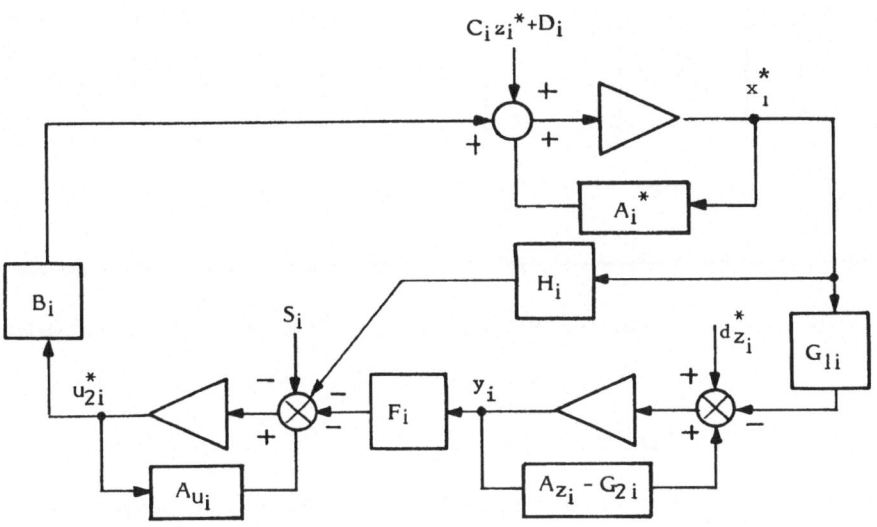

Fig. 6.5

6.7.2. - Model-following method (HAS-80) (CHE-81)

Let the optimization problem be :

$$\min J = 1/2 \sum_{i=1}^{S} \int_0^\infty (x_i' \, Q_i \, x_i + u_i' \, R_i \, u_i) \, dt$$

subject to :

$$\dot{x}_i = A_i \, x_i + B_i \, u_i + C_i \, z_i \qquad\qquad (6.7.16)$$
$$z_i = \sum_{j=1}^{S} L_{ij} \, x_j$$

In order to solve this problem, Hassan and Singh (HAS-80) use the following model of interactions :

$$\dot{z} = A_{zi} \, z_i$$

and the optimization problem can be rewritten as :

$$\min J = 1/2 \sum_{i=1}^{S} \int_0^\infty (y_i' \, \tilde{Q}_i \, y_i + u_i' \, R_i \, u_i) \, dt$$

subject to :

$$\dot{y}_i = \tilde{A}_i \, y_i + \tilde{B}_i \, u_i$$

where :

$$y_i = \begin{bmatrix} x_i \\ \\ z_i \end{bmatrix} \qquad \widetilde{A}_i = \begin{bmatrix} A_i & C_i \\ \\ 0 & A_{z_i} \end{bmatrix} \qquad \widetilde{B}_i = \begin{bmatrix} B_i \\ \\ 0 \end{bmatrix} \quad \text{and} \ \widetilde{Q}_i = \begin{bmatrix} Q_i & 0 \\ \\ 0 & 0 \end{bmatrix}$$

The optimal solution of this problem is given by :

$$u_i = - R_i^{-1} \, \widetilde{B}_i \, P_i' \, y_i$$

where P_i is the solution of the Riccati equation. This control can be written in the form :

$$u_i = - K_{1i} \, x_i - K_{2i} \, z_i \qquad\qquad (6.7.17)$$

Now, consider the subsystem model whose input $\hat{z}_i$ is provided by the inter-connection model :

$$\dot{\hat{x}}_i = A_i \, \hat{x}_i + B_i \, u_i + C_i \, \hat{z}_i \qquad\qquad (6.6.18)$$

where $\hat{x}_i$ is the state of subsystem model. Then, if we substitute u_i by its value (from (6.6.17), after replacing z par $\hat{z}$) in (6.6.16) and (6.6.18) we obtain :

$$\dot{\bar{x}}_i = \hat{A}_i \, \bar{x}_i + C_i \, z_i - B_i \, K_{2i} \, \hat{z}_i \qquad\qquad (6.6.19)$$

with

$$\hat{A}_i = A_i - B_i \, K_{1i}$$
$$\dot{\hat{x}}_i = \hat{A}_i \, \hat{x}_i + (C_i - B_i \, K_{2i}) \, \hat{z}_i \qquad\qquad (6.6.20)$$

where $\bar{x}_i$ is the ith subsystem state resulting from the use of $\hat{z}_i$ instead of z_i.

Substracting (6.6.20) from (6.6.19) we get :

$$\dot{\tilde{x}}_i = \widetilde{A}_i \, \tilde{x}_i + C_i \, \tilde{z}_i$$

where x_i and z_i are error vectors given by ;

$$\tilde{x}_i = \bar{x}_i - \hat{x}_i$$
$$\tilde{z}_i = z_i - \hat{z}_i$$

and the optimization problem is rewritten :

$$\min J = 1/2 \sum_{i=1}^{S} \int_{o}^{\infty} (\tilde{x}_i{'} H_i \tilde{x}_i + \tilde{z}_i{'} S_i \tilde{z}_i) \, dt$$

subject to :

$$\dot{\tilde{x}}_i = \tilde{A}_i \tilde{x}_i + C_i \tilde{z}_i$$

The optimal solution of this problem is given by :

$$\tilde{z}_i^* = - S_i^{-1} C_i{'} P_i \tilde{x}_i$$

where P_i is the solution of the decomposed Riccati equation. Then we get :

$$z_i^* = \hat{z}_i - S_i^{-1} C_i{'} P_i \tilde{x}_i$$

and the decentralized control is given by :

$$\dot{x}_i = (\hat{A}_i + B_i K_{2i} S_i^{-1} C_i{'} P_i) x_i + C_i z_i - B_i K_{2i} (\hat{z}_i + S_i^{-1} C_i{'} P_i \hat{x}_i)$$
$$\dot{\hat{x}}_i = \hat{A}_i \hat{x}_i + (C_i - B_i K_{2i}) \hat{z}_i$$
$$\dot{\hat{z}}_i = A_{zi} \hat{z}_i$$

Figure 6.6 illustrates the control structure.

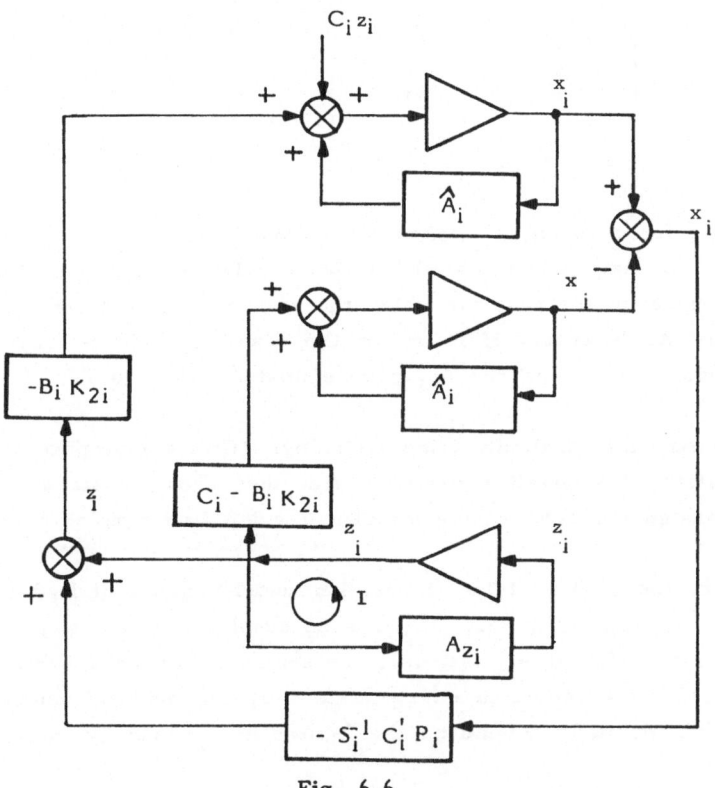

Fig. 6.6

Choice of the interaction model

It is clear that the choice of A_{zi} plays an important part in the existence of such controller. Hassan and Singh (HAS-80) proposed to choose A_{zi} as the block of the global matrix A corresponding to the vector z_i. To clarify this idea, consider the following example.

Let the overall system (with two subsystem) matrix be :

$$A + CL = \left[\begin{array}{ccc|ccc} a_{11} & a_{12} & 0 & 0 & a_{15} \\ a_{21} & a_{22} & a_{23} & 0 & 0 \\ \hline a_{31} & 0 & a_{33} & a_{34} & a_{35} \\ a_{41} & 0 & a_{43} & a_{44} & a_{45} \\ 0 & 0 & a_{53} & a_{54} & a_{55} \end{array}\right]$$

Then, according with the proposition of Hassan and Singh in (HAS-80), the interaction model is :

$$z_1 = \left[\begin{array}{c} x_5 \\ x_3 \end{array}\right] \qquad A_{z_1} = \left[\begin{array}{cc} a_{55} & a_{53} \\ a_{35} & a_{33} \end{array}\right]$$

$$z_2 = x_1 \quad , \quad A_{z2} = a_{11}$$

This rule for choosing A_{zi} takes into account the sparsity of the overall dynamic matrix by eliminating columns where the off-diagonal parts are zero. This provides a good model in the case for which the selected elements are the dominant ones in the rows of A. However, if it is not the case, this choice may mean a longer correction period on-line and the A_{zi} may be unstable (MOO-81).

To overcome this difficulty, Chen and Singh (CHE-81) propose to choose A_{zi} as a diagonal matrix with negative elements, i.e. they choose a stable A_{zi} or, in other words, they assign the poles of the model in the left half complex plane.

However, the choice of the interaction model requires indeed the judgment of the designer. As a matter of fact, there is no simple, infallible way of choosing A_{zi}. Anyway, for any choice of A_{zi} (stable), the errors between the system state and the model state will be corrected to a large extent by the feedback channel. Of course, the accuracy will partially depend on the figures in A_{zi}. Finally, note that the loop I (see Figure 6.6) can be viewed as a reference input. It is independent of any control that could be applied.

6.8. – DECENTRALIZED CONTROL FOR SYSTEMS
WITH OVERLAPPING INFORMATION SET

In the previous section, we have considered some synthesis methods for decentralized systems assuming that the actions of the controllers are carried out using disjoint information sets, i.e. there is no sharing of information among the local control stations. However, in practice, many system models, for example traffic regulation (ISA-73), power systems (SIL-78, CAL-78), economic systems (AOK-76), multiple control systems for reliability enhancement (SIL-80), are constituted of overlapping subsystems, and the sharing of information among the controllers is absolutely essential.

One way of designing a controller for this type of systems is to use the techniques of expansion and contraction. The idea is to expand (under certain conditions) the state space of the original system so that the expanded system contains all the necessary information about the original system and that the subsystems appear as disjoint. Then, conventional techniques for the design of a decentralized controller can be used for the expanded system and the resulting controller is contracted for implementation on the original system.

6.8.1. – Expansion, contraction, and inclusion

Consider the system S given by :

$$S : \dot{x} = Ax + Bu \qquad x(0) = x_0 \qquad (6.8.1)$$

where $x \in R^n$, $u \in R^m$ are the state and the input, and A, B are appropriate constant matrices. Associate the performance index :

$$J(x_0, u) = \int_0^\infty (x' \Omega x + u' Ru) \, dt$$

where $\Omega \geqslant 0$ and $R > 0$ are appropriate weighting matrices.

Associate the pair $(\widetilde{S}, \widetilde{J})$ with the pair (S, J), where the system $\widetilde{S}$ and its associated performance index are given by :

$$\tilde{S} : \dot{\tilde{x}} = \tilde{A}\tilde{x} + \tilde{B}u \qquad \tilde{x}(0) = \tilde{x}_0$$
$$\tilde{J}(\tilde{x}_0, u) = \int_0^\infty (\tilde{x}' \tilde{Q} \tilde{x} + u' \tilde{R}u) \, dt$$

$\tilde{Q} \geqslant 0$ and $\tilde{R} > 0$ are appropriate weighting matrices, and $\tilde{x} \in R^{\tilde{n}}$, $\tilde{n} \geqslant n$.

Introduce the linear transformation :

$$\tilde{x} = Tx \qquad\qquad\qquad\qquad\qquad (6.8.2)$$

where T is an $\tilde{n} \times n$ constant matrix with full column rank. Let $x(t, x_0, u)$ and $\tilde{x}(t, \tilde{x}_0, u)$ be the states of the systems S and $\tilde{S}$ corresponding to the initial conditions x_0, $\tilde{x}_0$ and the fixed input u. The linear transformation (6.8.2) can be used to relate the pairs (S, J) and $(\tilde{S}, \tilde{J})$ in the context of inclusion as follows :

<u>Definition 6.2</u> (IKE-81). The pair $(\tilde{S}, \tilde{J})$ includes the pair (S, J) if there exists a matrix T such that for any initial condition x_0 of S, the choice

$$\tilde{x}_0 = T \, x_0 \qquad\qquad\qquad\qquad\qquad (6.8.3)$$

of the initial state $\tilde{x}_0$ of $\tilde{S}$ implies that :

$$x(t; x_0, u) = T^I x(t; \tilde{x}_0, u) \qquad \text{for all } t \geqslant 0$$
$$J(x_0, u) = \tilde{J}(\tilde{x}_0, u)$$

for any input $u(t)$, where T^I is a generalized inverse of T.

If the pair $(\tilde{S}, \tilde{J})$ includes the pair (S, J), then $(\tilde{S}, \tilde{J})$ is said to be and <u>expansion</u> of (S, J), and (S, J) is called a <u>contraction</u> of $(\tilde{S}, \tilde{J})$. Note that if $(\tilde{S}, \tilde{J})$ includes (S, J), then the optimization problem corresponding to $(\tilde{S}, \tilde{J})$ is equivalent to the one for (S, J) provided that (6.8.3) holds.

Now, under the transformation (6.8.2) the matrices of the expanded system and the original system are related by ;

$$\tilde{A} = TAT^I + M_A, \qquad\qquad \tilde{B} = TB + N_B,$$

$$\tilde{Q} = (T^I)' Q \, T^I + M_Q \text{ and } \tilde{R} = R + N_R$$

where M_A, N_B, M_Q and N_R are constant complementary matrices of proper dimension.

Theorem 6.12 (IKE-81, SIL-82a). The pair $(\tilde{S},\tilde{J})$ includes the pair (S,J) if either :

(i) $M_A T = 0, \quad N_B = 0, \quad T^I M_Q T = 0$ and $N_R = 0$ \hfill (6.8.4)

or (ii) $M^I M_A^i T = 0, \ T M_A^{i-1} N_B = 0, \quad M_Q M_A^{i-1} T = 0$

$M_Q M^{i-1} N = 0$ and $N_R = 0 \quad (i=1,2,\ldots,\tilde{n})$ \hfill (6.8.5)

We point out that the conditions (6.8.4) and (6.8.5) do not imply each other, they are two different sets of conditions for contraction and expansion (IKE-81). Although the choice of matrices T, M_A, N_B is by no means unique (IKE-80), they must be chosen such that the expanded system is controllable and observable and such that it can be used for controller or observer design, (a detailed study about this choice is given by Malinowski and Singh (MAL-85)).

In the following, we consider the design problem and discuss the conditions under which the control $u = - \tilde{K}\tilde{x}$ can be contracted to $u = - Kx$ for implementation on the original system S.

Definition 6.3 (IKE-81, SIL-82a). The control $u = - \tilde{K}\tilde{x}$ of the expansion $\tilde{S}$ is contractible to the control $u = - Kx$ for the original system S if $\tilde{x}_0 = Tx$ implies that

$$Kx (t; x_0, u) = \tilde{K}\tilde{x} (t; \tilde{x}_0, u) \qquad \forall t \geqslant 0$$

for any fixed input u.

Theorem 6.13 (IKE-81). If

$$M_A T = 0 \text{ and } N_B = 0$$

then any control law $u = - \tilde{K}\tilde{x}$ is contractible to the control law $u = - Kx$, and K is given by :

$$K = \tilde{K} T$$

6.8.2. - Overlapping decomposition

Consider again the system (6.8.1), and assume that its state vector x is composed of three subvectors x_1, x_2 and x_3 of dimension n_1, n_2 and n_3 respectively, i.e. :

$$x = (x_1', x_2' x_3')'$$

and

$$n = n_1 + n_2 + n_3$$

Let the input be decomposed into two stations :

$$u = (u_1' u_2')'$$

where $u_1 \in R^{m_1}$, $u_2 \in R^{m_2}$ and $m = m_1 + m_2$.

With this representation the system S can be described as :

$$
\begin{bmatrix} \dot{x}_1 \\ \dot{x}_2 \\ \dot{x}_3 \end{bmatrix} =
\begin{bmatrix} A_{11} & A_{12} & A_{13} & x_1 \\ A_{21} & A_{22} & A_{23} & x_2 \\ A_{31} & A_{32} & A_{33} & x_3 \end{bmatrix}
\begin{bmatrix} B_{11} & B_{12} \\ B_{21} & B_{22} \\ B_{31} & B_{32} \end{bmatrix}
\begin{bmatrix} u_1 \\ u_2 \end{bmatrix}
\qquad (6.8.6)
$$

where the submatrices correspond to the components of the state and input vectors.

Let the state be decomposed into two overlapping components (the generalization to any number of components is obvious) :

$$\tilde{x}_1 = (x_1' x_2')'$$
$$\tilde{x}_2 = (x_2' x_3')'$$

such that the new state vector is :

$$\tilde{x} = (\tilde{x}_1' \tilde{x}_2')'$$

This vector is related to x by :

$$\tilde{x} = Tx$$

where $\tilde{x} \in R^{\tilde{n}}$ and $\tilde{n} = n_1 + 2n_2 + n_3$, and T is a $\tilde{n} \times n$ non square matrix defined by :

$$
T = \begin{bmatrix} I_1 & 0 & 0 \\ 0 & I_2 & 0 \\ 0 & I_2 & 0 \\ 0 & 0 & I_3 \end{bmatrix}
$$

where $I_i = i = 1,2,3$ are unity matrices of dimension $n_i \times n_i$ $i = 1,2,3$.

Define the expansion $\widetilde{S}$:

$$\widetilde{S} : \dot{\widetilde{x}} = \widetilde{A} \, \widetilde{x} + \widetilde{B} \, u$$

where $\qquad \widetilde{A} = TAT^I + M$, $\quad \widetilde{B} = TB + N$

Ikeda et al. (IKE-81) propose the choice (not unique) :

$$T^I = \begin{bmatrix} I_1 & 0 & 0 & 0 \\ 0 & \frac{1}{2} I_2 & \frac{1}{2} I_2 & 0 \\ 0 & 0 & 0 & I_3 \end{bmatrix} , \quad M = \begin{bmatrix} 0 & \frac{1}{2} A_{12} & -\frac{1}{2} A_{12} & 0 \\ 0 & \frac{1}{2} A_{22} & -\frac{1}{2} A_{22} & 0 \\ 0 & -\frac{1}{2} A_{22} & \frac{1}{2} A_{22} & 0 \\ 0 & -\frac{1}{2} A_{32} & \frac{1}{2} A_{32} & 0 \end{bmatrix} , \quad N = 0$$

which satisfies the conditions (given by part (i) of Theorem 6.12) for $\widetilde{S}$ to be an expansion of S. With the transformation (6.8.7) the expansion $\widetilde{S}$ is given by :

$$\widetilde{S} : \begin{bmatrix} \dot{\widetilde{x}}_1 \\ \dot{\widetilde{x}}_2 \end{bmatrix} \begin{bmatrix} A_{11} & A_{12} & 0 & A_{13} \\ A_{21} & A_{22} & 0 & A_{23} \\ A_{21} & 0 & A_{22} & A_{23} \\ A_{31} & 0 & A_{32} & A_{33} \end{bmatrix} \begin{bmatrix} \widetilde{x}_1 \\ \widetilde{x}_2 \end{bmatrix} \begin{bmatrix} B_{11} & B_{12} \\ B_{21} & B_{22} \\ B_{21} & B_{22} \\ B_{31} & B_{32} \end{bmatrix} \begin{bmatrix} u_1 \\ u_2 \end{bmatrix} \qquad (6.8.8)$$

By comparing the systems S and $\widetilde{S}$, we see that the overlapping decomposition of the system S results in a disjoint decomposition of the expansion $\widetilde{S}$ of S. Standard decentralized control techniques can thus be used to design a controller for $\widetilde{S}$.

The expansion $\widetilde{S}$ can be represented as two interconnected subsystems :

$$\widetilde{S}_1 : \dot{\widetilde{x}}_1 = \widetilde{A}_1 \, \widetilde{x}_1 + \widetilde{B}_1 \, u_1 + \widetilde{A}_{12} \, \widetilde{x}_2 + \widetilde{B}_{12} \, u_2$$
$$\widetilde{S}_2 : \dot{\widetilde{x}}_2 = \widetilde{A}_2 \, \widetilde{x}_2 + \widetilde{B}_2 \, u_2 + \widetilde{A}_{21} \, \widetilde{x}_1 + \widetilde{B}_{21} \, u_1$$

where the matrices of the decoupled subsystems are given by :

$$\tilde{S}_1^D : \dot{\tilde{x}}_1 = \tilde{A}_1 \tilde{x}_1 + \tilde{B}_1 u_1$$
$$\tilde{S}_2^D : \dot{\tilde{x}}_2 = \tilde{A}_2 \tilde{x}_2 + \tilde{B}_2 u_2$$

and :

$$\tilde{A}_1 = \begin{bmatrix} A_{11} & A_{12} \\ A_{21} & A_{22} \end{bmatrix} \quad \tilde{B}_1 = \begin{bmatrix} B_{11} \\ B_{12} \end{bmatrix}$$

$$\tilde{A}_2 = \begin{bmatrix} A_{22} & A_{23} \\ A_{32} & A_{33} \end{bmatrix} \text{ and } \tilde{B}_2 = \begin{bmatrix} B_{22} \\ B_{32} \end{bmatrix}$$

The interconnection matrices are given by :

$$\tilde{A}_{12} = \begin{bmatrix} 0 & A_{13} \\ 0 & A_{23} \end{bmatrix} \quad \tilde{A}_{21} = \begin{bmatrix} A_{21} & 0 \\ A_{31} & 0 \end{bmatrix} \quad \tilde{B}_{12} = \begin{bmatrix} B_{12} \\ B_{22} \end{bmatrix} \text{ and } \tilde{B}_{21} = \begin{bmatrix} B_{21} \\ B_{31} \end{bmatrix}$$

Let us associate the following performance indices with the decoupled subsystems :

$$\tilde{J}_1 (\tilde{x}_{10}, u_1) = \int_0^\infty (\tilde{x}_1', \tilde{Q}_1 \tilde{x}_1 + u_1' \tilde{R}_1 u_1) \, dt$$
$$\tilde{J}_2 (\tilde{x}_{20}, u_2) = \int_0^\infty (\tilde{x}_2', \tilde{Q}_2 \tilde{x}_2 + u_2' \tilde{R}_2 u_2) \, dt$$

where $\tilde{x}_{10}$ and $\tilde{x}_{20}$ are the initial states of $\tilde{S}_1^D$ and $\tilde{S}_2^D$ and $\tilde{Q}_1$, $\tilde{Q}_2$, $\tilde{R}_1$ and $\tilde{R}_2$ are weighting matrices of appropriate dimensions.

The global performance index can be written :

$$\tilde{J} (\tilde{x}_0, u) = \int_0^\infty (\tilde{x}' \tilde{Q} \tilde{x} + u' \tilde{R} u) \, dt$$

where

$$\tilde{Q} = \text{diag.} (\tilde{Q}_1, \tilde{Q}_2)$$
$$\tilde{R} = \text{diag.} (\tilde{R}_1, \tilde{R}_2)$$

In virtue of part (i) of Theorem 6.12, the performance index $\tilde{J} (\tilde{x}_0, u)$ is a expansion of :

$$J (x_0, u) = \int_0^\infty (x' Qx + u' Ru) \, dt$$

with

$$Q = T^I \tilde{Q} T$$
$$R = \tilde{R}$$

and J (x_0, u) is the performance index associated to the original system.

Now, the local control laws :

$$u_1 = - \tilde{\tilde{K}}_1 \, \tilde{x}_1$$
$$u_2 = - \tilde{K}_2 \, \tilde{x}_2$$

are calculated to optimize the local performance indices for the local subsystems $\tilde{S}_1^D$ and $\tilde{S}_2^D$. The global control is then written :

$$\tilde{K} = \begin{bmatrix} \tilde{K}_1 & \vdots & \\ --- & + & --- \\ & \vdots & \tilde{K}_2 \end{bmatrix} = \begin{bmatrix} \tilde{K}_{11} & \tilde{K}_{12} & \vdots & \\ ------ & + & ----- \\ & & \vdots & \tilde{K}_{23} & \tilde{K}_{24} \end{bmatrix}$$

and the control to be implemented on the original system is given by the contraction :

$$K = \tilde{K} T = \begin{bmatrix} \tilde{K}_{11} & | & \tilde{K}_{12} & | \\ ---- & + & ---- & --- \\ & | & \tilde{K}_{23} & \tilde{K}_{24} \end{bmatrix}$$

6.9. - CONCLUSION

Decentralized control and in general constrained structure control induce parametric optimization procedures for the synthesis of adequate control structures.

In this chapter, we have presented most of the available techniques for dealing with this problem. In order to give a more complete and realistic solution to this kind of problem, we have also considered that the system could be perturbed by small or large parameter variations or affected by external disturbances. The solution provides a robust decentralized control.

Since decentralized control remains an active research area with practical impact in many fields of application, the presented results should be completed soon by other efficient techniques and algorithms.

CHAPTER 7

STRUCTURAL ROBUSTNESS

7.1. - INTRODUCTION

The problem of designing an optimal control structure such that the system has no fixed modes and that the information transfer is minimized was considered in Chapter 5. Chapter 6 was then concerned with the parametric optimization technics which can be used to determine the gains of the structurally constrained controller. This problem was also considered with the addition of robustness constraints refering to external disturbances and variations of the system parameter values.

When dealing with large scale systems, it is also of a real practical importance to consider that the system can be subjected to structural perturbations. A recent number of studies have paid attention to the problem of synthesizing regulators which, in addition to provide some desirable properties to the controlled system when every component is properly working, preserve a certain number of these properties under the occurence of structural perturbations (SIL-75) (SIL-78) (DAV-81) (LOC-83) (ALB-83) (ACK-84). In presence of an operating controlled system, structural disturbances can affect the plant itself or the control system. A structural perturbation related to the first class is defined for composite systems as the possible disconnection of one or more subsystems. This concept was originally introduced by Siljak (SIL-75) (SIL-78) who considered the stability of a composite system when certain subsystems may be disconnected. The other class of structural perturbations is concerned with controller failures affecting either sensor measurements, actuator behaviour or line cuts.

In (DAV-81), the robust decentralized servomechanism problem (DAV-76c) is addressed with an additional reliability requirement. The problem consists in synthesizing a decentralized controller to solve the robust servomechanism problem for a S-decentralized plant such that if the j-th control agent fails, the resulting system still has the property that satisfactory robust tracking/regulation occurs for control agents $(1,2,\ldots,j-1)$, $\forall j \in \{1,2,\ldots,S\}$ (sequential reliability).

Also in (LOC-83), the problem of designing a completely decentralized regulator which performes robust zero-error regulation in normal operating conditions and preserves zero-regulation of all the error variables but the i-th one when a failure occurs in the i-th feedback loop (reliability) is discussed.

In the present chapter, the robustness constraints are specified in a different way and we are concerned with a pure structural problem. The study is focused on the consequences of structural perturbations on stabilizability or pole assignability using structurally constrained feedback control. Both structural perturbations affecting the plant (Section 7.2) and affecting the controller (Section 7.3) are considered. In the last section, the design is faced of an optimal control structure which, besides preventing some modes of the system (the unstable ones for example) to be fixed modes in normal operating conditions, guarantees that these modes (or another subset) are not fixed modes when the system is subjected to a prespecified class of structural perturbations (affecting the controller).

7.2. - STRUCTURAL PERTURBATIONS AFFECTING THE SYSTEM (OZG-82)

Consider the following system composed of S subsystems :

$$\dot{x} = Ax + Bu \qquad (7.2.1)$$

$$A = \begin{bmatrix} A_{11} & \text{-------} & A_{1S} \\ \vdots & & \vdots \\ A_{S1} & \text{-------} & A_{SS} \end{bmatrix} \qquad B = \begin{bmatrix} B_1 & & & \\ & B_2 & & 0 \\ & 0 & \ddots & \\ & & & B_S \end{bmatrix}$$

where $A_{ij} \in R^{n_i \times n_j}$ and $B_i \in R^{n_i \times m_i}$, (i,j=1,...,S). This system is controlled by decentralized state feedback in the form :

$$u = Kx \qquad K = \text{block-diag.} [K_1 \ K_2 \ ... \ K_S] \qquad (7.2.2)$$
where
$$K_i \in R^{m_i \times n_i} \ , \ (i=1,...,S).$$

We assume that (7.2.1) is subjected to structural perturbations which consist in the disconnection of one or more subsystems. The interconnection structure of the system (7.2.1) can be represented by a digraph G = (V,E). Each node $v_i \in V$ is associated to a subsystem Si, i $\in$ {1,...,S} and each edge $(e_i, e_j) \in E$ represents a nonzero interconnection from Si to Sj which implies that $A_{ji} \neq 0$ and vice versa (this digraph was also used in (COR-76b), see Chapter 2, Section 2.2.3.b). A subgraph $G_T = (V_T, E_T)$ of G is a collection of nodes $V_T \subseteq V$ and the set of edges $E_T \subseteq E$ connecting the nodes V_T in G. The matrices A_T and B_T are defined as the submatrices of A and B corresponding to the subsystems associated to V_T.

Then the following result provides sufficient conditions under which the system (7.2.1) is pole assignable using the control (7.1.2) under structural perturbations.

<u>Theorem 7.1</u> (OZG-82). If for every strongly connected subgraph (see Chapter 2, Section 2) $G_T = (V_T, E_T)$, the matrix pairs (A_T, B_T) are controllable, then the system (7.2.1) has no decentralized fixed modes under all structural perturbations.

<u>Remark 7.1</u>. The problem of existence of decentralized solutions to the servomechanism problem (DAV-76c) is also discussed in (OZG-82). The results are briefly presented in this remark. The model (7.2.1) is modified as follows :

$$\dot{x} = Ax + Bu + E\omega \qquad (7.2.3)$$
$$y = Cx$$

$$
E = \begin{bmatrix} E_1 \\ \vdots \\ \vdots \\ E_S \end{bmatrix} \quad
C = \begin{bmatrix} C_1 & & \\ & C_2 & 0 \\ & 0 & \ddots \\ & & & C_S \end{bmatrix}
$$

where A and B are the same as in (7.1.1), $E_i \in R^{n_i \times \Omega}$, and $C_i \in R^{p_i \times n_i}$, (i=1,...,S). y= $(y'_1 \; y'_2 \ldots y'_S)'$ is the output to be regulated so that the errors

$$e_i = y_i - y_i^{ref} \quad (i=1,...,S)$$

tend to zero as t -> ∞.

We assume that the disturbance vector w satisfies :

$$\dot{z}_1 = F_1 z_1$$
$$\omega = H_1 z_1 \qquad (7.2.4)$$

where $z_1 \in R^{\hat{n}_1}$ and where (H_1, F_1) is observable and $z_1(0)$ is not known. The reference input vector y^{ref} satisfies :

$$\dot{z}_2 = F_2 z_2$$

$$y^{ref} = H_2 \, z_2 \tag{7.2.5}$$

where $z_2 \in R^{\hat{n}_2}$ and where (H_2, F_2) is observable and y^{ref} is measurable. The minimal polynomials of F_1 and F_2 are denoted by $\Lambda_1 (p)$ and $\Lambda_2 (p)$ and their least common multiple by $\Lambda(p)$. Let the zeros of $\Lambda(p)$ (multiplicities included) be given by $\lambda_1, \lambda_2, \ldots, \lambda_q$.

A system is said to be decentrally retunable under structural perturbations if after any structural perturbation, decentralized controllers can be designed so as to solve the servomechanism problem for the perturbated system. We have the following result :

<u>Theorem 4.2</u> (OZG-82). If for every strongly connected subgraph $G_T = (V_T, E_T)$
(i) the matrix pairs (A_T, B_T) are controllable
(ii) the subsystems (C_T, A_T, B_T) have no transmission zero coinciding with λ_1, $\lambda_2, \ldots, \lambda_q$ (C_T is defined in a similar way as A_T and B_T)
then the system (7.2.3) is decentrally retunable under all structural perturbations.

7.3. - STRUCTURAL PERTURBATIONS AFFECTING THE CONTROL SYSTEM   (TRA-84b)

In this section, we are also concerned by the problem of pole assignability using structurally constrained controllers but structural perturbations are now supposed to affect the controller. The following subsection specifies the type of perturbations that we consider and provides a model for the perturbated controlled system.

7.3.1. - Structural perturbations characterization (TRA-84b)

Consider the class of linear time-invariant systems described by the following state-space representation :

$$\dot{x}(t) = Ax(t) + Bu(t)$$
$$y(t) = Cx(t) \tag{7.3.1a}$$

where $x(t) \in R^n$, $u(t) \in R^m$, $y(t) \in R^r$ are the state, input and output vectors, respectively and A, B, C are real matrices of appropriate dimensions.

Define $B = [b_1, \ldots, b_m]$
        $C = [c_1, \ldots, c_r]'$

so that the equivalent representation of the system in the frequency domain can be written :

$$y(p) = W(p) \, u(p) \qquad (7.3.1b)$$

with

$$y_j(p) = \sum_{i=1}^{m} w_{j,i}(p) \, u_i(p) \qquad (j=1,\ldots,r)$$

$$w_{ij}(p) = c_j(pI-A)^{-1}b_i$$

Consider the following feedback control law for system (7.3.1) :

$$u = K \, y \qquad (7.3.2)$$

whose structure is specified by the feedback matrix K :

$$K = (k_{ij})_{\substack{i=1,\ldots,m \\ j=1,\ldots,r}} \qquad \text{with some } k_{ij} \text{ constrained to be zero,}$$

We assume that the controlled system behaviour may be perturbed by failures of the controller components (sensors, actuators, lines). These failures are specified below :

Definition 7.1.

1. If the ith actuator, $i \in \{1,\ldots,m\}$ fails at time τ, then $u_i(t) = 0$, $t \geqslant \tau$

2. If the ith sensor, $i \in \{1,\ldots,p\}$ fails at time τ, then $y_i(t) = 0$, $t \geqslant \tau$

The behaviour of the ith actuator can be expressed by:

$$u_i(t) = \alpha_i \, \overline{u}_i(t) \qquad (i=1,\ldots,m) \qquad (7.3.3)$$

$$\alpha_i = \begin{cases} 1 \text{ if the actuator is properly working} \\ 0 \text{ if a failure occurs} \end{cases}$$

$\overline{u}_i(t)$ is the control that should be applied to the system and $u_i(t)$ is the control that is effectively applied.

Similarly, the behaviour of the ith sensor can be expressed by :

$$\overline{y}_i(t) = \beta_i \, y_i(t) \qquad (i=1,\ldots,r) \qquad (7.3.4)$$

$$\beta_i = \begin{cases} 1 \text{ if the sensor is properly working} \\ \\ 0 \text{ if a failure occurs} \end{cases}$$

$\bar{y}_i(t)$ is the measured value of the real output $y_i(t)$.

Line failures must be considered differently from actuator or sensor failures. Practical considerations lead to distinguish the two following situations :
1. The physical lines establishing the feedback from one output to one input are isolated one from another.
2. The lines establishing the feedback connections from a set of outputs to a set of inputs (corresponding each to a given geographical station) are put together in a unique physical line. This situation corresponds to geographically distributed systems for which there is a natural partitioning of inputs and outputs in several stations S_i, (i=1,...,S).

Definition 7.2.
1. If the line associated to the feedback connection between output y_j, $j \in \{1,...,r\}$ and input u_i, $i \in \{1,...,m\}$ fails at time τ, then $k_{ij} = 0$, $t \geqslant \tau$.
2. If the line associated to the feedback connection between station S_j and station S_i, $i,j \in \{1,...,S\}$ fails at time τ, then :

$$k_{sv} = 0 \text{ for all } s, v \text{ such that } u_s \in S_i \text{ and } y_v \in S_j.$$

(If a reordering of inputs and outputs according to the stations has been performed on the system model (7.3.1), this corresponds to setting to zero the whole block K_{ij} in the feedback matrix K defined in (7.3.2)).

In view of this definition, we can define a new feedback matrix $\bar{K}$ which takes into account the line failures :

1. $\bar{K} = (l_{ij} \ k_{ij}) i=1,...,m$

   $\hspace{3cm} j = 1,...,r$ $\hspace{4cm}$ (7.3.5a)

$$l_{ij} = \begin{cases} 0 \text{ if there is a break of the line j-i} \\ \\ 1 \text{ otherwise} \end{cases}$$

2. $\bar{K} = \text{block } (L_{ij} \ K_{ij}) \ i,j=1,...,S$

$$L_{ij} = \begin{cases} 0 \text{ if there is a break of the line between station } S_j \text{ and station } S_i \hspace{1cm} (7.3.5b) \\ \\ 1 \text{ otherwise} \end{cases}$$

If we define :

$$\alpha = \quad \text{diag.}[\ \alpha_1, \ldots, \ \alpha_m] \ \in \ R^{m \times m} \tag{7.3.6}$$
$$\beta = \quad \text{diag.}[\ \beta_1, \ldots, \beta_p] \ \in \ R^{r \times r}$$

we obtain the following model for the perturbated system :

PLANT

State space :
$$\dot{x}(t) = Ax(t) + B \ \alpha \ \bar{u}(t)$$
$$\bar{y}(t) = \beta C x(t) \tag{7.3.7a}$$

Frequency domain :
$$\bar{y}(p) = \beta W(p) \ \alpha \bar{u}(p) \tag{7.3.7b}$$

CONTROL

$$\bar{u}(t) = \bar{K} \ \bar{y}(t) \tag{7.3.8}$$

The closed-loop system (7.3.7) (7.3.8) is given by :
$$\dot{x}(t) = (A + B \alpha \bar{K} \beta \ C) x (t)$$
and illustrated by the following scheme :

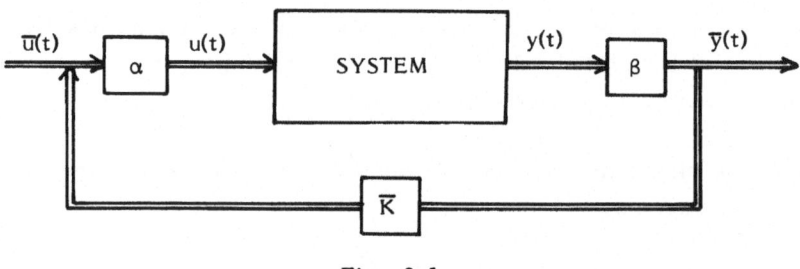

Fig. 3.1

Remark 7.2.
2. In the case of dynamic feedback control as :

$$\dot{z} = Sz + Ry$$
$$u = Qz + Ky + v \tag{7.3.9}$$

we suppose that the structure of the compensator is conditionned by the structure of the output feedback matrix K (S, R and Q have the same structure than K). The existence of a solution to the problem of stabilizability and pole assignability (existence of fixed modes) with structurally constrained dynamic compensation can thus be solved by considering the static feedback of the same structure ((WAN-73), see Chapter 2, Section 2.2.3a).

7.3.2. - Structural robustness

For a given controlled system, practical considerations (like sensor technology, line fiability...) make some structural perturbations more probable than others. Therefore, a designer generally wants to restrict the study by specifying a class of perturbations which he considers like the most probable.

Let $Fa = \{\alpha^1, \ldots, \alpha^a\}$, $Fs = \{\beta^1, \ldots, \beta^s\}$ and $F_L = \{\overline{K}^1, \ldots, \overline{K}^L\}$, where $\overline{K}$, α, and β are defined in (7.3.5) and (7.3.6), represent a class of actuator, sensor, and line failures. Then $P = \{F_a, F_s, F_L\}$ specifies a class of structural perturbations.

A controlled system (7.3.1) (7.3.9) is said to be <u>structurally robust with respect to P</u> if it remains pole assignable under all possible structural perturbations within the class P. In this case, the controller (7.3.9) itself is said to be structurally robust with respect to P.

To a class of structural perturbations $P = \{F_a, F_s, F_L\}$, we can associate a class of perturbated systems Σ_P composed by the set of perturbated systems (7.3.7) (7.3.8) such that $\alpha \in F_a$, $\beta \in F_s$, and $\overline{K} \in F_L$. Then from the definition of fixed modes (WAN-73) (see Chapter 2), it comes :

Proposition 7.1. The controlled system (7.3.1) (7.3.9) is structurally robust with respect to P if and only if no perturbated system within the class Σ_P has fixed modes.

Introduce the following definition :

Definition 7.3. Given the controlled system (7.3.1) (7.3.9), $\lambda_0 \in \sigma(A)$ is a <u>structurally robust mode</u> with respect to P if and only if no perturbated system within the class Σ_P has λ_0 as a fixed mode.

Using this definition, Proposition 7.1 can be rewritten as follows :

Corollary 7.1. The controlled system (7.3.1) (7.3.9) is structurally robust with respect to P if and only if all the modes of (7.3.1) $\in \sigma$ (A) are structurally robust with respect to P.

Remark 7.2. If we are interested by robustness with reference to the problem of stabilization, the necessary and sufficient condition of Corollary 7.1 must be replaced by "the set of no structurally robust modes of (7.3.1) (7.3.9) is stable".

7.3.3. - Characterization of structurally robust modes

In this section, we use the above definitions and the characterizations of fixed modes (see Chapter 3) to provide three characterizations of structurally robust modes. The first two characterizations are given in the context of decentralized control, i.e. the feedback matrix has a block-diagonal structure. The third one can be used for arbitrary feedback structures.

7.3.3.a. - In the state space

Consider that the system (7.3.1) is partitioned in S stations. If the corresponding reordering of inputs and outputs is performed, we obtain the following model :

$$
\begin{cases}
\dot{x} = Ax + \sum_{i=1}^{S} B_i \, u_i \\
y_i = C_i \, x \qquad (i=1,\ldots,S)
\end{cases}
\tag{7.3.10}
$$

with $B = [B_1, B_2, \ldots, B_S]$, $C = [C'_1, C'_2, \ldots C'_S]'$ and where $B_i \in R^{n \times m_i}$ and $C_i \in R^{r_i \times n}$.

The feedback structure is supposed to be decentralized :

$$
u_i(t) = K_{ii} \, y_i(t) \qquad (i=1,\ldots,S)
\tag{7.3.11}
$$

The matrices α and β (defined in (7.3.6)) specifying the actuator and sensor failures are partitionned in the same way :

$$
\begin{cases}
\alpha = \text{block-diag.}[\chi_1,\ldots,\chi_S] \\
\chi_i = \text{diag}[\alpha_{i_1},\ldots,\alpha_{i_{m_i}}] \qquad i=1,\ldots,S \\
\beta = \text{block-diag.}[\Gamma_1,\ldots,\Gamma_S] \\
\Gamma_i = \text{diag.}[\beta_{i_1},\ldots,\beta_{i_{r_i}}] \qquad i=1,\ldots,S
\end{cases}
\tag{7.3.12}
$$

First, note that in the context of decentralized control, the failure of the line associated to the feeback-loop at station i is equivalent to the elimination of station i in the system model (7.3.10) (indeed there is no more use made of $y_i(t)$, and $u_i(t)$ remains identically zero). A configuration of line failures $\overline{K}* \in F_L$ can therefore be represented by defining the set $\pi* = \{1,\ldots,S\} - \{i/L_{ii} = 0\}$, L_{ii} as defined in (7.3.5).

The following characterization is a straightforward extension of the result of Anderson and Clements (AND-82) (see Chapter 3, Section 3.3.1) :

Proposition 7.2. Given the decentralized controlled system (7.3.10) (7.3.11), $\lambda_0 \in \sigma(A)$ is a structurally robust mode with respect to $P = \{F_a, F_s, F_L\}$ if and only if :

$\forall \in F_a$, $\quad \forall \beta \in F_s$, $\quad \forall \pi*$ corresponding to $\overline{K}* \in F_L$,

$$
\text{rank} \begin{bmatrix} A - \lambda_0 I & B_K \alpha_K \\ \beta_{\pi*-K} \; C_{\pi*-K} & 0 \end{bmatrix} \geqslant n \tag{7.3.13}
$$

for all k such that $K = \{i_1,\ldots,i_k\} \subset \pi*$,
where :
$$B_K = [B_{i_1}, \ldots, B_{i_k}] \qquad C_{\pi*-K} = [C'_{i_{k+1}},\ldots,C'_{i_S}]'$$

$$\alpha_K = \text{Block-diag.} [\chi_{i_1}, \cdots, \chi_{i_k}]$$

$$\beta_{\pi*-K} = \text{block-diag.} [\Gamma_{i_{k+1}},\ldots, \Gamma_{i_S}]$$

Proposition 7.2 means that we use the matrix rank test (7.3.13) for every perturbated system within P in order to check whether or not λ_0 is a decentralized fixed mode for some of them. If we want to conclude whether a decentralized control is structurally robust, we must check all the modes of the system. This is obviously a laborious task. From a practical point of view, there is no doubt that the following characterization is more convenient since the whole set of non structurally robust modes (if any) is determined in one step.

7.3.3.b. - In the frequency domain

This characterization is based on the fixed mode characterization of Vidyasagar and Wiswanadham (VID-83) (see Chapter 3, Section 3.2.3).

It provides a direct determination of the non structurally robust polynomial, whose zeros are the non structurally robust modes of the system. The same notations as in Section 3.2.3 are used.

Consider the partitioned system (7.3.10) in a frequency domain representation :

$$
y = \begin{bmatrix} W_{11}(p) & \cdots & W_{1S}(p) \\ W_{S1}(p) & \cdots & W_{SS}(p) \end{bmatrix} u \tag{7.3.14}
$$

and the decentralized feedback control (7.3.11).

We recall that the fixed polynomial $\alpha(p)$ of the system (7.3.14), whose zeros are the decentralized fixed modes of (7.3.14), is given by the g.c.d. of the characteristic polynomial $\phi(p)$ of (7.3.14) and the minors $W\begin{bmatrix} I \\ J \end{bmatrix}$ of $W(p)$ corresponding to non singular square submatrices of K' (VID-83) :

$$
\alpha(p) = \text{g.c.d.} \quad \{ \ \phi(p), \quad W \begin{bmatrix} I_1 \cup I_2 \cup \cdots \cup I_S \\ J_1 \cup J_2 \cup \cdots \cup J_S \end{bmatrix} \}
$$

$$
I_i \subset R_i = \{ \sum_{j=1}^{i-1} r_j + 1, \ldots, \sum_{j=1}^{i-1} r_j + r_i \} \tag{7.3.15}
$$

$$
J_i \subset M_i = \{ \sum_{j=1}^{i-1} m_j + 1, \ldots, \sum_{j=1}^{i-1} m_j + m_i \} \qquad i = 1, \ldots, S
$$

$$
\| I_i \| = \| J_i \|
$$

Given a class of perturbations $P = \{ F_a, F_s, F_L \}$, it is clear from Definition 7.3 that the non structurally robust polynomial of (7.3.14) with respect to P is equal to the l.c.m. of the fixed polynomials of all the perturbated systems within Σ_P. Consider a structural perturbation $\alpha \in F_a$, $\beta \in F_s$, and $\overline{K} \in F_L$, (as defined in (7.3.6) and (7.3.5)) then the corresponding perturbated system is given by (7.3.7) (7.3.8) :

$$
\overline{y}(p) = \beta \ W(p) \ \alpha \overline{u}(p)
$$
$$
\overline{u} = \overline{K} \ \overline{y}
$$

The matrix $\beta\, W(p)\, \alpha$ is obtained from $W(p)$ by setting to zero any row i corresponding to $\beta_i = 0$, $i \in \{1,\ldots,r\}$ and any column j corresponding to $\alpha_j = 0$, $j \in \{1,\ldots,m\}$. It follows that the minors $[\beta\, W\, \alpha]\begin{bmatrix} I \\ J \end{bmatrix}$ such that $i \in I$ or $j \in J$ are equal to zero.

From another hand, $\overline{K}$ is obtained from K by setting to zero the diagonal blocks K_{ii} corresponding to $L_{ii} = 0$ (i.e. there is a failure of the line implementing the feedback loop at station i). It follows that the square submatrices

$$K' \begin{bmatrix} I = I_1 \cup \ldots \cup I_i \cup \ldots \cup I_S \\ J = J_1 \cup \ldots \cup J_i \cup \ldots \cup J_S \end{bmatrix}$$

such that $I_i \neq 0$, $J_i \neq 0$ are structurally singular.

The following result is straightforward from the above discussion.

<u>Proposition 7.3.</u> The non structurally robust polynomial of (7.3.14) (7.3.11) with respect to the structural perturbation α, β, $\overline{K}$ is given by :

$$\xi(p) = \text{g.c.d} \; \{ \; \phi(p) \; , \; W \begin{bmatrix} I' \\ J' \end{bmatrix} \; \}$$

(7.3.16)

$$I' = (I'_1 \cup I'_2 \cup \ldots \cup I'_S) - \{ I'_i \text{ such that } L_{ii} = 0 \}$$
$$J' = (J'_1 \cup J'_2 \cup \ldots \cup J'_S) - \{ J'_i \text{ such that } L_{ii} = 0 \}$$

$$I'_i \subset R'_i = R_i - \{ k \text{ such that } \beta_k = 0 \}$$
$$J'_i \subset M'_i = M_i - \{ k \text{ such that } \alpha_k = 0 \}$$

(i=1,...,S)

$$\| I'_i \| = \| J'_i \|$$

R_i and M_i are defined in (7.3.15)

Using the above proposition, we can consider the problem of robustness with respect to a class of perturbations as follows :

<u>Proposition 7.4.</u> The non structurally robust polynomial of (7.3.14) (7.3.15) with respect to the class of structural perturbations $P = \{ F_a, F_s, F_L \}$ is given by :

$$\xi_P(p) = \text{l.c.m} \; \{ \; \xi(p) \; \}$$
$$\Sigma_P$$

(7.3.17)

As an example, consider that we are concerned by the failure of one actuator, i.e. $P = \{F_a\}$, $F_a = \{ \alpha^1 = \text{block-diag.}[\,01\ldots1\,] \; , \; \ldots, \alpha^i = \text{block-diag.}[\,1..101..1\,]$,..., $\alpha^a = \text{block-diag.}[\,1\ldots10\,]\}$, then the non structurally robust polynomial is given by :

$$\xi_p(p) = \text{l.c.m.} \quad \{g.c.d. \quad \{ \phi(p), \ W \begin{bmatrix} I' \\ J' \end{bmatrix} \} \}$$
$$k=1,\ldots m$$

$$I' = I_1 \cup I_2 \cup \ldots \ I_S \qquad J' = J_1 \cup J_2 \cup \ldots \cup J_S$$
$$I'_i \subset R_i$$
$$J'_i \subset M'_i = M_i - \{ k \}$$
$$\| I'_i \| = \| J'_i \|$$

The problem of one sensor or one line failure can also be simply formulated in a similar way (TRA-84b). Although the complexity of the problem grows obviously with the number of perturbations that we consider, this characterization presents the interest to bring back all the calculations to the original non perturbated system.

7.3.3.c. – Graph-theoretic characterization

This characterization derives from the fixed modes graph-theoretic characterization of Locatelli et al. (LOC-77) (see Chapter 3, Section 3.6.2). The graph-theoretic framework allows to consider arbitrary feeback structures (not necessarily decentralized). The counterpart is that the approach is only applicable for systems with simple modes.

The same digraph $\Gamma_S = (V_S, \ L_S)$ as in Section 3.6.2 is associated to the system (7.3.1) and the same notations as in that section are used.

Refer to the fixed modes characterization of Theorem 3.30. Then λ_0 is a structurally robust mode with respect to some structural perturbation $(\alpha, \beta, \overline{K})$ if the perturbation does not result in the desappearance of all the elementary cycles of Γ_S for which λ_0 is a pole.

The perturbations can be easily integrated by modifying the digraph Γ_s associated to the original system. From Definitions 7.1 and 7.2, the ith actuator failure can be expressed by the elimination of the vertex $i \in V_{1S}$, the jth sensor failure can be expressed by the elimination of the vertex $(j+m) \in V_{2S}$, and the failure of the line supporting kij can be expressed by the elimination of the edge $(j+m, \ i) \in L_{2S}$. The following result comes.

Proposition 7.5

1. λ_0 is structurally robust with respect to the ith actuator (j^{th} sensor) failure if and only if λ_0 is a pole of some elementary cycle of Γ_S in which the vertex $i \in V_{1S}$ ($j+m \in V_{S2}$) is not involved.

2. λ_0 is structurally robust with respect to the failure of the line supporting k_{ij} if

and only if λ_0 is a pole of some elementary cycle of Γ_S in which the edge $(j+m,i) \in L_{S2}$ is not involved.

For a perturbation $(\alpha, \beta, \overline{K})$ involving several actuator, sensor, and line failures, it is clear that the set of conditions of robustness are given by the intersection of the conditions corresponding to every elementary perturbation. The same is true for a class of perturbations.

The following corollary provides some results refering to particular cases of practical interest :

Corollary 7.2.

1. λ_0 is structurally robust with respect to one actuator (sensor) failure if and only if λ_0 is a pole of at least two elementary cycles of Γ_S such that the sets of vertices $\in V_{1S}$ ($\in V_{2S}$) involved in each cycle are disjoint.

2. λ_0 is structurally robust with respect to one line failure of if and only if λ_0 is a pole of at least two elementary cycles of Γ_S such that the sets of edges $\in L_{2S}$ involved in each cycle are disjoint.

3. λ_0 is structurally robust with respect to one actuator, sensor or line failure (occuring one at a time) if and only if λ_0 is a pole of at least two elementary disjoint cycles of Γ_S.

Remark 7.3. This graph-theoretic approach allows the determination of the nature of the non structurally robust modes. We consider the digraph Γ'_S associated to the perturbated system (Γ'_S is obtained by removing the vertices and edges corresponding to the perturbation).

λ_0 is a non structurally fixed mode (SEZ-81a) for the perturbated system if some elementary cycles remain in Γ'_S for which λ_0 is not a pole due to a pole-zero cancellation in the cycle transmittances.

λ_0 is a structurally fixed mode (SEZ-81a) for the perturbated system if the absence, due to the failure, of elementary cycles for which λ_0 is not a pole is not a consequence of pole-zero cancellations. Nevertheless, some edges $\in L_{1S}$ for which λ_0 is a pole remain in Γ'_S.

λ_0 is an uncontrollable or inobservable mode for the perturbated system (only for actuator and sensor failures) if no edge $\in L_{1S}$ for which λ_0 is a pole remains in Γ'_S.

The following scheme illustrates the possible consequences of a perturbation :

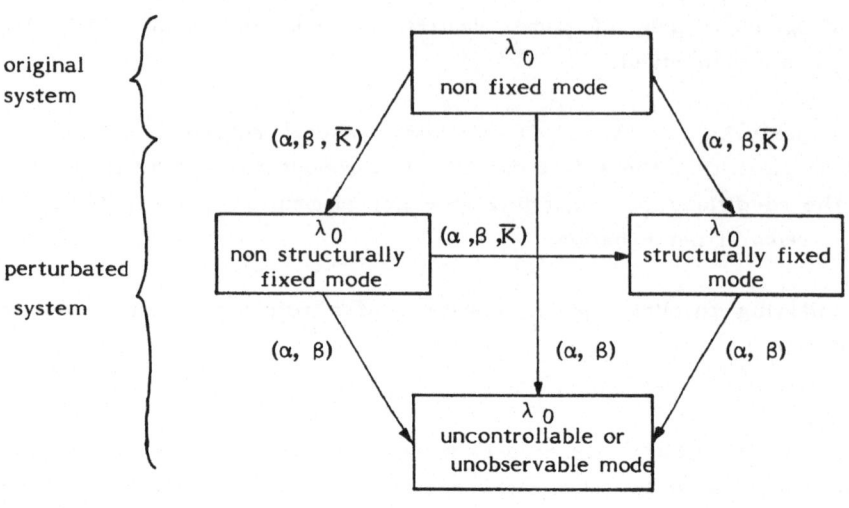

Fig. 7.2

7.3.4. - Example

Consider the 3-station system described by the following transfer matrix :

$$
W(p) = \begin{bmatrix} \dfrac{3}{p-2} & 0 & \dfrac{p+1}{p(p-2)} \\[3mm] \dfrac{1}{p-2} & \dfrac{1}{p+2} & \dfrac{1}{p(p-2)} \\[3mm] \dfrac{1}{p+1} & \dfrac{1}{p+2} & 0 \end{bmatrix}
$$

for which the characteristic polynomial is : $\phi(p) = p(p+1)(p+2)(p-2)$. Consider a decentralized feedback structure given by the feedback matrix :

$$K = \text{block-diag.} [\, k_{11}, \; k_{22}, \; k_{33} \,]$$

Using one or the other of the fixed mode characterizations given in Theorem 3.4 or Theorem 3.30, we can determine that this system has a non structurally fixed mode at $\lambda_0 = -1$.

Now let us determine, for example, the non structurally robust modes with respect to one actuator failure :

1. <u>Using the frequency domain characterization</u> (Proposition 7.4).

We have : $P_1 = \{1\}$ $P_2 = \{2\}$ $P_3 = \{3\}$
$M'_1 = \{1\} - \{k\}$ $M'_2 = \{2\} - \{k\}$ $M'_3 = \{3\} - \{k\}$

The non structurally robust polynomial is given by :

$$\xi_P(p) = \text{l.c.m.} \quad \{ \text{g.c.d.} \quad \{\phi(p), W \begin{bmatrix} I_1 & I_2 & I_3 \\ J_1 & J_2 & J_3 \end{bmatrix} \} \}$$

$$= \text{l.c.m.} \{ \text{g.c.d.} \{ \phi(p), W \begin{bmatrix} 2 \\ 2 \end{bmatrix} \quad W \begin{bmatrix} 3 \\ 3 \end{bmatrix} \quad W \begin{bmatrix} 2 & 3 \\ 2 & 3 \end{bmatrix} \}$$

$$\text{g.c.d.} \{ \phi(p), W \begin{bmatrix} 1 \\ 1 \end{bmatrix} \quad W \begin{bmatrix} 3 \\ 3 \end{bmatrix} \quad W \begin{bmatrix} 1 & 3 \\ 1 & 3 \end{bmatrix} \}$$

$$\text{g.c.d.} \{ \phi(p) \quad W \begin{bmatrix} 1 \\ 1 \end{bmatrix} \quad W \begin{bmatrix} 2 \\ 2 \end{bmatrix} \quad W \begin{bmatrix} 1 & 2 \\ 1 & 2 \end{bmatrix} \}\}$$

$$= \text{l.c.m.} \quad \{ \text{g.c.d.} \quad \{p(p+1)(p+2)(p-2); p(p+1)(p-2); 0; (p+1)\} ;$$
$$\text{g.c.d.} \{ p(p+1)(p+2)(p-2); 3p(p+1)(p-2); 0; (p+1)(p+2)\} ;$$
$$\text{g.c.d.} \{ p(p+1)(p+2)(p-2); 3p(p+1)(p-2); 0; (p+1)(p+2); 3p(p+1)\}\} ;$$

$$= \text{l.c.m.} \quad \{ (p+1) ; (p+1)(p+2) ; p(p+1)\}$$

$$\xi_P(p) = p(p+1)(p+2)$$

Therefore the system has three non structurally robust modes with respect to one actuator failure : $\lambda_0 = -1$ $\lambda_1 = 0$, $\lambda_2 = -2$.
Obviously, the fixed mode $\lambda_0 = -1$ appears also as a non structurally robust mode.

2. <u>Using the graph-theoretic characterization</u> (Corollary 7.2). The digraph Γ_S associated to the system is the following :

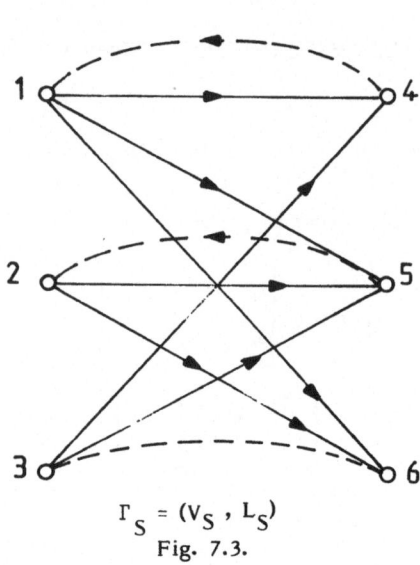

$$\Gamma_S = (V_S , L_S)$$

Fig. 7.3.

with $V_{1S} = \{1, 2, 3\}$
$V_{2S} = \{4, 5, 6\}$
$L_{1S} = \{(1,4),(1,5),(1,6),(2,5),(2,6),(3,4),(3,5)\}$
$L_{2S} = \{(4,1),(5,2),(6,3)\}$

Γ_S has five elementary cycles for which the transmittances are given below :

$$\begin{cases} \mathcal{C}_1 = (1,4,1) \\ \\ T_1(p) = \dfrac{3}{p-2} \end{cases} \qquad \begin{cases} \mathcal{C}_2 = (2,5,2) \\ \\ T_2(p) = \dfrac{1}{p+2} \end{cases} \qquad \begin{cases} \mathcal{C}_3 = (1,6,3,4,1) \\ \\ T_3(p) = \dfrac{p}{p+1} \cdot \dfrac{p+1}{p(p-2)} = \dfrac{1}{p(p-2)} \end{cases}$$

$$\begin{cases} \mathcal{C}_4 = (2,6,3,5,2) \\ \\ T_4(p) = \dfrac{1}{p(p-2)(p+2)} \end{cases} \qquad \begin{cases} \mathcal{C}_5 = (1,5,2,6,3,4,1) \\ \\ T_5(p) = \dfrac{p+1}{p(p+2)(p-2)^2} \end{cases}$$

It appears clearly that $\lambda_0 = -1$ is a non structurally fixed mode. Indeed, as a consequence of the pole-zero cancellation in $T_3(p)$, there is no cycle for which $\lambda_0 = -1$ is a pole.

Now, we use the first result in Corollary 7.2 to determine the non structurally robust modes with respect to one actuator failure. The only mode which is a pole of two elementary cycles such that the sets of vertices $\in V_{1s}$ involved in each cycle are disjoint is $\lambda = 2$. These two cycles are $\mathcal{C}_1$ and $\mathcal{C}_4$. Therefore, the non structurally robust modes with respect to one actuator failure are :

$$\lambda_0 = -1 \qquad \lambda_1 = 0 \qquad \lambda_2 = -2$$

We obtain the same result as using the frequency domain characterization with a significant lower number of calculations.

7.3.5. – Structurally robust control design
The choice of the information pattern

A significant advantage of the graph-theoretic characterization is that the robustness conditions of Proposition 7.5 and Corollary 7.2 can easily be expressed using binary variables associated to the components of the digraph. This provides a way for solving the problem of optimal structurally robust control design.

Let us consider the system (7.3.1) with the assumption that it has simple poles.

We study the same problem as the one solved by Locatelli et al. (LOC-77) which was presented in Section 5.3.3 but we add robustness constraints. The same notation as in Sections 3.6.2 and 5.3.3 are used.

The problem consists in determining a minimal set $S^* \subset S$ such that every pole of the system contained in the set $\Lambda^* = \{\lambda_1^*, \ldots \lambda_h^*\}$ is structurally robust. S is defined as in (3.6.1) by :

$$(j,i) \in S \text{ if } k_{i,j} \neq 0 \quad (i=1,\ldots,m) \quad ; \quad (j=1,\ldots,r)$$

The optimization criterion remains :

$$R(S^*) = \sum_{(i,j) \in S^*} r_{i,j}$$

where $r_{i,j}$ is a cost associated to the feeback connection from the output i to the input j.

Of course, the problem has a solution if and only if every element of Λ^* is structurally robust with respect to S.

If we want to determine a structurally robust control structure with respect to a unique perturbation, the simplest approach is to solve the problem presented in Section 5.3.3 for the perturbated system (7.3.7). The line failures are considered by eliminating the corresponding feedback connections from S.

Now, if we want to take into account a class of perturbations, the structure of the program remains the same but new constraints expressing the robustness requirement must be added.

Our study will be restricted to the following classes of perturbations :
- one actuator failure
- one sensor failure
- one line failure
- one actuator, sensor, or line failure (only one at a time)

which correspond to the cases considered in Corollary 7.2.

Consider first the class of perturbations specifying one actuator, sensor, or line failure (case 3 of Corollary 7.2). Then, the original program is easily modified. The constraint (C^g) is replaced by :

$$\sum_{(i,j) \in L_{1S}} v^g_{i,j} \quad z^g_{i,j} \geqslant 2$$

which assures that two edges (i,j) for which λ^*_g is a pole will be retained.

The two following constraints must be added :

$$(C^g_4) \quad \sum_{\substack{i \in V_{1S} / \\ (k,i) \in L_{2S}}} \sum_{(i,j) \in L_{1S}} (v^g_{i,j}, \; z^g_{i,j}) \; z^g_{k,i} \; \geqslant \; 2$$

which elimines the possibility of a unique cycle for which λ_g^* is a pole of order two.

$$(C^g_5) \quad \sum_{j/(i,j) \in L_S} z^g_{i,j} \leqslant 1 \qquad i \in V_S$$

which guarantees that the two cycles do not involve the same vertices. This is sufficient to asssure that the two cycles are disjoint because the boolean nature of the variables certifies that the two cycles are not composed by the same edges.

Remark 7.4. A significant advantage of this approach is that every mode is treated separately. Consequently, we can impose to a mode λ_i^* to be structurally robust whereas another mode $\lambda_j{}^*$ is required not to be fixed only (we modify C_1^i and add C_4^i, C_5^i only).

In the case for which we consider the class of perturbations specifying one actuator failure or the class of perturbations specifying one sensor failure (case 1 of Corollary 7.2), the corresponding programs are particular cases of the program established above. The two cycles are not required to be disjoint : for actuator (sensor) failure, they are not allowed to involve the same vertices of V_{1S} (V_{2S}) but some vertices of V_{2S} (V_{1S}) can be used twice. Therefore, the constraint (C_5^g) is relaxed as follows :

<table>
<tr><td>Actuator failure</td><td>Sensor failure</td></tr>
<tr><td>$\displaystyle\sum_{j/(i,j)\,\in\,L_S} z_{i,j}^g \leqslant 1 \qquad i \in V_{1S}$</td><td>$\displaystyle\sum_{j/(i,j)\,\in\,L_S} z_{i,j}^g \leqslant 1 \qquad i \in V_{2S}$</td></tr>
</table>

Finally, consider the case of the class of perturbations specifying one line failure (case 2 of Corollary 7.2).

The two cycles cannot be composed to be the same edges of L_{2S} but an edge from L_{1S} can belong to the two cycles.

This problem can be solved by considering some no boolean variables or, if we want to preserve the advantageous boolean nature of the program, by adding some redundant boolean variables.

1. <u>The variables $z_{i,j}^g$, associated with the edges $(i,j) \in L_{1S}$ are not boolean :</u>

$$(i,j) \in L_{1S}, \quad z_{i,j}^g = \begin{cases} 0 \text{ if } (i,j) \text{ does not belong to the retained cycles} \\ 1 \text{ if } (i,j) \text{ belongs to one retained cycle} \\ 2 \text{ if } (i,j) \text{ belongs to two retained cycles} \end{cases}$$

The program to be solved is the same as the one already established but the constraint (C_5^g) must be removed.

2. <u>Two boolean variables $z_{i,j}^g$ and $x_{i,j}^g$ are associated to each edge $(i,j) \in L_{1S}$</u>

The constraints C_1^g, C_2^g and C_4^g are modified as follows :

$$(C_1^g) \qquad \sum_{(i,j) \in L_{1S}} v_{i,j}^g (z_{i,j}^g + x_{i,j}^g) \geqslant 2$$

$$(C_2^g) \qquad \sum_{j/(i,j) \in L_S} z_{i,j}^g + x_{i,j}^g = \sum_{j/(j,i) \in L_S} z_{j,i}^g + x_{j,i}^g \qquad i \in V_S$$

$$(C_4^g) \qquad \sum_{\substack{(i,j) \in L_{1S} \\ i \in V_{1S}/(k,i) \in L_{2S}}} (z_{i,j}^g + x_{i,j}^g) \, v_{i,j}^g \, z_{k,i}^g \geqslant 2$$

Moreover, the constraint C_5^g is removed.

As an example, given the same system as in the example of Section 7.3.3, consider the following problem :

Find the feeback structure $S^* \subset S$ with a minimal number of feedback connections such that :

$-\lambda_1^* = -1$ is not a fixed mode
$-\lambda_2^* = 2$ is structurally robust with respect to the class of perturbations specifying one actuator, sensor, or line failure (case 3 of Corollary 7.2).

Only the feeback connections specified by $S = \{(1,1),(2,1),(3,1),(2,2),(2,3),(3,3)\}$ are allowed, and the associated costs are :

$$\begin{array}{l} r_{i,j} = 1 \\ (i,j) \in S \end{array}$$

The solution is obtained by solving the following boolean linear program :

$$\min \; w_{4,1} + w_{5,1} + w_{6,1} + w_{5,2} + w_{5,3} + w_{6,3}$$

$$(C_1^1) \quad z_{1,6}^1 \geqslant 1$$

$$(C_1^2) \quad z_{1,4}^2 + z_{1,5}^2 + z_{3,4}^2 + z_{3,5}^2 \geqslant 2$$

$$(C_4^2) \quad (z_{1,4}^2 + z_{1,5}^2)(z_{4,1}^2 + z_{5,1}^2 + z_{6,1}^2) + (z_{3,4}^2 + z_{3,5}^2)(z_{5,3}^2 + z_{6,3}^2) \geqslant 2$$

$$z_{1,4}^2 + z_{1,5}^2 + z_{1,6}^2 \leqslant 1$$

$$z_{2,5}^2 + z_{2,6}^2 \leqslant 1$$

$$z_{3,4}^2 + z_{3,5}^2 \leqslant 1$$

(C_5^2)

$$z_{4,1}^2 \leqslant 1$$

$$z_{5,1}^2 + z_{5,2}^2 + z_{5,3}^2 \leqslant 1$$

$$z_{6,1}^2 + z_{6,3}^2 \leqslant 1$$

and for $g = 1,2$

$$z_{1,4}^g + z_{1,5}^g + z_{1,6}^g = z_{4,1}^g + z_{5,1}^g + z_{6,1}^g$$

$$z_{2,5}^g + z_{2,6}^g = z_{5,2}^g$$

$$z_{3,4}^g + z_{3,5}^g = z_{5,3}^g + z_{6,3}^g$$

(C_2^g)

$$z_{4,1}^g = z_{3,4}^g + z_{1,4}^g$$

$$z_{5,1}^g + z_{5,2}^g + z_{5,3}^g = z_{1,5}^g + z_{2,5}^g + z_{3,5}^g$$

$$z_{6,1}^g + z_{6,3}^g = z_{1,6}^g + z_{2,6}^g$$

(C_3^g)

$$z_{4,1}^g \leqslant w_{4,1} \qquad\qquad z_{5,2}^g \leqslant w_{5,2}$$

$$z_{5,1}^g \leqslant w_{5,1} \qquad\qquad z_{5,3}^g \leqslant w_{5,3}$$

$$z_{6,1}^g \leqslant w_{6,1} \qquad\qquad z_{6,3}^g \leqslant w_{6,3}$$

There is a unique optimal solution :

$$S^* = \{(1,1),\ (2,3),\ (3,1)\}$$

corresponding to the following feedback structure :

$$K^* = \begin{bmatrix} k_{11} & 0 & k_{13} \\ 0 & 0 & 0 \\ 0 & k_{32} & 0 \end{bmatrix}$$

7.4. - CONCLUSION

When a controlled system is operating, it may happen that some component of the controller or of the system itself fails resulting in a structural modification of the controlled system. Such structural perturbations may be dangerously detrimental for a good pursuit of the operations. As an example, consider that the perturbated system is unstable.

Two approaches can be used to prevent such inacceptable situations. The first one consists in implementing a system for failure detection and diagnosis and then proceeding to a real time reconfiguration of the controller. Depending on the dynamics of the system, this solution, (which may require installing new components, proceeding to new measurements...) may be too much time consuming and therefore unefficient. The second approach consists in taking into account the eventuality of some structural perturbations in the design of the control system. The synthesis is then performed such that the controller preserves, under structural perturbations, some desirable properties to the controlled system.

Such controller is said to be structurally robust. In this chapter, the study focuses on the consequences of structural perturbations on stabilizability or pole assignability using structurally constrained feedback control. In Section 7.1, we consider structural perturbations affecting the plant in the sense that some subsystems may be disconnected. In Section 7.2, structural perturbations affect the controller. They stem from actuator, sensor, or line failures. The concept of structurally robust modes is introduced and some characterizations are provided. In Section 7.3, the design is faced of a robust control feedback structure which minimizes the cost associated to the information transfer.

APPENDIX 1

MULTIVARIABLE SYSTEM ZEROS

This appendix is concerned by the different types of zeros appearing in multivariable system. Each type of zero is defined and some relationships are outlined.

Consider the following time-invariant multivariable system :

$$\dot{x} = Ax + Bu$$
$$y = Cx + Du$$

where $x \in R^n$, $u \in R^m$, and $y \in R^r$ (max $(m,r) \leqslant n$) are the state, input, and output vectors, respectively. A, B, C, D are constant matrices of appropriate dimensions. The polynomial matrix :

$$P(p) = \begin{bmatrix} pI - A & B \\ C & D \end{bmatrix} \qquad (n+r, n+m)$$

is called the system matrix (ROS-70). If rank $P(p) = q$, then the Smith's form of $P(p)$ is given by :

$$S(p) = \begin{bmatrix} S^*(p)_{q,q} & 0_{q,n+m-q} \\ 0_{n+r-q,q} & 0_{n+r-q,n+m-q} \end{bmatrix}$$

where $S^*(p) = \text{diag}(s_1, s_2, \ldots, s_q)$ and s_i, $(i=1,\ldots,q)$ (s_i divides s_{i+1}), are the invariant polynomials of $P(p)$. If $M_j(p)$ denotes the greatest common divisor of all j^{th} order minors of $P(p)$, then the polynomial s_j is given by :

$$s_j(p) = \frac{M_j(p)}{M_{j-1}(p)} \qquad (j=1,2,\ldots,q)$$

with $M_0 = 1$.

The transfer function matrix of the system is :

$$G(p) = C (pI-A)^{-1} B + D = \frac{N(p)}{d(p)}$$

and its Smith-Mc Millan form is given by :

$$M(s) = \begin{bmatrix} M^*(p)_{qxq} & 0_{q,m-q} \\ 0_{r-q,q} & 0_{r-q,m-q} \end{bmatrix}$$

where

$$M^*(p) = \text{diag.}(\frac{\epsilon_1(p)}{\psi_1(p)} , ..., \frac{\epsilon_q(p)}{\psi_q(p)})$$

and $\frac{\epsilon_i}{\psi_i}$ is the ith invariant polynomial of $N(p)$ divided by the characteristic polynomial of $G(p)$, i.e. $\phi(p)$, and $q = \text{rank } G(p)$. Note that ϵ_i divides ϵ_{i+1} and ψ_{i+1} divides ψ_i.

The first clear classification of the zeros of linear multivariable systems was given by Rosenbrock (ROS-70). We find the following types of zeros :

Element Zeros (E.Z.) :

An element zero is any value of p for which the numerator of an element $g_{ij}(p)$ of $G(p)$ vanishes.

This type of zero has no special meaning in multivariable systems theory beyond its role in mono-variable system theory.

Decoupling Zeros (D.Z.) :

The decoupling zeros, introduced by Rosenbrock (ROS-70), are associated with the existence of uncoupled modes. They are defined as the values of p for which the matrices $(pI-A \quad B)$ and/or $\binom{pI-A}{C}$ are rank deficient.

These zeros are commonly known as the uncontrollable and/or unobservable modes. They are associated with a pole-zero cancellation and, as a consequence, they do not appear in the corresponding transfer function.

Three types of decoupling zeros can be defined :

- the input-decoupling zeros (I.D.Z.) which are the uncontrollable modes
- the output-decoupling zeros (O.D.Z.) which are the unobservable modes
- the input-output-decoupling zeros (I.O.D.Z.) which are the simultaneously uncontrollable and unobservable modes.

So, we have :

I.O.D.Z. = I.D.Z. ∩ O.D.Z.
D.Z.   = I.D.Z. ∪ O.D.Z.

Transmission Zeros (T.Z.) : (ROS-70)

These zeros are defined as the roots of the numerator polynomials of the Smith-Mc Millan form of $G(p)$. In terms of the minors of $G(p)$, they are the roots of the g.c.d. of the numerators of all the qth order minors of $G(p)$ (q = rank $G(p)$). Note that these minors must be adjusted to have $\phi(s)$ as their common denominator.

A transmission zero appears as a pole in some entries of $G(p)$ and as a zero in some others. The T.Z. are physically associated with the transmission-blocking properties of the system (see (MAC-76)). Note that Rosenbrock calls these zeros the zeros of the transfer matrix (ROS-70).

Invariant Zeros (I.Z.) :

The roots of the invariant polynomials of the system matrix $P(p)$ are called the invariant system zeros. In terms of the minors of $P(p)$, the invariant zeros are the roots of the monic g.c.d. of all the minors of $P(p)$ of maximum order.

Physically, the system invariant zeros are associated with the zero-output behaviour of the system. They correspond to the particular values of the complex frequency for which the system output is identically zero. In the general case of non-square systems, the set of invariant zeros is composed by the set of transmission zeros plus some decoupling zeros.

System Zeros (S.Z.) : (ROS-74)

The system zeros are the roots of the monic g.c.d. of all the minors of $P(p)$ of the form P_I^I, $I = \{1,2,\ldots,n, \, n+i_1, \, n+i_2, \, \ldots, \, n+i_k\}$, where P_I^I denotes the minor obtained by selecting the rows and columns corresponding to the set I in $P(p)$ and $0 \leqslant k \leqslant \min(m,r)$.

Roughly speaking, the set of system zeros is the union of the set of transmission zeros and the set of decoupling zeros :

$$S.Z. = T.Z. \cup D.Z.$$

Note also that the set of invariant zeros is included in the set of system zeros. These relationships are illustrated by Figure A1.1.

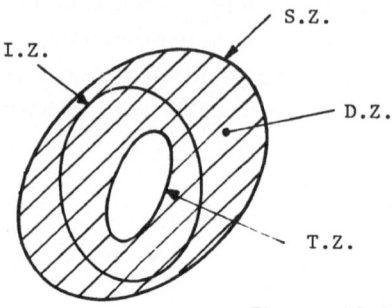

Figure A1.1.

When the system is completely controllable and completely observable, the sets of S.Z., I.Z. and T.Z. coincide because the set of decoupling zeros is empty.

So far, the various types of zeros have been defined in terms of the minors of $G(p)$ or $P(p)$. Some authors give equivalent definitions in terms of the particular frequency values for which $G(p)$ and $P(p)$ loose rank :

Z1 : Wolowich (WOL-73a)

The zeros of the controllable and observable system (A, B, C, D) are those p_0 such that rank $P(p_0)$ < rank $P(p)$.

Z2 : Davison and Wang (DAV-74 et 76e)

The transmission zeros of the system (A, B, C, D) are those complex numbers p_0 which satisfy the following inequality :

$$\text{rank } P(p_0) < n + \min(m,r)$$

Therefore, the transmission zeros (multiplicity included) are the roots of the g.c.d. of all the $(n+\min(m,r))^{\text{th}}$ order minors of $P(p)$. Note that if rank $P(s)$ < $n+\min(m,r)$, then every complex number is a transmission zero and the system is said to be degenerated (DAV-74).

The T.Z. as defined here are the roots (including multiplicities) of the polynomial obtained by multiplying all numerator polynomials of the Smith-McMillan form of $G(p)$ ((DAV-76e),Theorem 1). It is clear that the present definition coincides with Rosenbrock's definition. Note that for the special case of non degenerated systems with D = 0, the T.Z. are the roots of the transmission polynomials of the system (A, B, C) defined by Morse (MOR-73).

Z3 : Wolovich (WOL-73b), Desoer and Schulman (DES-74)

The transfer matrix of the system can be factorized as :

$$G(p) = C (pI-A)^{-1} B+D = V(p) T^{-1}(p) + D$$

where $V(p)$ and $T(p)$ are relatively right prime polynomial matrices. The zeros of the system (A, B, C, D) are those complex numbers p_0 such that

$$\text{rank } V(p_0) < \text{rank } V(p)$$

It is obvious that Z1 and Z3 are equivalent and, except for multiplicities, these definitions are equivalent to Rosenbrock's definition of transmission zeros.

APPENDIX 2

A FORTRAN SUBROUTINE TO EVALUATE THE FIXED MODES USING

OPEN-LOOP AND CLOSED-LOOP SYSTEM POLES

PURPOSE

The FORTRAN IV subroutine DATFM evaluates the set of fixed modes of a system described by :

$$\dot{x} = Ax + Bu$$
$$y = Cx$$

with respect to the output or state feedback control.

$$u = K y \qquad K \in K_F$$
$$\text{or} \quad u = K x \qquad K \in K_F$$

where K_F is the set of admissible feedback matrices specifying the feedback structure.

The subroutine DATFM uses the algorithm described in Section 2.4.1 which is based on the Definition 2.2 of fixed modes. It calculates the intersection between the set of open-loop system poles (eigenvalues of A) and the set of closed-loop system poles (eigenvalues of A+BKC) for three different values of K.

UTILIZATION

The subroutine statement is :
SUBROUTINE DATFM (N, M, L, A, B, C, AK, EPS, Z, JJ, AA, WK, ZF, ZV)

INPUT ARGUMENTS

N   Order of the system.

M   Number of system inputs.

L   Number of system outputs. This parameter is set equal to N in the case of state feedback.

A,B,C System matrices of dimension (N,N), (N,M) and (LXN), respectively.

AK Real matrix of dimension (MxN) describing the desired feedback structure, i.e. AK $\in$ K_F.

EPS Accuracy positive parameter used to compare the eigenvalues.

OUTPUT ARGUMENTS

JJ   Number of fixed modes. JJ=0 means that the system has no fixed modes with respect to the given feedback structure K_F and for an accuracy equal to EPS.

Z   Complex vector with N components containing the set of fixed modes (if any) in the first JJ positions.

WORK AREA ARGUMENTS

AA Work area of dimension (NxN).

WK Work area of dimension N.

ZF Complex work area of dimension N.

ZV Complex work area of dimension NxN.

CALLED SUBROUTINES

EIGRF Eigenvalues calculation subroutine, described in : "IMSL Library Manual", Edition 8, 1980.

GAUSS Standard IBM subroutine for random numbers generation described in : 1130 scientific subroutine Package (1130-Cm-02X), Programmer's Manual, IBM, Publ. IBM H20-0252-3, 1968.

MULT Two real matrices multiplication (see listing).

MULT3 Three real matrices multiplication (see listing).

```
        SUBROUTINE DATFM(N,M,L,A,B,C,AK,EPS,Z,JJ,AA,WK,ZF,ZV)

        IMPLICIT REAL*8 (A-H,O-Y),COMPLEX*16(Z),INTEGER(I-N)
        DIMENSION A(N,N),B(N,M),C(L,N),AK(M,N),Z(N),ZF(N)
        DIMENSION AA(N,N),WK(N),ZV(N,N)

        NIT=3
        IX=675543

        *** OPEN-LOOP POLES CALCULATION ***

        IJOB=0
        DO 26 I=1,N
        DO 26 J=1,N
26      AA(I,J)=A(I,J)
        CALL EIGRF(AA,N,N,IJOB,Z,ZV,N,WK,IER)

        DO 70 II=1,3
        IF(II.EQ.2) IX=975
        IF(II.EQ.3) IX=79861

        *** FEEDBACK MATRIX SELECTION ***

        DO 38 I=1,M
        DO 38 J=1,L
        IF(AK(I,J).EQ.0) GO TO 38
        CALL GAUSS(IX,0.33,0,V)
        AK(I,J)=V
38      CONTINUE

        *** CLOSED-LOOP POLES CALCULATION ***

        IF(L.NE.N) GO TO 44
        CALL MULT(B,N,M,AK,N,N,AA,N,N,N,M,L)
        GO TO 50
44      CALL MULT3(B,N,M,AK,M,N,C,L,N,N,M,L,N,AA,N)
50      DO 52 I=1,N
        DO 52 J=1,N
52      AA(I,J)=A(I,J)+AA(I,J)
        CALL EIGRF(AA,N,N,IJOB,ZF,ZV,N,WK,IER)

        *** INTERSECTION OF THE SET OF CLOSED-LOOP POLES AND ***
```

```
*** THE SET OF OPEN-LOOP POLES ***

      JJ=0
      NN=N
      NV=0
      IF(II.EQ.1)LI=N
      DO 68 I=1,LI
      IF (NV.EQ.0) GO TO 62
      NN=NN-1
      NV=0
62    DO 66 J=1,NN
      XX=REAL(Z(I))
      VV=REAL(ZF(J))
      IF(DABS(XX-VV).GT.EPS)GO TO 66
      YY=AIMAG(Z(I))
      WW=AIMAG(ZF(J))
      IF(DABS(YY-WW).GT.EPS)GO TO 66
      NV=1
      JJ=JJ+1
      Z(JJ)=Z(I)
      NNN=NN-1
      DO 64 K1=J,NNN
64    ZF(K1)=ZF(K1+1)
      GO TO 68
66    CONTINUE
68    CONTINUE
      IF(JJ.EQ.0)RETURN
      LI=JJ
70    CONTINUE
      RETURN
      END



      SUBROUTINE MULT(A,NA,MA,B,NB,MB,C,NC,MC,N,M,L)

      ***********************************************
      * TWO REAL MATRICES MULTIPLICATION      *
      *  C = A * B                            *
      *  A (N,M) < (NA,MA)                    *
      *  B (M,L) < (NB,MB)                    *
      *  C (N,L) < (NC,MC)                    *
      ***********************************************
```

```
        IMPLICIT REAL*8(A-H,O-Y), INTEGER(I-N)
        DIMENSION A(NA,MA),B(NB,MB),C(NC,MC)
        DO 1 I=1,N
        DO 1 J=1,L
        C(I,J)=0.D0
        DO 1 K=1,M
1       C(I,J)=C(I,J)+A(I,K)*B(K,J)
        RETURN
        END


        ***********************************************
        * THREE REAL MATRICES MULTIPLICATION   *
        *   QQ = A.B.C                         *
        *WITH                                  *
        *   A(N,M)   〈 (NA,NA)                 *
        *   B(M,L)   〈 (NB,MB)                 *
        *   C(L,K)   〈 (NC,MC)                 *
        *   QQ(N,K)  〈 (NQ,NQ)                 *
        ***********************************************


        SUBROUTINE MULT3(A,NA,MA,B,NB,MB,C,NC,MC,N,M,L,K,QQ,NQ)

        IMPLICIT REAL*8 (A-H,O-Y)
        DIMENSION A(NA,MA),B(NB,MB),C(NC,MC),QQ(NQ,NQ)

        DO 2 I=1,N
        DO 2 J=1,K
        S=O
        DO 1 II=1,M
        DO 1 JJ=1,L
1       S=S+A(I,II)*B(II,JJ)*C(JJ,J)
        QQ(I,J)=S
2       CONTINUE
        RETURN
        END

        SUBROUTINE GAUSS(IX,S,AM,V)

        ******************************************************
        * CALCUL DE DISTRIBUTION NORMALE V          *
        * DE VALEUR MOYENNE AM ET DE VARIANCE S     *
        ******************************************************
```

```
      A=0.0
      DO 1 I=1,12
      CALL RANGE(IX,IY,Y)
      IX=IY
1     A=A+Y
      V=(A-6.0)*S+AM
      RETURN
      END



      SUBROUTINE RANGE(IX,IY,YFL)

      *******************************************************
      * CALCULS OF A UNIFORM DISTRIBUTION          *
      * RANDOM VARIABLE BETWEEN 0 AND 1            *
      *******************************************************

      IY=IX*65539
      IF (IY)5,6,6
5     IY=IY+2147433647+1
6     YFL=IY
      YFL=YFL*0.4656613E-9
      RETURN
      END
```

APPENDIX 3
A FORTRAN ROUTINE TO
EVALUATE FIXED MODES USING THEIR SENSITIVITY

This appendix is concerned with the evaluation of fixed modes using the sensitivity approach (see § 2.4.2 and 3.5.4). The Appendix 3.1 gives a routine to compute the fixed modes of a system using variation calculus. This routine corresponds to the Algorithm 2.2 in Paragraph 2.4.2. The routine provided by Appendix 3.2 uses the eigenvalues gradient calculation approach and corresponds to the Algorithm 3.1 in Paragraphe 3.5.4.

Appendix 3.1

ROUTINE BASED ON VARIATION CALCULUS

The FORTRAN subroutine STFM1 computes the fixed modes of a system with simple modes. Based on the Algorithm 2.2, it computes the variations of the modes of the system resulting from changes in the feedback matrix. If the variation is zero, the corresponding mode is of course a fixed mode. The system is described by :

$$\dot{x} = Ax + Bu$$
$$y = Cx$$

and the feedback structure is described by the set of admissible matrices K_F.

UTILIZATION

The subroutine statement is

SUBROUTINE STFM1 (A,N,B,M,C,L,AK,EPS,MM,Z)

INPUT ARGUMENTS

N   Order of the system.

M   Number of system inputs.

L   Number of system outputs. This parameter is set equal to N in the case of state feedback.

A,B,C   System matrices of dimension (N,N),(N,M) and (L,N), respectively.

AK   Real feedback matrix of dimension (M,L) with admissible structure. The values of its entries are such that the closed-loop system has simple modes.

EPS   Accuracy positive small parameter.

OUTPUT ARGUMENTS

MM   Number of fixed modes (if any). MM=0 means that the system has no fixed modes with respect to the feedback structure K_F and for an accuracy equal to EPS.

Z   Complex vector of length N containing the set of fixed modes (if any) in the first MM positions.

REQUIRED MEMORY

If N > 20 the dimension statement must be modified according to :

DIMENSION WK(N), Z0(N,N), Z2(N,N), Z3(N,N), Z1(N,N,N)

CALLED SUBROUTINES

EIGRF   Eigenvalues calculation subroutine, described in : "IMSL Library Manual", Edition 8, 1980.

LISTING

```
      SUBROUTINE STFM1(A,N,B,M,C,L,AK,EPS,MM,Z)

      IMPLICIT REAL*8(A-H,O-Y),COMPLEX*16(Z)
      DIMENSION A(N,N),B(N,M),C(L,N),AK(M,L),Z(N)
      DIMENSION WK(20),Z1(20,20,20)
```

```
      DIMENSION Z0(20,20),Z2(20,20),Z3(20,20)

      NA=20
      IF(L.NE.N) GO TO 40
      DO 37 I=1,N
      DO 37 J=1,N
      DD=0
      DO 35 K=1,M
35    DD=DD+B(I,K)*AK(K,J)
      A(I,J)=A(I,J)+DD
      Z2(I,J)=A(I,J)+(0.,1.)*0.
37    Z3(I,J)=DD+(0.,1.)*0.
.     GO TO 48

40    DO 44 I=1,N
      DO 44 J=1,N
      DD=0
      DO 42 II=1,M
      DO 42 JJ=1,L
42    DD=DD+B(I,II)*AK(II,JJ)*C(JJ,J)
      A(I,J)=A(I,J)+DD
      Z2(I,J)=A(I,J)+(0.,1.)*0.
44    Z3(I,J)=DD+(0.,1.)*0.

48    IJOB=0
      CALL EIGRF(A,N,N,IJOB,Z,Z0,NA,WK,IER)
      MM=0
      DO 60 K=1,N
      DO 60 I=1,N
      DO 58 J=1,N
58    Z1(I,J,K)=Z2(I,J)
60    Z1(I,I,K)=Z2(I,I)-Z(K)
      DO 100 K=1,N
      DO 55 II=1,N
      DO 54 JJ=1,N
54    Z0(II,JJ)=(0.,0.)
55    ZO(II,II)=(1.,0.)
      DO 70 J=1,N
      IF(J.EQ.K) GO TO 70
      DO 65 I1=1,N
      DO 65 J2=1,N
```

```
65      Z2(I2,J2)=Z2(I2,J2)+Z0(I2,K2)*Z1(K2,J2,J)
        DO 67 I3=1,N
        DO 67 J3=1,N
67      Z0(I3,J3)=Z2(I3,J3)
70      CONTINUE

        ZZ=(0.,0.)
        DO 75 I4=1,N
        DO 75 J4=1,N
75      ZZ=ZZ+Z0(I4,J4)*Z3(J4,I4)

        RR=CDABS(ZZ)
        IF(RR.GT.EPS) GO TO 100
        MM=MM+1
        Z(MM)=Z(K)
100     CONTINUE
        RETURN
        END
```

Appendix 3.2

ROUTINE BASED ON GRADIENT CALCULATION

The FORTRAN routine STFM2 performes the same task as STFM1.

Based on the algorithm (3.1), it computes the gradient of the system modes with respect to the feedback matrix. If the gradient is zero (sufficiently small), then the corresponding mode is a fixed mode. In addition, STFM2 determines the type of the fixed modes by determining the structural sensitivity matrix (see § 3.5.4).

The data required by STFM2 are the same as the input arguments of STFM1 (see Appendix 3.1).

CALLED SUBROUTINES

EIGRF   See Appendix 3.1.
FIKT   Left eigenvector index search (see listing).
MODF   Calculation of the sensitivity with respect to the feedback matrix (see listing)
ECRIV   Writing of a complex vector (see listing).

LISTING

```
        PARAMETER (N=12,M=3,L=3)
        PARAMETER (IK=2*N+1,N2=N-2)
        IMPLICIT REAL*8(A-H,O-Y),COMPLEX*16(Z)
        DIMENSION ZM(N),Z(N,N),ZT(N,N)
        DIMENSION ZMT(N),ZMS(N)
        DIMENSION A(N,N),B(N,M),C(L,N),AK(M,N)
        DIMENSION WK(IK),AA(N,N),AS(N,N),AB(N,N)
        INTEGER S(N,N,N2),SS(N,N)

        READ(23,*)EPS
        DO 4 I=1,N
4       READ(23,*)(A(I,J),J=1,N)
        DO 5 I=1,N
5       READ(23,*)(B(I,J),J=1,M)
        IF(L.EQ.N) GO TO 7
        DO 6 I=1,L
6       READ(23,*)(C(I,J),J=1,N)
7       DO 8 I=1,M
8       READ(23,*)(AK(I,J),J=1,L)

        AL=0
        IF(L.EQ.N) GO TO 40
        DO 37 I=1,N
        DO 37 J=1,N
        AL=AL+0.05
        DD=0
        DO 35 K=1,M
35      DD=DD+B(I,K)*AK(K,J)
        AS(I,J)=AL*A(I,J)+DD
        AB(I,J)=A(I,J)+DD
37      AA(J,I)=AB(I,J)
        GO TO 48

40      DO 44 I=1,N
        DO 44 J=1,N
        AL=AL+0.05
        DD=0
        DO 42 II=1,M
        DO 42 JJ=1,L
```

```
42    DD=DD+B(I,II)*AK(II,JJ)*C(JJ,J)
      AS(I,J)=AL*A(I,J)+DD
      AB(I,J)=A(I,J)+DD
44    AA(J,I)=AB(I,J)

      *** CLOSED-LOOP EIGENVALUES AND ***
      *** EIGENVECTORS CALCULATION      ***

48    IJOB=1
      CALL EIGRF(AB,N,N,IJOB,ZM,Z,N,WK,IER)
      IJOB=1
      CALL EIGRF(AA,N,N,IJOB,ZMT,ZT,N,WK,IER)

      IR=0
      IR2=0
      DO 100 K=1,N
      CALL FIKT(ZM,ZMT,K,KT,N)
      CALL MODF(B,AK,C,N,M,L,Z,ZT,K,KT,SK)
      IF(SK.GT.EPS) GO TO 100
      IR=IR+1
      ZM(IR)=ZM(K)

      IS=0
      DO 70 II=1,N
      DO 70 JJ=1,N
      S(II,JJ,IR)=0
      IF(A(II,JJ).EQ.0) GO TO 70
      ZZ=ZT(II,KT)*Z(JJ,K)
      RS=CDABS(ZZ)
      IF(RS.LE.EPS) GO TO 70
      S(II,JJ,IR)=1
      IS=IS+1
70    CONTINUE

      IF(IS.GT.0) GO TO 100
      IR2=IR2+1
      ZMS(IR2)=ZM(K)
      IR=IR-1
100   CONTINUE
      IRT=IR+IR2
      IF(IRT.NE.0) GO TO 120
      WRITE(6,110)
```

```
110    FORMAT(/,5X,,'THE SYSTEM HAS NO FIXED MODES',//)
       STOP

120    WRITE(6,125)
125  · FORMAT(/,5X,'THE FIXED MODES ARE :',/)
       IRT=IR*IR2
       IF(IRT.EQ.0) GO TO 140
       CALL ECRIV(ZM,IR,N)
       CALL ECRIV(ZMS,IR2,N)
       WRITE(6,130)
130    FORMAT(/,5X,'THE STRUCTURALLY FIXED MODES OF TYPE II ARE :',/)
       CALL ECRIV(ZMS,IR2,N)
       GO TO 160
140    IF(IR.NE.0) GO TO 155
       CALL ECRIV(ZMS,IR2,N)
       WRITE(6,150)
150    FORMAT(/,5X,'ALL THE FIXED MODES ARE STRUCTURAL',//)
       'OF TYPE II',/)
       STOP

155    CALL ECRIV(ZM,IR,N)
160    CONTINUE

       *** CALCULATION OF CLOSED-LOOP EIGENVALUES   ***
       *** AND EIGENVECTORS OF AN EQUIVALENT SYSTEM ***

       DO 170 I=1,N
       DO 170 J=1,N
170    AA(J,I)=AS(I,J)
       IJOB=1
       CALL EIGRF(AS,N,N,IJOB,ZMS,Z,N,WK,IER)
       IJOB=1
       CALL EIGRF(AA,N,N,IJOB,ZMT,ZT,N,WK,IER)

       IS=0
       DO 230 K=1,N
       CALL FIKT(ZMS,ZMT,K,KT,N)
       CALL MODF(B,AK,C,N,M,L,Z,ZT,K,KT,SK)
       IF(SK.GT.EPS) GO TO 230
       JS=0
       DO 190 II=1,N
       DO 190 JJ=1,N
```

```
      SS(II,JJ)=0
      IF(A(II,JJ).EQ.0) GO TO 190
      ZZ=ZT(II,KT)*Z(JJ,K)
      RS=CDABS(ZZ)
      IF(RS.LE.EPS) GO TO 190
      SS(II,JJ)=1
      JS=JS+1
190   CONTINUE
      IF(JS.EQ.0) GO TO 230
      DO 220 KK=1,IR
      DO 200 II=1,N
      DO 200 JJ=1,N
      IF(SS(II,JJ).NE.S(II,JJ,KK)) GO TO 220
200   CONTINUE
      IS=IS+1
      ZMS(IS)=ZM(KK)
220   CONTINUE
230   CONTINUE
      IF(IS.EQ.0) GO TO 250
      WRITE(6,240)
240   FORMAT(/,5X,'THE STRUCTURALLY FIXED MODES OF TYPE I ARE :',/)
      CALL ECRIV(ZMS,IS,N)
      IF(IR.NE.IS) GO TO 250
      STOP

250   WRITE(6,260)
260   FORMAT(/,5X,'THE NON STRUCTURALLY FIXED MODES ARE :',/)
      IF(IS.NE.0) TO GO 265
      CALL ECRIV(ZM,IR,N)
      STOP

265   DO 300 I=1,IR
      DO 290 J=1,IS
      ZZ=ZM(I)-ZMS(J)
      RX=DREAL(ZZ)
      RX=DABS(RX)
      IF(RX.GT.EPS)GO TO 270
      RY=DIMAG(ZZ)
      RY=DABS(RY)
      IF(RY.LE.EPS) GO TO 290
270   WRITE(6,280)
```

```
280     FORMAT(5X,'(',F12.6,' +J',F12.6,' )')
        GO TO 300
290     CONTINUE
300     CONTINUE
        STOP
        END


        *** LEFT EIGENVECTOR INDEX SEARCH ***


        SUBROUTINE FIKT(ZA,ZB,K,KT,N)


        IMPLICIT REAL*8(A-H,O-Y),COMPLEX*16(Z)
        DIMENSION ZA(N),ZB(N)


        EPS=1.E-10
        DO 5 II=1,N
        ZZ=ZA(K)-ZB(II)
        VA=DREAL(ZZ)
        VB=DABS(VA)
        IF(VB.GT.EPS) GO TO 5
        WA=DIMAG(ZZ)
        WB=CDABS(ZZ)
        IF(WB.GT.EPS) GO TO 5
        KT=II
        RETURN
5       CONTINUE
        RETURN
        END


        ****************************************************
        * CALCULATION OF THE SENSITIVITY WITH       *
        * RESPECT TO THE FEEDBACK MATRIX            *
        ****************************************************


        SUBROUTINE MODF(B,AK,C,N,M,L,Z,ZT,K,KT,SK)


        IMPLICIT REAL*8(A-H,O-Y),COMPLEX*16(Z)
        DIMENSION B(N,M),C(L,N),AK(M,N)
        DIMENSION Z(N,N),ZT(N,N)


        SK=0
        DO 50 I=1,M
```

```
      DO 50 J=1,L
      ZSK=(0.,0.)
      IF(AK(I,J).EQ.0) GO TO 50
      XW=0
      YW=0
      DO 10 LL=1,N
      XW=XW+DREAL(ZT(LL,KT))*B(LL,I)
10    YW=YW+DIMAG(ZT(LL,KT))*B(LL,I)
      ZW=XW+(0.,1.)*YW
      IF(L.EQ.N) GO TO 30
      XV=0
      YV=0
      DO 20 LL=1,N
      XV=XV+C(J,LL)*DREAL(Z(LL,K))
20    YV=YV+C(J,LL)*DIMAG(Z(LL,K))
      ZV=XV+(0.,1.)*YV
      GO TO 40
30    ZV=Z(J,K)
40    ZSK=ZW*ZV
      SK1=CDABS(ZSK)
      SK=DMAX1(SK,SK1)
50    CONTINUE
      RETURN
      END




      ************************************
      * COMPLEX VECTOR WRITTING     *
      * Z (N)      N ≤ NMAX         *
      ************************************


      SUBROUTINE ECRIV(Z,N,NMAX)

      COMPLEX*16 Z(NMAX)

      DO 5 I=1,N
5     WRITE(6,10)Z(I)
10    FORMAT(15X,'(',F12.6,' +J',F12.6,' )')
      RETURN
      END
```

ANDERSON AND CLEMENTS TEST PACKAGE FOR REAL MODES

PURPOSE

The FORTRAN W subroutine ACTFM evaluates the set of real decentralized fixed modes of a NS-station system described by :

$$\dot{x} = Ax + \sum_{i=1}^{NS} B_i u_i$$

$$y_i = C_i x \qquad (i=1,\dots,NS) \tag{A4.1}$$

Defining :

$$B = (B_1,\dots, B_{NS})$$

$$C = (C',\dots, C'_{NS})' \tag{A4.2}$$

the system can be written :

$$\dot{x} = Ax + Bu \tag{A4.3}$$
$$y = Cx$$

The subroutine ACTFM uses the algebraic characterization of fixed modes described in section 3.3.1. Note that ACTFM examines real poles only.

UTILIZATION

The subroutine statement is :
SUBROUTINE ACTFM (A,N,B,M,C,L,NS,IM,IR,EPS,Z,JJ)

INPUT ARGUMENTS

N   Order of the system.

M   Number of the system inputs.

L   Number of the system outputs.

A,B,C   System matrices of dimension (N,N), (N,M) and (L,N), respectively.

NS   Number of the control stations.

IM   Integer vector of dimension NS containing the number of inputs of the ith station in the ith position.

IR   Integer vector of dimension NS containing the number of outputs of the ith station in the ith position.

EPS   Small positive real number defining the zero accuracy.

IT   Option parameter :

         IT = 1 writting of the open-loop poles.

         IT $\neq$ 1 no writting.

OUTPUT ARGUMENTS

JJ   Number of real decentralized fixed modes of the system. JJ=0 means that the system has not real decentralized fixed modes.

XX   Real vector of length N containing the real decentralized fixed modes (if any) in the first JJ positions.

CALLED SUBROUTINES

EIGRF   See Appendix 2.

DSVD   Subroutine computing the Singular Value Decomposition of an arbitrary real rectangular matrix, described in :

         GARBOW B.S., J.M. BOYLE, J.J. DONGARRA, C.B. MOLER

         "Matrix Eigensystem Routines - EISPACK Guide Extension".

         Lecture notes in computer science n° 51, Springer-Verlag, New-York, 1977.

RANK   Subroutine determining the rank of a real matrix of dimension (M,N) by calling DSVD.

REQUIRED MEMORY

If NT=N+max(M,L)>20 or NS>5, then the dimension statement must be changed according to :

```
      DIMENSION II(NS), ZV(NT,NT), AA(NT,NT)
      DIMENSION WK(NT), U(NT,NT), V(NT,NT), RV1(NT)
```

LISTING

```
      SUBROUTINE ACTFM(A,N,B,M,C,L,NS,IM,IR,EPS,Z,JJ,IT)

      IMPLICIT REAL*8(A-H,O-Y),INTEGER(I-N),COMPLEX*16(Z)
      DIMENSION A(N,N),B(N,M),C(L,N),IM(NS),IR(NS),Z(N)
      DIMENSION II(5),ZV(20,20),AA(20,20)
      DIMENSION WK(20),U(20,20),V(20,20),RV1(20)

      NT=20

      *** OPEN-LOOP POLES CALCULATION ***

15    IJOB=1
      DO 16 I=1,N
      DO 16 J=1,N
16    AA(I,J)=A(I,J)
      CALL EIGRF(AA,N,NT,IJOB,Z,ZV,NT,WK,IER)

      IF(IT.NE.1) GO TO 25
      WRITE(6,18)
18    FORMAT(5X,'THE OPEN-LOOP POLES ARE ',/)
      DO 20 I=1,N
20    WRITE(6,22) Z(I)
22    FORMAT(5X,E12.6,2X,E12.6,/)

25    JJ=0
      EPS1=10D-8
      DO 90 JI=1,N
      YY=AIMAG(Z(JI))
      IF(DABS(YY).GT.EPS1) GO TO 90
      XX=REAL(Z(JI))
      IF(JI.EQ.1) GO TO 40
      JII=JI-1
      DO 30 I=1,JII
      YY=AIMAG(Z(I))
      IF(DABS(YY).GT.EPS1) GO TO 30
      XX1=REAL(Z(I))
```

```
         IF(DABS(XX-XX1).LT.EPS1) GO TO 90
30       CONTINUE

         *** BUILDING OF THE MATRIX BLOCK-DIAG.((A-SI),O) ***

40       DO 44 I=1,N
         DO 42 J=1,N
42       AA(I,J)=A(I,J)
44       AA(I,I)=AA(I,I)-XX
         N1=N+1
         IK=N+M
         IK1=N+L
         DO 46 I=N1,IK1
         DO 46 J=N1,IK
46       AA(I,J)=O.

         *** COMPLEMENTARY SUBSYSTEMS SEARCH ***

         MAX=2**NS-2
         DO 90 NB=1,MAX
         I1=O
         NU=NB
         DO 48 I=1,NS
         ND=NU/2
         NR=NU-ND*2
         IF (NR.EQ.0) GO TO 48
         I1=I1+1
         II(I1)=I
48       NU=ND

         *** BUILDING OF THE TESTED MATRIX ***

56       I1=1
         IP=N
         JP=N
         DO 68 IS=1,NS
         IC=0
         JC=0
         IK2=IS-1
         IF (IK2.LT.1) GO TO 60
         DO 58 K=1,IK2
         IC=IC+IR(K)
```

```
58    JC=JC+IM(K)
60    IF(IS.EQ.II(I1)) GO TO 64
      IK3=IP+1
      IK4=IP+IR(IS)
      DO 62 I=IK3,IK4
      IC=IC+1
      DO 62 J=1,N
62    AA(I,J)=C(IC,J)
      IP=IP+IR(IS)
      GO TO 68
64    IK5=JP+1
      IK6=JP+IM(IS)
      DO 66 J=IK5,IK6
      JC=JC+1
      DO 66 I=1,N
66    AA(I,J)=B(I,JC)
      JP=JP+IM(IS)
      I1=I1+1
68    CONTINUE
      CALL RANK(AA,IP,JP,NT,WK,U,V,RV1,IRANK,EPS)
      IF(IRANK.LT.N) GO TO 78
      GO TO 90
78    JJ=JJ+1
      Z(JJ)=Z(JI)
90    CONTINUE
      RETURN
      END


      SUBROUTINE RANK(A,M,N,NM,W,U,V,RV1,IRANK,EPS)


      ************************************************************
      * DETERMINATION OF THE RANK OF A REAL MATRIX    *
      * A : MxN                                       *
      * NM = A ⩾ M,⩾ N                                 *
      * REAL MATRICES :                               *
      * W(NM),U(NM,NM),V(NM,NM),RV1(NM)               *
      * RESULT IN IRANK                               *
      ************************************************************
      REAL*8 A(NM,N),W(N),U(NM,N),V(NM,N),RV1(N)
      DO 1 I=1,NM
1     W(I)=O.
      CALL DSVD(NM,N,M,A,W,.FALSE.,U,.FALSE.,V,IERR,RV1)
```

```
      IRANK=0
      MM=N
      IF(M.GE.N) GO TO 2
      MM=M
2     DO 3 I=1,MM
      IF (W(I).LT.EPS) GO TO 3
      IRANK=IRANK+1
3     CONTINUE
      RETURN
      END
```

DETERMINATION OF THE GRADIENT MATRIX OF THE PERFORMANCE INDEX
BY USING VARIATION CALCULUS

This appendix uses variation calculus (LEV-70) to determine the derivatives of the performance index.

A5.1. Preliminaries

The three following results are useful in the subsequent development.

Theorem A5.1 (BEL-70). If the integral : $x = - \int_0^\infty e^{At} C e^{Bt}$ exists for all C, it represents the unique solution of $Ax + xB = C$.

Theorem A5.2 (BEL-70). Consider the matrix $e^{(A + \epsilon B)t}$, where ϵ is a small real number. It can be approximated to the first order in ϵ as :

$$e^{(A + \epsilon B)t} = e^{At} + \epsilon \int_0^t e^{A(t-s)} B e^{As} ds$$

Kleinman's lemma (KLE-66). Let $f(x)$ be a trace function. If, for all x and for $\epsilon \longrightarrow 0$, we can write :

$$f(x + \epsilon \Delta x) = f(x) + \text{Trace } M(x) - \Delta x$$

$$\text{then} : \frac{df(x)}{dx} = M'(x)$$

A5.2. Development (TAR-85)

Consider the system :

$$\dot{x}(t) = A x(t) + B u(t)$$
$$y(t) = C x(t)$$

and the performance index :

$$J = \int_0^\infty (x'Q\ x + u'\ Ru)\ dt.$$

Let the control be given by u = - Ky, so that the closed-loop system is :

$$\dot{x} = (A - BKC)\ x = Dx$$

which has the solution : $x = e^{Dt}\ x_0$, $x_0 = x(0)$

Substituting in the expression of y and using the trace function properties, we get :

$$J(D) = Tr\ [\ \int_0^\infty e^{D'\ t}\ Q_1(K,C)\ e^{Dt}\ dt\ X_0\ ]$$

where

$$X_0 = E\ [\ x(0)\ x(0)'\ ]$$
$$Q_1(K) = Q + C'\ K'\ R\ KC$$

Suppose that the system matrices are subjected to small perturbations, i.e. :

A changes to $A + \varepsilon_A$. ΔA
B changes to $B + \varepsilon_B$. ΔB
C changes to $C + \varepsilon_C$. ΔC
D changes to $D + \varepsilon_D$. ΔD

where ε_A, ε_B, ε_C and ε_D are small real numbers of the same order, i.e. $\varepsilon_A \simeq \varepsilon_B \simeq \varepsilon_C \simeq \varepsilon_D$. Then the closep-loop system becomes :

$$\dot{x}(t) = (D + \varepsilon.\ \Delta D)\ x(t)$$

with $\Delta D = \Delta A - \Delta B.\ KC - B\ .\ \Delta K.C - B.K.\ \Delta C$
and the criterion :

$$J(D + \varepsilon \Delta D) = Tr\ [\ \int_0^\infty e^{(D+\varepsilon \Delta D)'t}\ \ Q_1(K+\varepsilon \Delta K,\ C+\varepsilon \Delta C)\ e^{(D+\varepsilon \Delta D)t}\ dt\ X_0\ ]$$

By developing this expression to the first order, using Theorems A5.1 and A5.2, and the trace function property $Tr(AB) = Tr(BA) = Tr(A'B') = Tr(B'A')$, one obtains :

$$J(D+\varepsilon \Delta D) = J(D) + \varepsilon\ Tr\ [2S(C'K'R-PB).\Delta\ (KC) + 2SP.\Delta\ (A,B)\ ]$$

with

$$D'P + PD + Q + C'K'R\ KC = 0$$

$$DS + SD + X_0 = 0$$
$$\Delta D = \Delta(A,B) - B. \Delta(KC)$$
$$\Delta(KC) = \Delta K.C + K.\Delta C$$
$$\Delta(A,B) = \Delta A - \Delta B. KC$$

Two cases of variations are considered :

1 - Simultaneous variations on A and B

In this case, we have :
$$\Delta(KC) = 0 \text{ and } \Delta D = \Delta(A,B)$$
Then one gets :
$$J[D+\epsilon. \Delta(A,B)] = J(D) + \epsilon \text{ Tr } [2SP. \Delta(A,B)]$$
and by application of Kleinman's lemma :

$$\frac{\partial J}{\partial(A,B)} = 2 PS$$

This case combines two situations :

. variations on A only, we get $\dfrac{\partial J}{\partial A} = 2 PS.$

. variations on B only, we get $\dfrac{\partial J}{\partial B} = - PSC'K'$

2 - Simultaneous variations on K and C

In this case, we have :
$$\Delta(A,B) = 0 \quad \text{and} \quad \Delta D = - B \ \Delta(KC)$$
The criterion becomes :

$$J[D - \epsilon. B. \Delta(KC)] = J(D) + \epsilon \text{ Tr } [(2SC'K'R - 2SPB)] \quad (KC)$$

and by application of Kleinman's lemma :

$$\frac{\partial J}{\partial(KC)} = 2 (RKC - B'P) S$$

This case combines two situations :

. variations on K only, we get $\dfrac{\partial J}{\partial C} = 2 K'(RKC - B'P)S$

. variations on C only, we get $\dfrac{\partial J}{\partial K} = 2 (RKC - B'P) SC'$

***

APPENDIX 6

A FORTRAN ROUTINE TO DETERMINE AN OPTIMAL CONSTRAINED FEEDBACK MATRIX WITH POSSIBLE ROBUSTNESS REQUIREMENTS

This appendix provides a routine for the determination of an optimal (local) constrained feedback matrix $K \in K_F$ for the linear system :

$$\dot{x} = Ax + Bu \qquad x \in R^n, \quad u \in R^m$$
$$y = Cx \qquad y \in R^r$$

The optimization is performed with respect to the classical quadratic criterion (see § 6.3.1) and its sensitivities (see § 6.4.2). The routine considers two cases :

. State feedback : $u = Kx$

. Output feedback : $u = KCx = Ky$

The optimization problem is :

$$\min_{K \in K_F} J_3(K) = Tr(P V_0) + Tr(SPLPS) + Tr[(RKC-B'P)SFS(RKC-B'P)']$$

subject to $D'S + SD + Q + C'K'RKC = 0$
$$DP + PD' + V_0 = 0$$

where R, Q, L, F are weighting matrices of appropriate dimension.

The solution of this problem is given by :

$$\frac{\partial J_3}{\partial K} = 2 \ [(RKC + B'S)\mu + B' \lambda P + R(RKC+B'S)PFP] \ C'$$

with $D'S + SD + F_1(K) = 0$
$$DP + PD' + V_0 = 0$$
$$D\mu + \mu D' + F_2(K,P,S) = 0$$
$$D' \lambda + \lambda D + F_3(K,P,S) = 0$$
where $D = A + BKC$

$$F_1(K) = Q + C'K'RKC$$
$$F_2(K,P,S) = V_0 + PPSL + LSPP + B(RKC+B'S) PFP + PFP(RKC+B'S)'B'$$
$$F_3(K,P,S) = PSLS + SLSP + (RKC + B'S)' (RKC+B'S)PF +$$
$$+ FP(RKC+B'S)'(RKC+B'S)$$

This problem is solved by using the feasible direction method (gradient projection) to satisfy the structural constraints and an adaptative step size as given in (6.3.6).

The required data are :

SYSTEM DATA

N	Order of the system.
M	Number of inputs.
L	Number of outputs.
A,B,C	System matrices of dimension (N,N), (N,M) and (L,N), respectively.
IES	Option parameter :

IES = 0 if state feedback

IES $\neq$ 0 if output feedback.

OPTIMIZATION DATA

AL	Initial step size.
PI	Real positive number superior to 1.
ANU	Real positive number such that 0 < ANU < 1.
EPS	Accuracy positive small number considered as zero.
NI	Allowed maximum number of iterations.
IT	Option parameter :

IT = 1, writing of the intermediate results :

II iteration number

F gradient norm

CR criterion value

AL step size

VPD dominant closed-loop eigenvalue of the matrix S.

IT $\neq$ 1, no writting.

IGB Option parameter :

IGB=1 for minimizing the reduced criterion : $J_3 = Tr [PV_0]$.

This case corresponds to the algorithm of Geromel and Bernussou (see § 6.3.1)

IGB=2 for minimizing the criterion : $J_3 = \text{Tr} \, [\, PV_0] + \text{Tr} \, [\text{SPLPS}\,]$

IGB≠1 and IGB≠2 for solving the general problem described above.

AK   Initial stabilizing feedback matrix such that AK $\in K_F$.

V0   Initial state condition matrix V0 = E [x(0) x(0)'] .

Q   State weighting matrix of dimension (N,N)

R   Input weighting matrix of dimension (M,M).

PL   Weighting matrix of dimension (NxN) for the sensitivity with respect to A and B. This matrix is necessary if IGB≠1.

PF   Weighting matrix of dimension (N,N) for the sensitivity with respect to K and C.

CALLED SUBROUTINES

MULT   Two real matrices multiplication (see listing).

MULT3   Three real matrices multiplication (see listing).

F1KC   Calculation of $F_1(K)$ (see listing).

F2KSPV Calculation of $F_2(K,P,S)$ (see listing).

F3KSPV Calculation of $F_3(K,P,S)$ (see listing).

RKCBS Calculation of QQ = (R.AA+B'S) and of AB=QQ.P. State feedback : AA=K, output feedback : AA = KC (see listing).

MST   Calculation of F - (A+A') (see listing).

LYAPUN Lyapunov equation solving (see listing).

MATBF   Closed-loop matrix calculation (see listing).

PGV   Determination of the smallest (or biggest) element of a real vector (see listing).

MEV   Determination of the smallest real part of the eigenvalues of a real matrix (see listing).

CRIT   Criterion calculation (see listing).

GRADK   Gradient calculation (see listing).

EIGRF   Eigenvalue calculation subroutine, described in : "IMSL Library Manual", Edition 8, 1980.

LINV2F Real matrix inversion subroutine, described in "IMSL Library Manual", Edition 8, 1980.

PRINT   Writing of a real matrix (see listing).

REQUIRED MEMORY

The dimension statements must be modified if N > 15, M > 5, or L > 5 according to :

```
      DIMENSION A(N,N), B(N,M), C(L,N), AK(M,N), WK(IK), R(M,M)
```

The same for all the other matrices of dimension (N,N). Also the parameters NA,MA,LA and IK must be set to :

NA=N, MA=M, LA=L and IK $\geqslant$ N^2 + 3N

LISTING

```
      IMPLICIT REAL*8 (A-H,O-Y),COMPLEX*16(Z)
      COMMON /MATSS/ A(15,15),B(15,5),C(5,15),AK(5,15)
      COMMON /MATP1/ Q(15,15),R(5,5)
      COMMON /MATP2/ PL(15,15),PF(15,15)
      COMMON /MATL1/ S(15,15),P(15,15),VO(15,15)
      COMMON /MATL2/ PMU(15,15),SLA(15,15)
      COMMON /GRADS/ GA(15,15),GK(15,15)
      DIMENSION AF(15,15),AFF(15,15),G(15,15),D(15,15)
      DIMENSION E(15,15),AA(15,15),AC(15,15),WK(270)
      DIMENSION ZV(15),Z(15,15)


      NA=15
      MA=5
      LA=5
      IK=270

      *** SYSTEM MATRICES READING ***

      READ(19,*) N,M,L,IES
      DO 12 I=1,N
12    READ(19,*)(A(I,J),J=1,N)
      DO 17 I=1,N
17    READ(19,*)(B(I,J),J=1,M)
      NL=N
      IF(IES.EQ.0) GO TO 24
      NL=L
      DO 20 I=1,L
20    READ(19,*)(C(I,J),J=1,N)

      *** OPTIMIZATION DATA READING ***

24    READ(19,*)AL,PI,ANU,EPS,NI,IT,IGB
```

```
      DO 26 I=1,M
26    READ(19,*)(AK(I,J),J=1,NL)
      DO 28 I=1,N
28    READ(19,*)(VO(I,J),J=1,N)
      DO 32 I=1,N
32    READ(19,*)(Q(I,J),J=1,N)
      DO 36 I=1,M
36    READ(19,*)(R(I,J),J=1,M)
      IF(IGB.EQ.1) GO TO 52

      DO 44 I=1,N
44    READ(19,*)(PL(I,J),J=1,N)
      IF(IGB.EQ.2) GO TO 52

      DO 48 I=1,N
48    READ(19,*)(PF(I,J),J=1,N)

      *** ITERATIONS BEGINING ***

52    IF(IT.NE.1) GO TO 58
      WRITE(6,56)
56    FORMAT(6X,'II',10X,'F',10X,'CR',10X,'AL',10X,'VPD',/)

58    II=0
60    CONTINUE

      *** CLOSED-LOOP MATRIX DETERMINATION   ***

      CALL MATBF(N,M,NL,NA,MA,LA,AF)

      *** SOLVING THE 1ST LYAPUNOV EQUATION ***
      ***        (CALCULATION OF S)          ***
      DO 65 I=1,N
      DO 65 J=1,N
55    AFF(I,J)=-AF(I,J)
      CALL F1KC(N,NA,M,MA,NL,LA,AA,AC)
      CALL LYAPUN(AFF,N,NA,AC,NA,S,NA,P,G,ZV,Z,WK,IK)

      IF(IGB.NE.1) GO TO 72
      CALL CRIT(N,M,IGB,CR)
      IF(II.NE.0) GO TO 72
      WRITE(6,71) CR
```

```
71      FORMAT(5X,'INITIAL CRITERION VALUE=',E12.6,/)
72      IF(II.EQ.0) GO TO 75
        IF(IGB.NE.1) GO TO 73
        IF(CR.GE.Y) GO TO 135
73      CALL MEV(S,N,NA,G,ZV,Z,WK,IK,VPD)
        IF(VPD.LE.0) GO TO 135

        *** SOLVING THE 2ND LYAPUNOV EQUATION   ***
        ***           (CALCULATION OF P)        ***

75      DO 76 I=1,N
        DO 76 J=1,N
        AFF(I,J)=-AF(J,I)
76      E(I,J)=-VO(I,J)
        CALL LYAPUN(AFF,N,NA,E,NA,P,NA,AC,G,ZV,Z,WK,IK)


        CALL RKCBS(AA,M,N,NA,MA,LA,E,GK)
        IF(IGB.NE.1) GO TO 90

        IF(IES.EQ.0) GO TO 83
        DO 80 I=1,N
        DO 80 J=1,L
80      AC(I,J)=C(J,I)
        CALL MULT(GK,NA,NA,AC,NA,NA,G,NA,NA,M,N,L)
        GO TO 110
83      DO 85 I=1,M
        DO 85 J=1,NL
85      G(I,J)=GK(I,J)
        GO TO 110

90      CALL MULT(S,NA,NA,P,NA,NA,GA,NA,NA,N,N,N)

        *** SOLVING THE 3RD LYAPUNOV EQUATION ***
        ***           (CALCULATION OF MU)       ***

95      CALL F2KSPV(M,N,IGB,G,AFF,AA,NA,MA,LA)
        DO 97 I=1,N
        DO 97 J=1,N
97      AFF(I,J)=-AF(J,I)
        CALL LYAPUN(AFF,N,NA,AA,NA,PMU,NA,G,AC,ZV,Z,WK,IK)
```

```
*** SOLVING THE 4TH LYPAUNOV EQUATION ***
***        (CALCULATION OF LAMDA)        ***

      CALL F3KSP(E,M,N,IGB,AFF,G,AA,NA,MA,LA)
      DO 103 I=1,N
      DO 103 J=1,N
103   AFF(I,J)=-AF(I,J)
      CALL LYAPUN(AFF,N,NA,AA,NA,SLA,NA,G,AC,ZV,Z,WK,IK)

*** CRITERION CALCUATION ***

      CALL CRIT(N,M,IG,CR)

      IF(II.NE.0) GO TO 107
      WRITE(6,105) CR
105   FORMAT(5X,'INITIAL CRITERION VALUE=',E12.6,/)
      GO TO 109
107   IF(CR.GT.Y) GO TO 135

109   CALL GRADK(E,AA,AC,N,M,NL,NA,MA,LA,G)

*** GRADIENT PROJECTION ***

110   F=0
      DO 115 I=1,M
      DO 115 J=1,NL
      D(I,J)=0
      IF(AK(I,J).EQ.0) GO TO 115
      D(I,J)=G(I,J)
      Fl=DABS(G(I,J))
      F=DMAX1(F,Fl)
115   CONTINUE

      Y=CR
      II=II+1

      IF(IT.NE.1) GO TO 118
      WRITE(6,117) II,F,CR,AL,VPD
117   FORMAT(5X,I3,1X,E12.6,1X,E12.6,1X,E14.8,1X,E12.6,/)

118   IF(F.LT.EPS) GO TO 140
      IF(II.GT.NI) GO TO 160
```

```
      AL=PI*AL
      KL=1
120   DO 122 I=1,M
      DO 122 J=1,NL
122   AK(I,J)=AK(I,J)-AL*D(I,J)
      GO TO 60
135   CONTINUE

      IF(IT.NE.1) GO TO 138
      WRITE(6,136)CR,AL,VPD
136   FORMAT(22X,E12.6,1X,E14.8,1X,E12.6,/)

138   DO 139 I=1,M
      DO 139 J=1,NL
139   AK(I,J)=AK(I,J)+AL*D(I,J)
      IF(KL.EQ.1) AL=AL/PI
      AL=AL*ANU
      IF(AL.LT.1E-10) GO TO 160
      KL=0
      GO TO 120

140   WRITE(6,150) II
150   FORMAT(5X,'THE CONVERGENCE IS OBTAINED AFTER',1X,I4,1X,
      'ITERATIONS',/,5X,'THE OBTAINED FEEDBACK MATRIX IS',/)
      GO TO 170

160   WRITE(6,165)II
165   FORMAT(5X,'THE CONVERGENCE IS NOT OBTAINED AFTER',I4
      ,'ITERATIONS',/,5X,'THE OBTAINED FEEDBACK MATRIX IS',/)

170   CALL PRINT(AK,M,NL,MA,1)
      WRITE(6,180) Y
180   FORMAT(5X,'THE CRITERION VALUE= ',E12.6,//)
      WRITE(6,190) F
190   FORMAT(5X,'THE GRADIENT NORM VALUE EPS = ',E12.6,/)
      STOP
      END

      ************************************************
      * CLOSED-LOOP MATRIX DETERMINATION      *
      *  AA=A+B.AK.C IF OUTPUT FEEDBACK       *
```

```
*   A=A+B.AK    IF STATE FEEDBACK            *
*************************************************
    SUBROUTINE MATFB(N,M,NL,NA,MA,LA,AA)

    IMPLICIT REAL*8 (A-H,O-Y)
    COMMON /MATSS/ A(15,15),B(15,5),C(5,15),AK(5,15)
    DIMENSION AA(NA,NA)

    IF(NL.NE.N) GO TO 4
    CALL MULT(B,NA,MA,AK,MA,LA,AA,NA,NA,N,M,NL)
    GO TO 6
4   CALL MULT3(B,NA,MA,AK,MA,LA,C,LA,NA,N,M,NL,N,AA,NA)
6   DO 8 I=1,N
    DO 8 J=1,N
8   AA(I,J)=A(I,J)+AA(I,J)
    RETURN
    END


    ************************************************************
    * CALCULATION OF THE MATRIX QQ :                          *
    * QQ = -(Q+C'K'RKC) IF OUTPUT FEEDBACK (NL=L)    *
    * QQ = -(Q+K'RK)     IF STATE FEEDBACK(NL=N)       *
    ************************************************************
    SUBROUTINE F1KC(N,NA,M,MA,NL,LA,BB,QQ)

    IMPLICIT REAL*8 (A-H,O-Y)
    COMMON /MATSS/ A(15,15),B(15,5),C(5,15),AK(5,15)
    COMMON /MATP1/ Q(15,15),R(5,5)
    DIMENSION QQ(NA,NA),BB(NA,NA)

    IF(NL.EQ.N) GO TO 2
    CALL MULT(AK,MA,LA,C,LA,NA,BB,NA,NA,M,NL,N)
    GO TO 6
2   DO 4 I=1,M
    DO 4 J=1,NL
4   BB(I,J)=AK(I,J)
6   DO 10 I=1,N
    DO 10 J=1,N
    S=0
    DO 8 II=1,M
    DO 8 JJ=1,M
8   S=S+BB(II,I)*R(II,JJ)*BB(JJ,J)
10  QQ(I,J)=-S-Q(I,J)
```

```
      RETURN
      END


      ************************************************************
      *SOLVING THE MATRIX EQUATION :                          *
      *  A'Q+QA+C =0                                           *
      *INPUT ARGUMENTS :                                      *
      *  REAL A(N,N)   N ≤ NA                                 *
      *       C(N,N)   N ≤ NC                                 *
      *       R1(N,N)  N ≤ NA                                 *
      *       R2(N,N)  N ≤ NA                                 *
      *       WK(IK)   IK ≥ N*N+3*N                           *
      *  COMPLEX Z(N,N),Z1(N) N ≤ NA                          *
      * OUTPUT ARGEMENTS :                                    *
      *  REAL Q(N,N) N ≤ NQ                                   *
      *                                                       *
      * REFERENCE :                                           *
      *  "THE NUMERICAL SOLUTION OF  A'Q+QA=-C"               *
      *  W.D. HOSKINS, D.S. MEEK AND D.J. WALTON              *
      *  IEEE TRANS. AUT. CONT., VOL. AC-22,                  *
      *  N. 5, OCT. 1977, 882-883.                            *
      ************************************************************
      SUBROUTINE LYAPUN(A,N,NA,C,NC,Q,NQ,R1,R2,Z1,Z,WK,IK)

      IMPLICIT REAL*8 (A-H,O-Y)
      DIMENSION A(NA,NA),C(NC,NC),Q(NQ,NQ)
      DIMENSION R1(NA,NA),R2(NA,NA),WK(IK)
      COMPLEX*16 Z1(NA),Z(NA,NA)

      K=0
1     K=K+1
      IF(K.GT.100) GO TO 30
      DO 3 I=1,N
      DO 3 J=1,N
3     R1(I,J)=A(I,J)
      IJOB=0
      CALL EIGRF(R1,N,NA,IJOB,Z1,Z,NA,WK,IER)
      DO 5 I=1,N
5     WK(I)=REAL(Z1(I))
      II=0
      CALL PGV(WK,N,IK,PV,II)
      IF(PV.GT.0) GO TO 12
```

```
      DO 10 I=1,N
      I1=I+1
      DO 9 J=I1,N
      Q(J,I)=0
9     Q(I,J)=0
10    Q(I,I)=-1
      RETURN
12    II=1
      CALL PGV(WK,N,IK,GV,II)

      XX=(PV*GV)**0.5
      X=1/(PV+XX)**2
      ALPHA=2*PV*X
      BETA=PV*GV*ALPHA
      EPSI=X*(PV-XX)**2
      DO 14 I=1,N
      DO 14 J=1,N
14    R2(I,J)=A(I,J)
      IDGT=0
      CALL LINV2F(R2,N,NA,R1,IDGT,WK,IER)
      DO 17 I=1,N
      DO 17 J=1,N
17    Q(I,J)=R1(J,I)
      DO 20 I=1,N
      DO 20 J=1,N
20    A(I,J)=ALPHA*A(I,J)+BETA*R1(I,J)
      CALL MULT3(Q,NQ,NQ,C,NC,NC,R1,NA,NA,N,N,N,N,R2,NA)
22    DO 24 I=1,N
      DO 24 J=1,N
24    C(I,J)=ALPHA*C(I,J)+BETA*R2(I,J)
      IF(EPSI.GE.1.E-07) GO TO 1
      DO 25 I=1,N
      DO 25 J=1,N
25    Q(I,J)=-0.5*C(I,J)
      RETURN
30    WRITE(6,32) II
32    FORMAT(RX,'ITERATIONS NUMBER(LYAPUNOV)=',I4,//)
      RETURN
      END
```

```
***********************************************************
* FINDING THE SMALLEST/BIGGEST ELEMENT X OF      *
* THE REAL VECTOR Y OF DIMENSION N < NMAX.       *
* INPUT ARGEMENTS :                              *
*  Y REAL VECTOR                                 *
*  K OPTION PARAMETER :                          *
*    K=0     FOR SMALLEST ELEMENT                *
*    K≠0  FOR BIGGEST ELEMENT                    *
* USED AS AN OUTPUT, K IS THE INDEX OF THE       *
* DESIRED ELEMENT, .I.E. X=Y(K)                  *
***********************************************************


      SUBROUTINE PGV(Y,N,NMAX,X,K)

      IMPLICIT REAL*8 (A-H,O-Z)
      DIMENSION Y(NMAX)

      IF(K.NE.0) GO TO 2
      K=1
      X=Y(1)
      DO 1 I=2,N
      IF(Y(I).GE.X) GO TO 1
      X=Y(I)
      K=I
1     CONTINUE
      RETURN
2     K=1
      X=Y(1)
      DO 3 I=2,N
      IF(X.GE.Y(I)) GO TO 3
      X=Y(I)
      K=I
3     CONTINUE
      RETURN
      END


***********************************************************
* FINDING THE SMALLEST REAL PART VPD OF THE      *
* EIGENVALUES OF A REAL MATRIX A(N,N).           *
* WORKING AREA :                                 *
*  REAL*8 AA(N,N),WK(IK)                          *
```

```
*  COMPLEX*16 ZV(N),Z(N,N)                              *
*  N ≤ NA,   IK ≥ N                                     *
***********************************************************
SUBROUTINE MEV(A,N,NA,AA,ZV,Z,WK,IK,VPD)

IMPLICIT REAL*8 (A-H,O-Y)
COMPLEX*16 ZV(15),Z(15,15)
DIMENSION A(NA,NA),AA(NA,NA),WK(IK)

      DO 2 I=1,N
      DO 2 J=1,N
2     AA(I,J)=A(I,J)
      IJOB=0
      CALL EIGRF(AA,N,NA,IJOB,ZV,Z,NA,WK,IER)
      DO 4 I=1,N
4     WK(I)=REAL(ZV(I))
      II=0
      CALL PGV(WK,N,IK,VPD,II)
      RETURN
      END


*** CRITERION CALCULATION ***

SUBROUTINE CRIT(N,M,IGB,CR)

IMPLICIT REAL*8 (A-H,O-Y)
COMMON /MATP2/ PL(15,15),PF(15,15)
COMMON /MATL1/ S(15,15),P(15,15),V0(15,15)
COMMON /GRADS/ GA(15,15),GK(15,15)

      CR=0
      DO 2 I=1,N
      DO 2 J=1,N
2     CR=CR+S(I,J)*V0(J,I)
      IF(IGB.EQ.1) RETURN
      TR=0
      DO 6 I=1,N
      SD=0
      DO 4 II=1,N
      DO 4 JJ=1,N
```

```
4       SD=SD+GA(I,II)*PL(II,JJ)*GA(I,JJ)
6       TR=TR+SD
        CR=CR+TR
        IF(IGB.EQ.2) RETURN
        TR=0
        DO 10 I=1,N
        SD=0
        DO 8 II=1,M
        DO 8 JJ=1,M
8       SD=SD+GK(II,I)*PF(II,JJ)*GK(JJ,I)
10      TR=TR+SD
        CR=CR+TR
        RETURN
        END


*************************************************************
* CALCULATION OF THE MATRICES :                            *
*   QQ = R.AA + B'.S                                        *
*   AB = (R.AA+B'.S).P                                      *
* AA= K.C       IF OUTPUT FEEDBACK (NL=L)                  *
* AA= K         IF STATE FEEDBACK (NL=N)                   *
* AA(M,N)                                                   *
* QQ(M,N), N ⩽ NA, M ⩽ NA                                  *
* AB(M,N)                                                   *
*************************************************************

        SUBROUTINE RKCBS(AA,M,N,NA,MA,LA,QQ,AB)
        IMPLICIT REAL*8 (A-H,O-Y)
        COMMON /MATSS/ A(15,15),B(15,15),C(5,15),AK(5,15)
        COMMON /MATP1/ Q(15,15),R(5,5)
        COMMON /MALT1/ S(15,15),P(15,15),V0(15,15)
        DIMENSION QQ(NA,NA),AA(NA,NA),AB(NA,NA)

        DO 12 I=1,M
        DO 8 J=1,N
        SD=0
        DO 4 II=1,M
4       SD=SD+R(I,II)*AA(II,J)
        SS=0
        DO 6 K=1,N
6       SS=SS+B(K,I)*S(K,J)
8       QQ(I,J)=SS+SD
```

```
      DO 10 J=1,N
      SS=0
      DO 10 K=1,N
      SS=SS+QQ(I,K)*P(K,J)
10    AB(I,J)=SS
12    CONTINUE
      RETURN
      END



      *** CALCULATION OF THE MATRIX -F2(K,S,P) ***

      SUBROUTINE F2KSPV(M,N,IGB,G,AFF,F2,NA,MA,LA)

      IMPLICIT REAL*8 (A-H,O-Y)
      COMMON /MATSS/ A(15,15),B(15,5),C(5,15),AK(5,15)
      COMMON /MATP2/ PL(15,15),PF(15,15)
      COMMON /MATL1/ S(15,15),P(15,15),V0(15,15)
      COMMON /GRADS/ GA(15,15),GK(15,15)
      DIMENSION F2(NA,NA),AF(NA,NA),G(NA,NA)

      CALL MULT3(PL,NA,NA,GA,NA,NA,P,NA,NA,N,N,N,N,AFF,NA)
      DO 2 I=1,N
      DO 2 J=1,N
2     F2(I,J)=-V0(I,J)
      CALL MST(AFF,F2,N,NA)
      IF(IGB.EQ.2) RETURN
      CALL MULT(B,MA,NA,GK,NA,NA,G,NA,NA,N,M,N)
      CALL MULT3(G,NA,NA,PF,NA,NA,P,NA,NA,N,N,N,N,AFF,NA)
      CALL MST(AFF,F2,N,NA)
      RETURN
      END

      ************************************
      * CALCULATION OF F= F-(A+A')       *
      *   WHERE                          *
      *   A(N,N), F (N,N), N ≤ NA        *
      ************************************
      SUBROUTINE MST(A,F,N,NA)

      IMPLICIT REAL*8 (A-H,O-Y)
      DIMENSION A(NA,NA),F(NA,NA)
```

```
      DO 2 I=1,N
      DO 2 J=1,N
      A(I,J)=A(I,J)+A(J,I)
2     A(J,I)=A(I,J)
      DO 4 I=1,N
      DO 4 J=1,N
4     F(I,J)=F(I,J)-A(I,J)
      RETURN
      END



      *** CALCULATION OF -F3(K,S,P) ***

      SUBROUTINE F3KSP(E,M,N,IGB,AFF,D,F3,NA,MA,LA)

      IMPLICIT REAL*8 (A-H,O-Y)
      COMMON /MATSS/ A(15,15),B(15,5),C(5,15),AK(5,15)
      COMMON /MATP2/ PL(15,15),PF(15,15)
      COMMON /MATL1/ S(15,15),P(15,15),V0(15,15)
      COMMON /GRADS/ GA(15,15),GK(15,15)
      DIMENSION AFF(NA,NA),E(NA,NA),D(NA,NA),F3(NA,NA)

      CALL MULT3(S,NA,NA,PL,NA,NA,GA,NA,NA,N,N,N,N,AFF,NA)
      DO 2 I=1,N
      DO 2 J=1,N
2     F3(I,J)=0
      CALL MST(AFF,F3,N,NA)
      IF(IGB.EQ.2) RETURN

      DO 4 I=1,N
      DO 4 J=1,M
4     D(I,J)=E(J,I)
      CALL MULT3(D,NA,NA,GK,NA,NA,PF,NA,NA,N,M,N,N,AFF,NA)
      CALL MST(AFF,F3,N,NA)
      RETURN
      END



      *** GRADIENT CALCULATION ***

      SUBROUTINE GRADK(E,D,AC,N,M,NL,NA,MA,LA,G)
```

```
      IMPLICIT REAL*8 (A-H,O-Y)
      COMMON /MATSS/ A(15,15),B(15,5),C(5,15),AK(5,15)
      COMMON /MATP1/ Q(15,15),R(5,5)
      COMMON /MATP2/ PL(15,15),PF(15,15)
      COMMON /MATL1/ S(15,15),P(15,15),V0(15,15)
      COMMON /MATL2/ PMU(15,15),SLA(15,15)
      COMMON /GRADS/ GA(15,15),GK(15,15)
      DIMENSION AC(NA,NA),E(NA,NA),G(NA,NA),D(NA,NA)


      CALL MULT(E,NA,NA,PMU,NA,NA,G,NA,NA,M,N,N)
      DO 2 I=1,M
      DO 2 J=1,N
2     AC(I,J)=B(J,I)
      CALL MULT3(AC,NA,NA,SLA,NA,NA,P,NA,NA,M,N,N,N,D,NA)
      DO 4 I=1,M
      DO 4 J=1,N
4     G(I,J)=G(I,J)+D(I,J)
      IF(IGB.EQ.2) GO TO 6
      CALL MULT3(R,MA,MA,GK,NA,NA,PF,NA,NA,M,M,N,N,AC,NA)
      CALL MULT(AC,NA,NA,P,NA,NA,D,NA,NA,M,N,N)
      DO 5 I=1,M
      DO 5 J=1,N
5     G(I,J)=G(I,J)+D(I,J)
6     IF(NL.EQ.N) RETURN
      DO 8 I=1,N
      DO 8 J=1,NL
8     AC(I,J)=C(J,I)
      CALL MULT(G,NA,NA,AC,NA,NA,D,NA,NA,M,N,NL)
      DO 10 I=1,M
      DO 10 J=1,NL
10    G(I,J)=D(I,J)
      RETURN
      END



      SUBROUTINE MULT(A,NA,MA,B,NB,MB,C,NC,MC,N,M,L)

      ************************************************************
      * TWO REAL MATRICES MULTIPLICATION : C=A.B       *
      *  A (N,M), N ≤ NA, M ≤ MA                        *
```

```
*  B (M,L), M ≤ NB, L ≤ MB                             *
*  C (N,L), N ≤ NC, L ≤ MC                             *
******************************************************

   IMPLICIT REAL*8 (A-H,O-Y)
   DIMENSION A(NA,NA),B(NB,NB),C(NC,NC)

   DO 1 I=1,N
   DO 1 J=1,L
   C(I,J)=0.D0
   DO 1 K=1,M
 1 C(I,J)=C(I,J)+A(I,K)*B(K,J)
   RETURN
   END


*******************************************************************
* THREE REAL MATRICES MULTIPLICATION : Q = A.B.C     *
* THE MATRICE DIMENSION ARE :                        *
*  A(N,M), N, M ≤ MA                                 *
*  B(M,L), M, L ≤ MB                                 *
*  C(L,K), L ≤ NC, K ≤ MC                            *
*  QQ(N,K), N ≤ NQ, K ≤NQ                            *
*******************************************************************

   SUBROUTINE MULT3(A,NA,MA,B,NB,MB,C,NC,MC,N,M,L,K,QQ,NQ)

   IMPLICIT REAL*8 (A-H,O-Y)
   DIMENSION A(NA,MA),B(NB,MB),C(NC,MC),QQ(NQ,NQ)

   DO 2 I=1,N
   DO 2 J=1,K
   S=0
   DO 1 II=1,M
   DO 1 JJ=1,L
 1 S=S+A(I,II)*B(II,JJ)*C(JJ,J)
   QQ(I,J)=S
 2 CONTINUE
   RETURN
   END

   SUBROUTINE PRINT(A,M,N,MMAX,IT)
```

```
***********************************************************
* REAL MATRIX WRITING SUBROUTINE                         *
*  A:MxN    M<MMAX                                        *
* MMAX Maximal dimension of the rows of the matrix        *
* A as specified in the dimension statement of            *
* the calling program.                                    *
*  IT=1  FOR WRITING                                      *
***********************************************************
      REAL*8 A(MMAX,1)

      IF (IT.NE.1) GO TO 6
      K=(N-1)/10+1
      DO 3 KK=1,K
      NN=10
      DO 2 I=1,M
      IF (N.GT.10*KK) GO TO 1
      NN=N-10*(KK-1)
1     CONTINUE
2     WRITE(6,5)(A(I,(KK-1)*10+J),J=1,NN)
3     WRITE(6,4)
4     FORMAT(///)
5     FORMAT(10(D12.4))
6     RETURN
      END
```

REFERENCES

(ACK-84) ACKERMANN, J. (1984) Robustness against sensor failures, _Automatica_, Vol. 20, no. 2, 211-215.

(ALB-83) ALBERT, J. ALOS (1983) Stabilization of a class of plants with possible loss of outputs or actuator failures, _IEEE Trans. Aut. Cont._, Vol. AC-28, no. 2, 231-233.

(AND-71) ANDERSON, B.O.D. and MOORE, J. (1971) _Linear optimal control_, Prentice Hall.

(AND-81a) ANDERSON, B.O.D. and CLEMENTS, D.J. (1981) Algebraic characterization of fixed modes in decentralized control, _Automatica, Vol. 17_, no. 5, 703-712.

(AND-81b) ANDERSON, B.O.D. and MOORE, J. (1981) Time-varying feedback laws for decentralized control, _IEEE Trans. Aut. Cont._, Vol. AC-26, no. 5, 1133-1139.

(AND-82) ANDERSON, B.O.D. (1982) Transfer function matrix description of decentralized fixed modes, _IEEE Trans. Aut. Cont._, Vol. AC-27, no. 6, 1176-1182.

(AND-84) ANDERSON, B.O.D. and LINNEMANN, A. (1984) Spreading the control complexity in decentralized control of interconnected systems, _Systems & Control Letters_, Vol. 5, 1-8.

(AOK-72) AOKI, M. (1972) On feedback stabilizability of decentralized dynamic systems, _Automatica, Vol. 8_, no. 2, 163-173.

(AOK-73) AOKI, M., LI, M.T. (1973) Controllability and stabilizability of decentralized dynamic systems, _Proc. of Joint Automatic Control Conf. (JACC)_, Ohio State University, U.S.A., 278-286.

(AOK-76) AOKI, M. (1976), On decentralized stabilization and dynamic assignment problems, _J. of Int. Economics_, Vol. 6, 143-171.

(ARM-81) ARMENTANO, V.A. and SINGH, M.G. (1981) A new approach to the decentralized controller initialisation problem, _8th World Congress of IFAC_, Paper 41.2, Kyoto, Japan.

(ARM-82) ARMENTANO, V.A. and SINGH, M.G. (1982) A procedure to eliminate decentralized fixed modes with reduced information exchange, _IEEE Trans. Aut. Cont._, Vol. AC-27, no. 1, 258-260.

(BAR-72) BARTELS, R.H. and STEWART, G.W. (1972) Solution of matrix equation AX + XB = C, _Commun. ACM_, Vol. 15, no. 9, 820.

(BEL-70) BELLMAN, R. (1970) Introduction to matrix analysis, Mc Graw-Hill, 2nd edition.

(BER-81) BERNUSSOU, J. and GEROMEL, J.C. (1981) An easy way to find gradient matrix of composite matricial functions, IEEE Trans. Aut. Cont., Vol. AC-26, no. 2, 538-540.

(BER-82) BERNUSSOU, J. and TITLI, A. with the collaboration of AUTHIE, G. and CALVET, J.L. (1982) Interconnected dynamical systems : Stability, Decomposition and Decentralization, North-Holland.

(BIN-78) BINGULAC, S.P. (1978) Calculation of derivatives of characteristic polynomials, IEEE Trans. Aut. Cont., Vol. AC-23, no. 4, 751-753.

(BOG-61) BOGLIUBOV, N.N. and MITROPOLSKY, Yu.A. (1961) Asymptotic methods in the theory of non-linear oscillations, New York : Gordon Breach.

(BOW-76) BOWIE, W.S. (1976) Application of graph theory in computer systems, Int. J. Com. and Inf. Sci., Vol. 5, 9-31.

(BRA-70) BRASCH, F.M. and PEARSON, J.B. (1970) Pole placement using dynamic compensators ,IEEE Trans. Aut. Cont., Vol. AC-15, 34-43.

(BUR-81) BURROWS, C.R. and SAHINKAYA M.N. (1981) A new algorithm for determining structural controllability, Int. J. Cont., Vol. 33, no. 2, 379-392.

(CAL-78) CALOVIC, M., DJOROVIC, M. and SILJAK, D.D. (1978) Decentralized approach to automatic generation control of interconnected power systems, International Conference on Large High-voltage Electric Systems (CIGRE), Paris, France.

(CHE-81) CHEN, Y. and SINGH, M.G. (1981) Certain practical consideration in model-following method of decentralized control, Proc. IEE, Vol. 128, part D., no. 4, 149-155.

(CHE-84) CHEN, Y., MAHMOUD, M.S. and SINGH, M.G. (1984) An iterative block-diagonalization procedure for decentralized optimal control, Int. J. Syst. Sci., Vol. 15, no. 5, 563-573.

(COR-76a) CORFMAT, J.P. and MORSE, A.S. (1976) Control of linear systems through specified input channels, SIAM J. Cont. and Opt., Vol. 14, no. 1, 163-175.

(COR-76b) CORFMAT, J.P. and MORSE, A.S. (1976) Decentralized control of linear multivariable systems, Automatica, Vol. 12, no. 5, 479-495.

(DAV-68) DAVISON, E.J. and MAN, F.T. (1973) The numerical solution of A'Q+QA =-C, IEEE Trans. Aut. Cont., Vol. AC-13, 448-449.

(DAV-73) DAVISON, E.J., RAU, N.S. and PALMAY F.V. (1973) The optimal decentralized control of a power system consisting of a number of interconnected synchronous machines, Int. J. Cont., Vol. 18, no. 6, 1313-1328.

(DAV-74) DAVISON, E.J. and WANG, W.H. (1974) Properties and calculation of transmission zeros of linear multivariable time-invariant systems, Automatica, Vol. 10, 643-658.

(DAV-75) DAVISON, E.J. and GOLDENBERG, A. (1975) The robust control of a general servomechanism problem : the servo-compensator, Automatica, Vol. 11, 461-471.

(DAV-76a) DAVISON, E.J. (1976) Decentralized stabilization and regulation in large multivariable systems, In Direction in decentralized control, Many-person optimization and large scale systems (editors : HO, Y.C. and MITTER S.), Plenum Press, 303-323.

(DAV-76b) DAVISON, E.J. (1976) Multivariable tuning regulators : the feed-forward and robust control of a general servomechanism problem, IEEE Trans. Aut. Cont., Vol. AC-21, 35-47.

(DAV-76c) DAVISON, E.J. (1976) The robust decentralized control of a general servomechanism problem, IEEE Trans. Aut. Cont., Vol. AC-21, no. 1, 14-24.

(DAV-76d) DAVISON, E.J. (1976) The robust control of a servomechanism problem for linear time-invariant multivariable systems, IEEE Trans. Aut. Cont., Vol. AC-21, 25-34.

(DAV-76e) DAVISON, E.J. and WANG, W.H. (1976) Remark on multiple transmission zeros of a system, Automatica, Vol. 12, 195.

(DAV-77a) DAVISON, E.J. (1977) Connectability and structural controllability of composite systems, Automatica, Vol. 13, 109-123.

(DAV-77b) DAVISON, E.J. (1977) The robust decentralized servomechanism problem with extra stabilizing control agents, IEEE Trans. Aut. Cont., Vol. AC-22, no. 2, 256-259.

(DAV-78a) DAVISON, E.J. (1978) Decentralized robust control of unknown systems using tuning regulators, IEEE Trans. Aut. Cont., Vol. AC-23, no. 2, 276-289.

(DAV-78b) DAVISON, E.J., GESING, W. and WANG, W.H. (1978) An algorithm for obtaining the minimal realization of a linear time-invariant system and determining if a system is stabilizable-detectable, IEEE Trans. Aut. Cont., Vol. AC-23, no. 6, 1048-1054.

(DAV-79a) DAVISON, E.J. (1979) The robust decentralized control of a servomechanism problem for composite systems with input-output interconnections, IEEE Trans. Aut. Cont., Vol. 24, no. 2, 325-327.

(DAV-79b) DAVISON, E.J. and GESING, W. (1979) Sequential stability and optimization of large-scale decentralized systems, Automatica, Vol. 15, 307-324.

(DAV-81) DAVISON, E.J. and FERGUSON, I. (1981) The design of controllers for the multivable robust servomechanism problem using parameter optimization methods, IEEE Trans. Aut. Cont., Vol. AC-26, no. 1, 93-110.

(DAV-82) DAVISON, E.J. and CHANG, T. (1982) The design of decentralized controllers for the robust servomechanism problem using parameter optimization methods, American Automatic Control Conference, Arlington, Virginia, U.S.A., 905-909.

(DAV-83) DAVISON, E.J. and OZGUNER, U. (1983) Characterization of decentralized fixed modes for interconnected systems, Automatica, Vol. 19, no. 2, 169-182.

(DAV-85) DAVISON, E.J. and WANG, S.H. (1985) A characterization of decentralized fixed modes in terms of transmission zeros, IEEE Trans. Aut. Cont., Vol. AC-30, no. 1, 81-82.

(DES-74) DESOER, C.A. and SCHULMAN, D. (1974) Zeros and poles of matrix transfer function and their dynamical interpretation, IEEE Trans. Circ. Syst., Vol. CAS-21, 3-8.

(EVA-84) EVANS, F.J. and KRUSER, M. (1984) Pole assignment in decentralized systems : A structural approach, Proc. IEE, Vol. 131, part D, no. 6, 229-232.

(FAD-63) FADDEEV, D.K. and FADDEEVA, V.N. (1963) Computational methods of linear algebra, Freeman, 288.

(FES-79) FESSAS, P.S. (1979) A note on 'An example in decentralized control systems', IEEE Trans. Aut. Cont., Vol. AC-24, 669.

(FES-80) FESSAS, P.S. (1980) Decentralized control of linear dynamical systems via polynomial matrix methods : I - Two interconnected scalar systems, Int. J. Cont., Vol. 30, no. 2, 259-276. II-Arbitrary interconnected systems, Int. J. Cont., Vol. 32, no. 1, 127-147.

(FLE-63) FLETCHER, R. and POWELL, M.J.D. (1963) A rapidly convergent descend method for minimization, Computer J., Vol. 6, 163-663.

(FOS-77) FOSSARD, A. (1977) Multivariable systems control, North-Holland.

(GAN-79) GANTMACHER, F.R. (1979) The theory of matrices, Chelsea, New-York.

(GER-79a) GEROMEL, J.C. and BERNUSSOU, J. (1979) An algorithm for optimal decentralized regulation of linear quadratic interconnected systems, Automatica, Vol. 14, 489-491.

(GER-79b) GEROMEL, J.C. (1979) Contribution à l'étude des systèmes dynamiques interconnectés : Aspects de décentralisation, Thèse de Doctorat d'Etat, Université Paul Sabatier, Toulouse, France.

(GER-82) GEROMEL, J.C. and BERNUSSOU, J. (1982) Optimal decentralized control of dynamic systems, Automatica, Vol. 13, no. 5, 545-557.

(GER-84) GEROMEL, J.C. and PERES, P.L.D. (1985) Decentralized load-frequency control, IEE Proceedings, Vol. 132, Pt. D, n° 5, 225-230.

(GLO-76) GLOVER, K. and SILVERMAN, L.M. (1976) Characterization of structural controllability, IEEE Trans. Aut. Cont., Vol. AC-21, 534-537.

(HAR-65) HARARY, F., NORMAN, R.Z. and CARTWRIGHT, D. (1965) Structural models : An introduction to the theory of directed graphs, Wiley, New-York.

(HAS-78a) HASSAN, M.F. and SINGH, M.G. (1978) Robust decentralized controller for linear interconnected dynamical systems, Proc. IEE, Vol. 125, no. 5, 429-432.

(HAS-78b) HASSAN, M.F., SINGH, M.G. and TITLI, A. (1979) Near optimal decentralized control with a pre-specified degree of stability, Automatica, Vol. 15, 483-488.

(HAS-80) HASSAN, M.F. and SINGH, M.G. (1980) Decentralized controller with online interaction trajectory improvement, Proc. IEE, Vol. 127, part D, no. 3, 142-148.

(HOS-77) HOSKINS, W.D., MEEK, D.S. and WALTON, D.J. (1977) The numerical solution of $A^T Q + QA = - C$, IEEE Trans. Aut. Cont., Vol. AC-15, no. 5, 881-885.

(HUJ-84) HU, Y.Z. and JIANG, W.S. (1984) New characterization of decentralized fixed modes and their application, 9th Congress of I.F.A.C., Budapest, Hungary).

(IKE-79) IKEDA, M. and SILJAK, D.D. (1979) Counter examples to Fessas conjecture, IEEE Trans. Aut. cont., Vol. AC-24, no. 4, 670.

(IKE-80) IKEDA, M. and SILJAK, D.D. (1980) Overlapping decompositions, expansions and contraction of dynamic systems, Large Scale systems, Vol. 1, 29-38.

(IKE-81) IKEDA, M., SILJAK, D.D. and WHITE, D.E. (1981) Decentralized control with overlapping information sets, JOTA, Vol. 34, no. 2, 279-310.

(ISA-73) ISAKSEN, L. and PAYNE, H.J. (1973) Suboptimal control of linear systems by augmentation with application to freeway traffic regulation, IEEE Trans. Aut. Cont., Vol. AC-18, 210-219.

(JAM-83) JAMSHIDI, M. (1983) Large-Scale systems : Modeling and control, North-Holland, New-York.

(JOH-84) JOHNSTON, R.D., BARTON, G.W. and BRISK, M.L. (1984) Determination of the generic rank of structural matrices, Int. J. Cont., Vol. 40, no. 2, 257-264.

(KAI-80) KAILATH, T. (1980) Linear systems, Prentice-Hall.

(KAL-62) KALMAN, R.E. (1962) Canonical structure of dynamical systems, Proc. Nat. Acad. Sci. U.S.A., Vol. 48, 596-600.

(KAT-81) KATTI, S.K. (1981) Comments on Decentralized control of linear multivariable systems, Automatica, Vol. 17, no. 4, 665.

(KAU-68) KAUFMANN, A. (1968) Introduction à la combinatorique en vue des applications, Dunod, Paris, France.

(KAW-81) KAWASAKI, N. and SHIMEMURA, S. (1981) A method of deciding weighting matrices in an LQ-problem to locate all poles in the specified region, 8th world congress of I.F.A.C., Kyoto, Japan.

(KEV-75) KEVORKIAN, A.K. (1975) Structural aspects of large dynamic systems, 6th world congress of I.F.A.C., part III. A, paper 19.3, Boston, U.S.A.

(KLE-66) KLEINMAN, D.L. (1966) On the linear regulator problem and the matrix Riccati equation, MIT Electronic Systems Lab., Mass. Tech. Rept. ESL-R271 Cambridge, U.K.

(KOB-78) KOBAYASHI, H., HANAFUSA, H. and YOSHIKAWA, T. (1978) Controllability under decentralized information structure, IEEE Trans. Aut. Cont., Vol. AC-23, no. 2, 182-188.

(KOB-82) KOBAYASHI, H. and YOSHIKAWA, T. (1982) Graph-Thoeretic approach to controllability and localizability of decentralized control, IEEE Trans. Aut. Cont., Vol. AC-27, no. 5, 1096-1108.

(KOK-76) KOKOTOVIC, P.V., O'MALLEY, R.E. and SANNUTI, P. (1976) Singular perturbation and order reduction in control theory : An overview, Automatica, Vol. 12, 123-132.

(KRO-67) KROFT, D. (1967) All paths through a maze, Proc. IEEE, Vol. 55, 88-90.

(KWA-72) KWAKERNAAK, H. and SIVAN, R. (1972) Linear optimal control systems Wiley Interscience, New-York.

(LAN-64) LANCASTER, P. (1964) On eigenvalues of matrices dependent on a parameter, Numerische Mathematik, Vol. 6, 377-387.

(LEV-70) LEVINE, W.S. and ATHANS, M. (1970) On the determination of the optimal constant output feedback gains for linear multivariable systems, IEEE Trans. Aut. Cont., Vol. AC-15, no. 1, 44-48.

(LIA-69) LIN, P.M. and ALDERSON, G.E. (1969) 7th Alberton Conf. Circuit and System Theory, Urbana, U.S.A., 196.

(LIE-83) LINNEMANN, A. (1983) Fixed modes in parametrized systems, Int. J. Cont., Vol. 38, no. 2, 319-335.

(LIE-84) LINNEMANN, A. (1984) Decentralized control of dynamically interconnected system, IEEE Trans. Aut. Cont., Vol. AC-29, no. 11, 1052-1054.

(LIN-74) LIN, C.T. (1974) Structural controllability, IEEE Trans. Aut. Cont., Vol. AC-19, no. 3, 201-208.

(LIU-68) LIU, C.L. (1968) Introduction to combinatorial mathematics, Chapter 11, Mc Graw-Hill, New-York.

(LOC-77) LOCATELLI, A., SCHIAVONI, N. and TARANTINI, A. (1977) Pole placement : role and choice of the underlying information pattern, Ricerche di Automatica, Vol. 18, no. 1, 107-126.

(LOC-83) LOCATELLI, A., SCATTOLINI, R. and SCHIAVONI, N. (1983) Two problems in the design of discrete decentralized control systems subject to finite perturbations, 4th IMA Conference, Cambridge, England.

(MAC-76) MACFARLANE, A.G.J. and KARCANIAS, N. (1976) Poles and zeros of linear multivariable systems : a survey of the algebraic, geometric and complex-variable theory, Int. J. Cont., Vol. 24, no. 1, 33-74.

(MAL-85) MALINOWSKI, K. and SINGH, M.G. (1985) Controllability and observability of expanded system with overlapping decompositions, Automatica, Vol. 21, no. 2, 303-308.

(MAS-56) MASON, S.J. (1956) Feedback theory-further properties of signal flow graphs, Proc. Inst. Radio Engr., no. 44, 920-926.

(MEE-73) MEERKOV, S.M. (1973) Vibrational control, Automation and Remote control, Vol. 31, 201-209.

(MEE-80) MEERKOV, S.M. (1980) Principle of vibrational control : Theory and application, IEEE Trans. Aut. cont., Vol. AC-25, no. 4, 755-762.

(MIC-78) MICHEL, A.N., MILLER, R.K. and TANG, W. (1978) Lyapunov stability of interconnected systems : decomposition into strongly connected subsystems, IEEE Trans. Circ. Syst., Vol. CAS-25, 799-809.

(MOM-83) MOMEN, S. and EVANS, F.J. (1983) Structurally fixed modes in decentralized systems, Part I : Two control stations, Part II : General case, Proc. IEE, Vol. 130, part D., no. 6, 313-327.

(MOO-81) MOORE, B.C. (1981) Principal component analysis in linear systems : Controllability, observability and model reduction, IEEE Trans. Aut. cont., Vol. AC-26, no. 1, 17-32.

(MOG-66) MORGAN, B.S. (1966) Computational procedure for the sensitivity of an eigenvalue, Electronics Letters, Vol. 26, 197-198.

(MOR-73) MORSE, A.S. (1973) Structural invariants of linear multivariable systems, SIAM J. Control, Vol. 11, no. 3, 446-465.

(MOA-80) MORARI, M. and STEPHANOPOULOS, G. (1980) Part II : Structural aspects and the synthesis of alternative feasible control shemes, A.I. Ch. E. Jl, Vol. 26, 245.

(OZG-82) OZGUNER, U. and DAVISON, E.J. (1982) Decentralized retunability under structural perturbations, American Control Conference, Arlington, VA, U.S.A.

(OZG-85) OZGUNER, U. and DAVISON, E.J. (1985) Sampling and decentralized fixed modes, American Control Conference, Boston, MA, U.S.A.

(PAP-84) PAPADIMITRIOU, H. and TSITSIKLIS, J. (1984), A simple criterion for structurally fixed modes, Systems & Control Letters, Vol. 4, 333-337.

(PAR-74) PARASKEVOPOULOS, P.N., TSONIS, C.A. and TZAFESTAS, S.G. (1974) Eigenvalue sensitivity of linear time-invariant control systems with repeated eigenvalues, IEEE Trans. Aut. Cont., Vol. AC-19, 610-612.

(PEK-79) PETKOVSKI, D.B., and RAKIC, M. (1979) A series solution of feedback gains for output-constrained regulators, Int. J. Cont., Vol. 30, no. 4, 661-668.

(PEK-83) PETKOVSKI, D.B. (1983) Robust decentralized control system designs with application to DC/AC power systems, IFAC/IFORS Symposium on Large-Scale Systems : Theory and Application, Warsaw, Poland, 11-15.

(PEK-84a) PETKOVSKI, D.B. (1984) Robustness of decentralized control subject to linear perturbation in the system dynamics, Problems of control and Information Theory, Vol. 13, no. 1, 3-12.

(PEK-84b) PETKOVSKI, D.B. (1984) Robustunss of control systems subject to modelling uncertainties, R.A.I.R.O. Automatique/Systems Analysis and Control, Vol. 18, no. 3, 315-327.

(PET-84) PETEL, R.V. and MISRA, P. (1984) A numerical test for transmission zeros with application in characterizing decentralized fixed modes, 23rd Conference on Decision and Control (C.D.C.), Las Vegas, U.S.A.

(PIC-83a) PICHAI, V., SEZER, M.E. and SILJAK, D.D. (1983) A graphical test for structurally fixed modes, Mathematical Modelling, vol. 4, 339-348.

(PIC-83b) PICHAI, V., SEZER, M.E. and SILJAK, D.D. (1983) A graph-theoretic algorithm for hierarchical decomposition of dynamic systems with application to estimation and control, IEEE Trans. Syst. Man. & Cyb., Vol. SMC-13, no. 3, 197–207.

(PIC-84) PICHAI, V., SEZER, M. and SILJAK, D.D. (1984) A graph-theoretic characterization of structurally fixed modes, Automatica, Vol. 20, no. 2, 247–250.

(POT-79) POTTER, J.M., ANDERSON, B.O. and MORSE, A.S. (1979) Single-channel control of a two-channel system, IEEE Trans. Aut. Cont., Vol. AC-24, 491–492.

(PUR-82) PURVIANCE, J.E. and TYLEE, J.L. (1982) Scalar sinusoïdal feedback laws in decentralized control, 21th IEEE Conference on Decision and Control, Florida, U.S.A.

(PRE-81) PRESCOTT, R. and PEARSON, J.B. (1981) Private communication, Rice University, Houston, TX.

(RAO-69) RAO, V.V.B. and MURTI, V.G.K. (1969) Proc. Inst. Elec. Electron. Engrs., Vol. 57, 700.

(REI-81) REINSCHKE, K.J. (1981) Structurally complete systems with minimal input and output vectors, Large Scale Systems, Vol. 2, 235–242.

(REI-83) REINSHKE, K.J. (1983) Graph-theoretic approach to control systems, Third Conference on System Science, Lerchendal, Norway.

(REI-84a) REINSCHKE, K.J. (1984) Graph-theoretic characterization of fixed modes in centralized and decentralized control, Int. J. Cont., Vol. 39, no. 4, 715–729.

(REI-84b) REINSCHKE, K.J. (1984) Graph-theoretic characterization of structural properties of paths and cycle families, 9th world congress of I.F.A.C., Budapest, Hungary.

(ROS-65a) ROSENBROCK, H.H. (1965) Transfer matrix of linear dynamic system, Electronics Letters, Vol. 1, no. 4, 95–96.

(ROS-65b) ROSENBROCK, H.H. (1965) Sensitivity of an eigenvalue to changes in the matrix, Electronics Letters, Vol. 1, no. 10, 278–279.

(ROS-70) ROSENBROCK, H.H. (1970) State space and multivariable theory, Nelson, London.

(ROS-74) ROSENBROCK, H.H. (1973) The zeros of a systems, Int. J. Cont., Vol. 18, no. 2, 297–299. (1974) Correction to 'The zeros of a system', Int. J. Cont., Vol. 20, 525.

(ROY-70) ROY, B. (1970) Algèbre moderne et théorie des graphes, Dunod, Paris, France.

(RUN-85) RUNOLFSSON, T. and MEERKOV S.M. (1985) Vibrational-feedback control of decentralized systems : a design algorithm, IFAC Workshop on Modelling Errors, Boston, MA, U.S.A.

(SAE-79) SAEKS, R. (1979) On the decentralized control of interconnected dynamical systems, IEEE Trans. Aut. cont., Vol. AC-24, no. 2, 269-271.

(SEN-79) SENNING, M.F. (1979) Feasibly decentrlaized control, Ph. D. Thesis, ETH, Zurich, Switzerland.

(SER-82) SERAJI, H. (1982) On fixed modes in decentralized control systems, Int. J. Cont., Vol. 35, no. 5, 775-784.

(SEZ-81a) SEZER, M.E. and SILJAK, D.D. (1981) Structurally fixed modes, Systems & Control Letters, Vol. 1, no. 1, 60-64.

(SEZ-81b) SEZER, M.E. and HUSEYIN, O. (1981) Comment on 'Decentralized state feedback stabilization', IEEE Trans. Aut. Cont., Vol. AC-26, no. 2, 547-549.

(SEZ-81c) SEZER, M.E. and SILJAK, D.D. (1981) on structural decomposition and stabilization of large scale systems, IEEE Trans. Aut. cont., Vol. AC-26, 439-444.

(SEZ-83) SEZER, M.E. (1983) Minimal essential feedback patterns for pole assignment using dynamic compensation, 22th IEEE Conference on Decision and control.

(SHI-76) SHIELDS, R.W. and PEARSON, J.B. (1976) Structural controllability of multi-input linear systems, IEEE Trans. Aut. Cont., Vol. AC-21, no. 2, 203-212.

(SIL-73) SILJAK, D.D. (1973) On stability of large-scale system under structural perturbations, IEEE Trans. Syst. Man & Cyb., 415-147.

(SIL-75) SILJAK, D.D. (1975) Connective stability of competitive equilibrium, Automatica, Vol. 11, 389.

(SIL-76) SILJAK, D.D. and SUNDARESHAN, S.K. (1976) A multilevel optimization of large-scale dynamic systems, IEEE Trans. Aut. Cont., Vol. AC-21, 80.

(SIL-77) SILJAK, D.D. (1977) On reachability of dynamic systems, Int. J. Cont., Vol. 8, 321-338.

(SIL-78) SILJAK, D.D. (1978) Large scale systems : Stability and structure, North-Holland, New-York, U.S.A.

(SIL-80) SILJAK D.D. (1980) Reliable control using multiple control systems, Int. J. Cont., Vol. 31, 303-329.

(SIL-82a) SILJAK, D.D. and al. (1982) The inclusion principle for dynamic systems, Final report DE-ACO37 ET 29138-34, Santa Clara University, U.S.A.

(SIL-82b) SILJAK, D.D., PICHAI, V. and SEZER, M.E. (1982) Graph-theoretic analysis of dynamic systems, Report DE-ACO37 ET 29138-35, Santa Clara University, U.S.A.

(SIN-76) SINGH, M.G., HASSAN, M. and TITLI, A. (1976) A feedback solution for large interconnected dynamical systems using the prediction principle, IEEE Trans. Syst. Man & Cyb., Vol. SMC-6, 223-239.

(TAO-84) TAROKH, M. (1984) Fixed modes in decentralized control systems, First European Workshop on Real Time Control of Large Scale Systems, University of Patras, Greece.

(TAR-84) TARRAS, A.M. and TITLI, A. (1984) On a new algebraic characterization of decentralized fixed modes, IEEE International Conference on Computers, Systems & Signal Processing, Bangalore, India.

(TAR-85) TARRAS, A.M. (1985) Commande décentralisée des grands systèmes : modes fixes, sensibilité et robustesse paramétrique (in French), Doctorat thesis no. 5, Université Paul Sabatier, Toulouse, France.

(THI-71) THIRIEZ, M. (1971) The set covering problem : A group theoretic approach, R.A.I.R.O., Vol. 3, 84-103.

(TRA-84a) TRAVE L., TARRAS A.M., TITLI A., "Some problems in decentralized control in presence of fixed modes", 9th world congress of I.F.A.C., Budapest (Hungary), July 1984.

(TRA-84b) TRAVE L., "Commande par retour de sortie à structure décentralisée : Notion de modes fixes", Thèse de Docteur-Ingénieur, I.N.S.A. Toulouse (France), 1984.

(TRA-85) TRAVE L., TARRAS A.M., TITLI A., "An application of vibrational control to cancel unstable decentralized fixed modes", IEEE Trans. Auto. Contr., vol. AC-30, n° 3, pp. 283-286, 1985.

(TRA-86a) TRAVE, L. and TITLI, A. (1986) A sequencial algorithm to conclude on structural controllability of large scale systems, Internal report 85312, LAAS du CNRS, Toulouse, France.

(TRA-86b) TRAVE, L. and TITLI, A. (1986) Sequencial determination of minimal feedback patterns avoiding structurally fixed modes, 25th IEEE Conference on Decision and Control, Athens, Greece.

(TRA-87) TRAVE, L., TARRAS, A.M. and TITLI, A. (1987) Minimal feedback structure avoiding structurally fixed modes, Int. J. Control., Vol. 46, no. 1, 313-325.

(VAN-68) VAN TRESS, H.L. (1968) Detection, estimation and modulation theory, Part I, John Wiley & Sons, New-York, U.S.A.

(VID-83) VIDYASAGAR, M. and VISWANADHAM, N. (1983) Algebraic characterization of decentralized fixed modes, Systems & Control Letters, Vol. 3, 69-73.

(WAN-73a) WANG, S.H. and DAVISON, E.D. (1973) Properties of linear time-invariant multivariable system subject to arbitrary output and state feedback, IEEE Trans. Aut. Cont., Vol. AC-18, 24-32.

(WAN-73b) WANG, S.H. and DAVISON, E.D. (1973) On the stabilization of decentralized control systems, IEEE Trans. Aut. Cont., Vol. AC-18, 437-478.

(WAN-78a) WANG, S.H. and DAVISON, E.D. (1978) Minimization of transmission cost in decentralized control systems, Int. J. Cont., Vol. 28, no. 6, 889-896.

(WAN-78b) WANG, S.H. (1978) An example in decentralized control, IEEE Trans. Aut. Cont., Vol. AC-23, no. 5, 938.

(WAN-82) WANG, S.H. (1982) Stabilization of decentralized control systems via time-varying controllers, IEEE Trans. Aut. Cont., Vol. AC-27, no. 3, 741-744,.

(WIL-70) WILLEMS, J.L. (1970) Stability theory of dynamical systems, Nelson, London.

(WOL-73a) WOLOVICH, W.A. (1973) On determining the zeros of state-space systems, IEEE Trans. Aut. Cont., Vol. AC-18, 542-544.

(WOL-73b) WOLOVICH, W.A. (1973) On the numerators and zeros of rational transfer matrices, IEEE Trans. Aut. Cont. Vol. AC-18, 544-546.

(WOL-74) WOLOWICH, W.A. (1974) Linear multivariable systems, New-York : Springer.

(WON-67) WONHAM, W.M. (1967) On pole assignment in multi-input controllable linear systems, IEEE Trans. Aut. Cont., Vol. AC-12, 660-665.

(WON-74) WONHAM, W.M. (1974) Linear multivariable control : A geometric approach, Springer-Verlag, Berlin.

(XIN-82) XINOGALAS, T.C., MAHMOUD, M.S. and SINGH, M.G. (1982) Hierarchical computation of decentralized gain for interconnected systems, Automatica, Vol. 18, no. 4, 473-478.

(YAH-77) YAHAGI, T. (1977) Optimal output feedback control with reduced performance index sensitivity, Int. J. Cont., Vol. 25, no. 5, 769-783.

(ZHE-84) ZHENG, Y.F. (1984) The study of local controllability and observability of decentralized systems via polynomial models, Internal Report 84-002, Department of Mathematics, East China Normal University, Shanghai, China.

(ZOU-70) ZOUTENDIJK, G. (1970) Some algorithms based on the principle of feasible directions, in Non-linear Programing (Edited by ROSEN, J.B., MANGASARIAN J.L. and RITTER), Academic Press.

AUTHOR INDEX

SUBJECT INDEX

Lecture Notes in Control and Information Sciences

Edited by M. Thoma and A. Wyner

Lecture Notes in Control and Information Sciences

Edited by M. Thoma and A. Wyner

Lecture Notes in Control and Information Sciences

Edited by M. Thoma and A. Wyner